Lecture Notes in Control and Information Sciences 228

Editor: M. Thoma

Springer-Verlag London Ltd.

L. Dugard and E.I. Verriest (Eds)

Stability and Control of Time-delay Systems

Springer

Editors

L. Dugard
Laboratoire d'Automatique de Grenoble
UMR CNRS-INPG UJF, ENSIEG BP 46, 38402 St Martin d'Hères Cedex, France

E.I. Verriest
School of Electrical and Computer Engineering
Georgia Institute of Technology, Atlanta, GA 30332-0250, USA

ISBN 978-3-540-76193-8

British Library Cataloguing in Publication Data
Stability and control of time-delay systems. - (Lecture
notes in control and information sciences ; 228)
1.Delay lines 2.Delay lines - Stability
I.Dugard, L. II.Verriest, Erik Isidoor
003.8
ISBN 978-3-540-76193-8 ISBN 978-3-540-40935-9 (eBook)
DOI 10.1007/978-3-540-40935-9

Library of Congress Cataloging-in-Publication Data
Stability and control of time-delay systems / L. Dugard and
E.I. Verriest (eds.).
p. cm. - - (Lecture notes in control and information sciences : 228)
Includes bibliographical references.
ISBN 978-3-540-76193-8
1. Control theory. 2. Delay differential equations.
3. Stability. I. Dugard, L. II. Verriest, Erik I., 1950-
III. Series.
QA402.3.S75 1997 97-29314
629.8'312- -dc21 CIP

Originally published by Springer-Verlag London Limited in 1998

Typesetting: Camera ready by editors

69/3830-543210 Printed on acid-free paper

Foreword

These notes are primarily concerned with stability and control of linear delay differential equations. This is an active area of research with most of the recent results available only through journals. These notes are an attempt to remedy this situation.

The editors have chosen a variety of experts representing a broad spectrum of techniques. The authors of the chapters not only have given a good overall view of their subject but have also included an extensive bibliography as well as new and original results. Emphasis is placed on presenting the material in such a way that it can be directly applied to specific problems. Some attention also is paid to numerical schemes.

This is a welcome addition to the subject and should be useful to theoreticians as well as practitioners.

JACK HALE
4/9/1997

Introduction

The idea of editing a book on the stability and stabilization of time-delay systems emerged in the spring of 1996.

The two editors, Luc Dugard and Erik I. Verriest, participated in a French colloquium on "Analyse et commande des systèmes avec retard" organized in Nantes in the framework of GDR CNRS "Automatique". Many other authors contributing to this book also participated to this colloquium and the audience was enlarged with researchers from different countries. It is noticeable that most authors also participate to the 4th European Control Conference, Brussels, July 1-4, 1997, with an invited session dedicated to the stability and stabilization of continuous time delay systems.

The study of continuous-time delay systems has known a growing interest, in the past decade, in the automatic control community. Time-delay systems can be "tackled" from many points of view. In particular, the models of such systems can be considered as evolution in abstract systems, differential equations on rings or modulus, or as functional differential equations.

Surprisingly, only few books and monographies are dedicated to this subject. Motivated and encouraged by the enthusiasm of Silviu-Iulian Niculescu, who completed his Ph.D. dissertation on the stability and stabilization of continuous-time delay systems in early 1996, we decided to appeal to some specialists, recognized in the field, to edit a book on the subject, restricted to well-defined points. The book finds a niche in the time-delay system literature and should allow the interested reader to acquire a general idea of the problems posed by the stability and the stabilization of time-delay systems, as well as the various approaches and tools used to study and solve these problems. The book is characterized by a well defined spectrum. Stability analysis is studied in the first several chapters. Numerical aspects follow in the next ones. The stabilization problems are examined later, with some extensions to robustness issues and nonlinear aspects. This provides a coherent unity to the book.

The book consists of 14 chapters. The first chapter entitled "Stability and Robust Stability of Time-Delay Systems: A Guided Tour" is written by S.-I. Niculescu, E. I. Verriest, L. Dugard and J.-M. Dion. This chapter is intended to provide the reader with basic ideas on the various approaches to study the stability of time-delay systems. In particular, frequency domain and time domain approaches yield stability criteria which can be delay dependent or delay independent. Some extensions are given for the robust stability of uncertain time-delay systems. The stabilization aspects are not directly tackled in this chapter, but the results can be used for the study of the stability of the closed-loop time-delay systems.

The second chapter entitled "Convex Directions for Stable Polynomials and Quasipolynomials: A Survey of Recent Results" is written by L. Atanassova, D. Hinrichsen and V. L. Kharitonov. This chapter presents very recent results on robust stability of time-delay systems in the frequency framework. The notion of convex directions for stable polynomials and quasipolynomials is largely used here and applied to time-delay systems. This is a nice extension of some previous results on the stability of a polytope of polynomials or quasipolynomials, using the Edge Theorem.

The third chapter entitled "Delay-Independent Stability of Linear Neutral Systems: A Riccati Equation Approach" is written by E. I. Verriest and S. I. Niculescu. The linear neutral time-delay systems form a particular class of time-delay systems. The derivative of the delayed state appears in the system equation. Delay independent stability conditions are given in terms of some appropriate Riccati matrix equation coupled with a Lyapunov equation. The existence of solutions to these equations can be expressed in terms of feasibility of linear matrix inequalities (LMI).

The fourth chapter entitled "Robust Stability and Stabilization of Time-Delay Systems via Integral Quadratic Constraint Approach" is written by M. Fu, H. Li and S.-I. Niculescu. This chapter is devoted to robustness aspects using the integral quadratic criterion (IQC) approach. The stability conditions are expressed in terms of linear matrix inequalities. Based on these results, design procedures are given for the robust stabilization problem and explicit controller formulas are provided for static state feedback.

The fifth chapter entitled "Graphical Test for Robust Stability with Distributed Delayed Feedback" is written by E. I. Verriest. This chapter analyzes the performance degradation and stability margins for delay perturbations of a nominal state feedback control. Some conditions for stability of time-delay systems are re-interpreted as robust stability conditions and give frequency domain criteria. These criteria lead to interesting graphical methods that allow to derive stability margins in a straightforward way, in the spirit of the Nyquist criterion.

The sixth chapter entitled "Numerics of the Stability Exponent and Eigenvalue Abscissas of a Matrix Delay System" is written by J. Louisell. In this chapter, a method is presented, which is based on the analysis of the endpoint values of the solution to a functional equation occurring in the Lyapunov theory of delay equations. The existence of the solution to this functional equation is investigated in depth. A computational method is provided, that allows a very accurate determination of the system stability exponent.

The seventh chapter entitled "Moving Average for Period Delay Differential and Difference Equations" is written by B. Lehman and S. Weibel. This chapter extends some seminal works on the theory of averaging. In particular, it is shown that the delay must not be neglected in the averaged system to better approximate the dynamics of the original system. Two simple applications validate the developed theory on periodic delay differential and delay difference equations.

The eighth chapter entitled "On Rational Stabilizing Controllers for Interval Delay Systems" is written by L. Naimark, J. Kogan, A. Leizarowitz and E. Zeheb. This chapter is mainly concerned by the question of stabilizability of systems with an interval delay and fixed coefficients by rational controllers and intends to explain how to design these stabilizing controllers. Robustness aspects are also considered, when uncertainty is assumed on the coefficients of the rational transfer function.

The ninth chapter entitled "Stabilization of Linear and Nonlinear Systems with Time Delay" is written by W. M. Haddad, V. Kapila and C. Abdallah. This chapter is concerned with the design of fixed order dynamic feedback compensators for the linear case . Static full state feedback controllers are obtained for the nonlinear systems under sufficient conditions. In both cases, delay independent conditions are provided.

The tenth chapter entitled "Nonlinear Time Delay Systems: Tools for a Quantitative Approach to Stabilization" is written by J. P. Richard, A. Goubet-Bartholomeüs, P. A. Tchangani and M. Dambrine. In this chapter, a fairly general study is made. Qualitative and quantitative stability results are provided. The use of the comparison approach linked with vector Lyapunov functions appears as a simple and powerful tool for the study of nonlinear systems.

The eleventh chapter entitled "Output Feedback Stabilization of Linear Time-Delay Systems" is written by X. Li and C. E. de Souza. A delay dependent method is developed for designing linear dynamic output feedback controllers. Robustness considerations are given for uncertain polytopic systems and efficient numerical procedures are provided.

The twelfth chapter, entitled "Robust Control of Systems with A Single Input Lag" is written by G. Tadmor. This chapter develops a state-space design methodology for H_∞ problems and gap optimization in systems with a single input lag.

The thirteenth chapter entitled "Robust Guaranteed Cost Control for Uncertain Linear Time-Delay Systems" is written by H. Li, S.-I. Niculescu, L. Dugard and J.-M. Dion. The problem of stabilization of time-delay systems with linear fractional uncertainty is studied using the linear matrix inequality techniques. Some specific problems are considered, in particular, the case of mixed state and input delays. The guaranteed cost control problem is ensured through the feasibility of LMIs.

The fourteenth chapter entitled "Local Stabilization of Continuous Time-Delay Systems with Bounded Inputs" is written by S. Tarbouriech. The objective is to determine some domains of safe admissible states for which the stability of the saturated closed-loop system is guaranteed. The domains are obtained from an optimization linear program. Conditions are given in terms of solutions of finite dimensional algebraic Riccati equations.

This book gathers a fairly wide number of approaches, methods and tools for the (robust) stability analysis and the (robust) stabilization of time-delay systems. It makes the state of the art in the field and provides the readers with implementation and numerical issues as well as worked examples. This book is then a valuable tool for the control community and for the engineer who wants to acquire both the basic notions on the subject and some more advanced stability and stabilization results. Each chapter contains its own notations and definitions, and is provided with a rich bibliography which allows the readers to examine thoroughly a particular point.

Acknowledgement

The editors would like to acknowledge those who participated in this adventure. First the book would not have been possible without the enthusiasm and the excellent work of the twenty-seven authors involved in the fourteen chapters. They took much of their time to offer high quality contributions, witnessing their expertise in the field. The editors are also indebted to Springer for having made it possible to publish the book. In particular, Professor M. Thoma answered positively to our proposal for this book and Michael Jones, Editorial Assistant, Springer-Verlag London Ltd., gave excellent assistance for the preparation of the book. We thank very sincerely Professor J. K. Hale for his foreword, and for the stimulating discussions and many suggestions he gave - on the Georgia Institute of Technology side - that helped shape the theory. Thanks also to the many participants of the delay system workshop organized at the Center for Dynamical Systems and Nonlinear Studies of the Georgia Institute of Technology 1993-94.

Special thanks to Dr. Silviu-Iulian Niculescu and Dr. Huaizhong Li, postdoc at LAG, Grenoble. Silviu-Iulian played a major role in the completion of the book. His influence can be felt in many chapters. Huaizhong not only contributed to the scientific aspects but has also spent valuable time and energy putting the final manuscript in the right format with exemplary efficiency and LaTeX know how.

LUC DUGARD AND ERIK. I. VERRIEST

List of Contributors

C. T. Abdallah	Department of Electrical and Computer Engineering University of New Mexico Albuquerque, NM 87131, USA
L. Atanassova	Institut für Dynamische Systeme Universität Bremen D-28334 Bremen
M. Dambrine	LAIL, URA CNRS 1440 Ecole Centrale de Lille BP 48, 59651 Villeneuve d'Ascq Cedex, France
C. E. de Souza	National Laboratory for Scientific Computing LNCC/CNPq, Rua Lauro Müller 455, 22290-160 Rio de Janeiro, RJ, Brazil
J.-M. Dion	Laboratoire d'Automatique de Grenoble ENSIEG, BP 46 38402 Saint Martin d'Hères, France
L. Dugard	Laboratoire d'Automatique de Grenoble ENSIEG, BP 46 38402 Saint Martin d'Hères, France
M. Fu	Department of Electrical & Computer Engineering The University of Newcastle NSW 2308 Australia
A. Goubet -Bartholomeüs	LAIL, URA CNRS 1440 Ecole Centrale de Lille BP 48, 59651 Villeneuve d'Ascq Cedex, France
W. M. Haddad	School of Aerospace Engineering Georgia Institute of Technology, Atlanta GA 30332-0150, USA
D. Hinrichsen	Institut für Dynamische Systeme Universität Bremen D-28334 Bremen, Germany
V. Kapila	Dept. of Mech., Aerospace, and Manufacturing Engg. Polytechnic University Brooklyn, NY 11201, USA
V.L. Kharitonov	CINVESTAV-IPN Control Automatico A.P. 14-740 07000 Mexico D.F, Mexico

J. Kogan	Department of Mathematics and Statistics University of Maryland Baltimore, Maryland 21228, USA
B. Lehman	Dept. of Electrical and Computer Engineering Northeastern University Boston, MA 02115, USA
A. Leizarowitz	Department of Mathematics Technion - Israel Institute of Technology Haifa 32000, Israel
H. Li	Laboratoire d'Automatique de Grenoble ENSIEG, BP 46 38402 Saint Martin d'Hères, France
X. Li	Department of Electrical and Computer Engineering The University of Newcastle NSW 2308, Australia
J. Louisell	Department of Mathematics University of Southern Colorado Pueblo, CO 81001, USA
L. Naimark	Department of Electrical Engineering Technion - Israel Institute of Technology Haifa 32000, Israel
S.-I. Niculescu	Applied Math. Laboratory, ENSTA 32, boulevard Victor 75739, Paris, Cedex 15, France
J. P. Richard	LAIL, URA CNRS 1440 Ecole Centrale de Lille BP 48, 59651 Villeneuve d'Ascq Cedex, France
G. Tadmor	ECE Department Northeastern University Boston, MA 02115, USA
S. Tarbouriech	L.A.A.S - C.N.R.S. 7, Avenue du Colonel Roche 31077 Toulouse Cedex 4, France
P. A. Tchangani	LAIL, URA CNRS 1440 Ecole Centrale de Lille BP 48, 59651 Villeneuve d'Ascq Cedex, France
S. Weibel	Dept. of Electrical and Computer Engineering Northeastern University Boston, MA 02115, USA

E. I. Verriest School of Electrical and Computer Engineering
Georgia Institute of Technology, Atlanta
GA 30332-0250, USA

E. Zeheb Department of Electrical Engineering
Technion - Israel Institute of Technology
Haifa 32000, Israel

Table of Contents

Chapter 14 Local Stabilization of Continuous-time Delay Systems with Bounded Input 302

Stability and Robust Stability of Time-Delay Systems: A Guided Tour

Silviu-Iulian Niculescu* [1], Erik I. Verriest[2], Luc Dugard[3] and Jean-Michel Dion[3]

[1] Applied Math. Laboratory, ENSTA, 32, boulevard Victor, 75739, Paris, Cedex 15, France
silviu@ensta.fr

[2] School of Electrical and Computer Engineering, Georgia Institute of Technology, Atlanta, GA 30332-0250, USA
erik.verriest@ee.gatech.edu

[3] Laboratoire d'Automatique de Grenoble, ENSIEG, BP 46, 38402 Saint Martin d'Hères, France
dugard, dion@lag.ensieg.inpg.fr

Abstract. In this chapter, some recent stability and robust stability results on linear time-delay systems are outlined. The goal of this guided tour is to give (without entering the details) a wide overview of the state of the art of the techniques encountered in time-delay system stability problems. In particular, two specific stability problems with respect to delay (*delay-independent* and respectively *delay-dependent*) are analyzed and some references where the reader can find more details and proofs are pointed out. The references list is not intended to give a complete literature survey, but rather to be a source for a more complete bibliography. In order to simplify the presentation several examples have been considered.

1 Introduction

1.1 Basic ideas

In the mathematical description of a physical process, one generally assumes that the behaviour of the considered process depends only on the *present* (in the usual sense) state, asumption which is verified for a large class of dynamical systems.

However, there exist situations (for example, material or information transport), where this assumption is not satisfied and the use of a "classical" model in systems analysis and design may lead to poor performance. Moreover, small delays may lead to destabilization [73]. In such cases, it is better to consider that the system's behaviour includes also information on the *former* states. These systems are called *time-delay systems*.

---

* On leave from Laboratoire d'Automatique de Grenoble (France); Also with the Department of Automatic Control, University "Politehnica" Bucharest (Romania)

Following [110], the existence of a delay in a system model could have several causes, as, for example: the measure of a system variable, the physical nature of a system's composant or a signal transmission. A classification of *delays* with respect to the physical systems where they are encountered could be (see [93]): technological, transmission or information delay, respectively. Without discussing these causes and classifications, natural questions arise: *How to model? How to analyze the stability?* or *How to control such systems?*

Delay systems representations There are mainly *three ways* to model such systems: as evolutions in abstract spaces (infinite dimensional systems), as functional differential equations or as differential equations over a ring or module.

Evolutions in abstract spaces In this case, the delay system class is embedded in a larger class of linear systems for which the evolution is described by appropriate (bounded or not bounded) operators in infinite dimensional spaces [35, 36, 85] (see also [69] for a geometric theory or [44] for an operator theory framework). From a system theory point of view, this approach needs the introduction of appropriate concepts of stabilizability, observability, detectability, etc. Although this way is very general, the corresponding methods are not always easy to apply for specific problems. For further remarks and comments see also [13].

Functional differential equations In this case, we may have two different ways to consider a delay system, as evolutions in a *finite-dimensional* space [70, 93], or in a *function space* [70]. Some remarks on the effect of a delay on the boundedness, stability, continuation, integrability or oscillations can be found in [23]. From a system theory point of view, one can use classical concepts specific to "finite-dimensional" linear systems, or introduce "new" concepts more appropriate to a function space interpretation (see, for example, [111, 112, 159]). One of the possible advantages of such a modelling way lies in its facility to treat "infinite-dimensional" problems using "finite-dimensional" tools, with a *trade-off* to be paid on the conservatism of the obtained results.

Differential equations over rings or modules In this case, we have interesting "structural" properties, as stabilizability and observability [86, 166, 51, 65, 127]. In our opinion, these interpretations are better adapted for the cases when explicit information on the delay size is not needed. Further remarks, comparisons and examples can be found in [128, 161, 153].

Notice that, in some cases, classes of infinite-dimensional systems have nice representations (simpler) as functional differential equations (see, for example, Hale and Lunel [70] and the references therein).

Each described way has some advantages or inconvenients depending on the considered problem to be handled. Thus, for example, if we are interested in the *stabilization* problem, it is of some interest to know *whether* or *not* a *finite-dimensional* controller is *sufficient* to stabilize a given delay system.

Infinite-dimensional versus finite-dimensional Since the delay systems are in the *infinite dimensional* system class, we have two ways to analyze them, using tools specific to finite-dimensional systems:

- **Finite-dimensional approximations**, as for example: Padé approximants [98], Fourier-Laguerre series [151] or optimal Hankel rational approximants [56]. Specific problems for such approximants are the choice of their dimension and the stability [158, 64]. This approach will not be considered explicitly in this chapter.
- **Finite-dimensional interpretations**. As mentioned before (in the functional differential equation framework), a delay system can be described either as an evolution in a finite-dimensional space or in a function space. In each case and under appropriate assumptions, the considered "infinite-dimensional" problem may be "transformed" into a finite-dimensional one. It will be shown further how these transformations are done and what, in some cases, their restrictions may be. We do not claim here that the proposed approaches are the best, but emphasize that they are only alternative "ways" to analyze complex problems. Notice however that in some cases, the presented results are *necessary and sufficient* conditions.

Functional differential equations. A "short" historical perspective The study of functional differential equations started long before 1900 (see the works of Bernoulli, Euler, Condorcet or Volterra), but the basics and the mathematical formulation were developed in the 20th century. Thus, the notion of a *functional differential equation* was introduced by Myshkis [129] in 1949 as a *differential equation involving the function "$x(t)$" and its derivatives not only in the argument "t," (called time) but in several values of "t."* Thus, a classification of such differential equations includes: retarded, and neutral equations with point or distributed delays.

Without being exhaustive we cite some of the books which have marked the study of such systems (in the last 40 years) and which could be seen as basic works for this framework: Bellman and Cooke [12] (frequency based approach, entire functions), Krasovskii [95] (time-domain approach, extension of the Lyapunov second method to functional differential equations), Halanay [67] (extension of the Popov theory to time-delay systems), Răsvan [154] (absolute stability of time-delay systems), Lakshmikantam and Leela [97] (differential inequalities and comparison theorems), Burton [23] (refinements of the Lyapunov-Krasovskii theory and periodic solutions), Kolmanovskii and Nosov [93] (a good introduction to the stability of functional differential equations and a lot of examples) or Diekman *et al.* [44] ("small" solutions, operator theory approach). Some recent comprehensive introductions are Górecki, Fuksa, Gabrowski and Korytowski [60], Marshall et al. [118], Kolmanovskii and Myshkis [94] and Hale and Lunel [70]. The first volume of Bensoussan et al. [13] presents a detailed account of the *product space approach*, and the construction of the *structural operators*.

Delay as a parameter In the sequel, we consider a class of time-delay systems described by *linear delay-differential* equations including "point delays". The *delays* are seen as parameters of the system, and we are interested in analyzing the *stability property* with respect to them. The idea is to give *characterizations* of the *stability regions* for linear systems with delayed state in terms of delays. To the best authors' knowledge, this problem is *still open* (see [44] and the references therein), but in some cases, complete characterizations can be given. In this sense, two notions of *stability* are introduced: *delay-independent* and *delay-dependent* stability, respectively. The time-varying as well as the multiple delays cases are also considered, using some appropriate delay-independent / delay-dependent notions. Several examples including the scalar case are also presented.

Notations The following notations will be used throughout the chapter. $\mathbb{R}$ ($\mathbb{C}$) denotes the set of real (complex) numbers, $\mathbb{R}^* = \mathbb{R} - \{0\}$, $\mathbb{C}^* = \mathbb{C} - \{0\}$, $\mathcal{C}(0,1)$ denotes the unit circle in the complex plane. For a complex number $z \in \mathbb{C}$, $\bar{z}$ denotes its complex conjugate. $\mathbb{R}^+$ is the set of non-negative real numbers, $j\mathbb{R}$ denotes the imaginary axis of the complex plane, $j\mathbb{R}^* = j\mathbb{R} - \{0\}$, $\mathbb{R}^n$ denotes the n dimensional Euclidean space, and $\mathbb{R}^{n\times m}$ ($\mathbb{C}^{n\times m}$) denotes the set of all $n \times m$ real (complex) matrices. $\Lambda(M)$ represents the set of eigenvalues (spectrum) of the complex matrix $M \in \mathbb{C}^{n\times n}$. $diag(A, B)$ denotes the matrix $\begin{bmatrix} A & 0 \\ 0 & B \end{bmatrix}$, where the zero blocks have appropriate dimensions for the matrices $A \in \mathbb{C}^{m_1\times n_1}$, $B \in \mathbb{C}^{m_2\times n_2}$. $\mathbb{C}^+$ ($\mathbb{C}^-$) denotes the open right (left) half complex plan. $In(M) = (\pi(M), \nu(M), \delta(M))$ is the inertia of the complex matrix $M \in \mathbb{C}^{n\times n}$, where $\pi(M)$, $\nu(M)$ and $\delta(M)$ denote the number of eigenvalues with negative ($\mathbb{C}^-$), positive ($\mathbb{C}^+$) and zero real parts ($j\mathbb{R}$). $\mu(A)$ with $A \in \mathbb{R}^{n\times n}$ denotes the matrix measure of A given by: $\mu(A) = \lim_{h\to O+} \frac{\|I_n + hA\| - 1}{h}$. $\mathcal{C}_{n,\tau} = \mathcal{C}([-\tau, 0], \mathbb{R}^n)$ denotes the Banach space of continuous vector functions mapping the interval $[-\tau, 0]$ into $\mathbb{R}^n$ with the topology of uniform convergence. The following norms will be used: $\|\cdot\|$ refers to the Euclidean vector norm; $\| \phi \|_c = \sup_{-\tau\leq t\leq 0} \| \phi(t) \|$ stands for the norm of a function $\phi \in \mathcal{C}_{n,\tau}$. Moreover, we denote by $\mathcal{C}^v_{n,\tau}$ the set defined by $\mathcal{C}^v_{n,\tau} = \{\phi \in \mathcal{C}_{n,\tau} : \| \phi \|_c < v\}$, where v is a positive real number.

1.2 Linear delay systems class

In this chapter, a linear time-delay system is described by a functional differential equation of the form:

$$\dot{x}(t) = Ax(t) + \sum_{i=1}^{n_d} A_{di}x(t - \tau_i), \tag{1.1}$$

with an appropriate initial condition of the form (if the delays are supposed to be constant):

$$x(t_0+\theta) = \phi(\theta), \quad \theta \in [-\bar{\tau}, 0], \quad \bar{\tau} = \max_{i=\overline{1,n_d}} \tau_i, \quad (t_0, \phi) \in \mathbb{R}^+ \times C^v_{n,\bar{\tau}} \quad (1.2)$$

For such systems, one may associate a *triplet* Σ of the form $\Sigma = (A, A_d, \tau)$, where:

$$\begin{cases} A_d = [A_{d1}, \ldots, A_{dn_d}], \\ \tau = [\tau_1, \ldots \tau_{n_d}], \end{cases}$$

which allows the description of the considered system in the parameter space (A and A_d correspond to the *present* and respectively, *former* state). Throughout this chapter, Σ asymptotically stable means that the system (1.1)-(1.2) is asymptotically stable.

The case $n_d = 1$ is known as the *single delay* case. For $n_d \geq 2$, we shall also consider the *delay parameter space* $(\tau_1, \ldots \tau_{n_d}) \in \mathbb{R}^{n_d}$. If one has *multiple delays* ($n_d \geq 2$), we may have a particular situation — the *commensurable* case (i.e. there exists a delay value τ, such that all the delays τ_i are rational "multipliers" of τ). We shall see later that there are some similarities between this case and the single delay case.

Further specifications are given when the delays are continuous (or piecewise continuous) time-varying, but bounded functions. Thus, for the *single delay* case, we may have two different situations:

- continuous (or piecewise continuous) bounded time-varying delay function $\tau : \mathbb{R}^+ \mapsto \mathbb{R}$, $\tau(t) \leq \bar{\tau}$ for any $t \in \mathbb{R}^+$; in this case, the initial condition (1.2) becomes:

  $$x(t_0+\theta) = \phi(\theta), \quad \theta \in \mathcal{E}_{t_0}, \quad (1.3)$$

  where $\phi : \mathcal{E}_{t_0} \mapsto \mathbb{R}^n$ is a continuous and bounded vector-valued function and the definition domain $\mathcal{E}_{t_0}$ is given by:

  $$\mathcal{E}_{t_0} = \{t \in \mathbb{R} : t = \eta - \tau(\eta) \leq t_0, \eta \geq 0\}. \quad (1.4)$$

- continuous with bounded derivatives time-varying delay function, i.e. one needs the following natural supplementary condition:

  $$\dot{\tau}(t) \leq \beta < 1. \quad (1.5)$$

Uncertain linear delay systems As written in [206], "the term uncertainty refers to the differences or errors between models and reality." In this chapter we consider either, only *parameter uncertainty*, i.e. when the parameters (A, A_d) of the system are *not well known*, or parameter and delay time uncertainty, when (A, A_d, τ) are imprecise.

In this context, an *uncertain linear delay system* can be defined as a *triplet* $(\Sigma, \mathcal{D}, \Phi)$, where

- Σ is the *nominal* system (free of uncertainties), i.e. the triplet (A, A_d, τ) and
- the *pair* $(\mathcal{D}, \Phi)$ describes the uncertainty, where $\mathcal{D}$ is the *perturbation set*, i.e. a domain in which the uncertainty (physical parameters in our case) is known to lie and Φ is a *mapping* taking values from $\mathcal{D}$ which describes the way the uncertainty "acts" on the system's parameters.

This description allows a general framework for robustness stability issues. Notice that this definition is more general than the cases treated here, but we do not intend to review all the robustness problems in this framework.

We shall give also some classifications of delay systems involving parameter uncertainty. Notice that we make a *distinction* between the delay as a parameter and the others parameters (the pair (A, A_d)) of the systems. This aspect will be clarified next.

1.3 Delay-independent versus delay-dependent stability

In order to better fix the notions that will be used in this chapter, we consider the following two different cases (some stability notions and general results can be found in Appendix):

Single delay Following Mori [123], we have two different kinds of asymptotic stability for systems of the form (1.1)-(1.2), depending on the information on the delay size in the property:

- **Delay-independent**, i.e. the property holds for all positive (and finite) values of the delays. Hence this automatically implies robustness with respect to the delay time.
- **Delay-dependent**, i.e. the stability is preserved for some values of delays and the system is unstable for other values.

If the *delay-independent* notion is clear, the *delay-dependent* case has to be better specified. For the sake of simplicity and in order to have no ambiguity, we introduce the following *assumption*:

Assumption 1 *The system (1.1)-(1.2) free of delay ($\tau \equiv 0$) is asymptotically stable.*

With this assumption, we shall say that the system is *delay-dependent* stable, if it satisfies Assumption 1 and is unstable for some values of $\tau > 0$. It is easy to see that these notions are complementary one to the other. Thus, the *problem* considered here is of the form:

Problem 1 *Determine if a delay system of the form (1.1)-(1.2) satisfying Assumption 1 is delay-independent asymptotically stable or not. If not, find an optimal (sub-optimal) bound on the delay size which still ensures the stability property.*

Notice that we only consider here, in the *delay-dependent* case, the interval containing $\tau = 0$, i.e. of the form $[0, \tau^*)$, independently of possible other intervals (for general "delay-intervals" of the form $(\underline{\tau}, \overline{\tau})$, with $\underline{\tau} > 0$ see also [144]).

Remark 1. A natural question arises here: *Is the considered problem well posed or not?* Indeed, the system free of delay is finite-dimensional and the associated characteristic equation has a finite number of eigenvalues in the complex plane. The system with delayed state is an infinite-dimensional one and its characteristic equation has an infinite number of eigenvalues [165]. The answer is positive and we shall see later why.

For the time-varying delay case, *delay-independent* stability means that the stability property holds for any continuous (or piece-wise continuous) and bounded time-varying delay function, with any positive and finite bound in the specified class. The *delay-dependent* case could be defined by analogy.

Multiple delays The *delay-independent* and *delay-dependent* stability notions can be easily extended to this case, by taking into account the behaviour with respect to each delay. We have a particular "mixed" case, which could be called *delay-independent / delay-dependent*: delay-dependent stability in one delay (or several) and delay-independent stability in others (at least one) and all the possible combinations.

If the problem is posed in the *delay-parameter space*, it is clear that we may have two different *delays sets*:

Unbounded sets including the delays-independent and delay-independent / delay-dependent cases;

Bounded sets including only the delays-dependent case.

In conclusion, if a system is not *delays-independent* stable, two situations may occur: there exists at least one delay in which the system is *delay-independent*, and *delay-dependent* in all the others (the so-called "mixed" case in the unbounded sets class), or the stability is of *delay-dependent* type in each delay (the bounded sets class). Using the same formalism as in the single delay case, we can consider the following problem:

Problem 2 *Determine if a delay system of the form (1.1)-(1.2) satisfying Assumption 1 is delays-independent asymptotically stable or not. If not, find an optimal (sub-optimal, convex or not) region in the delay-parameter space which still ensures the stability property.*

If the delays are *commensurable*, we have the same *delay-independent* and *delay-dependent* notions as in the single delay case.

The *time-varying delay* cases can be defined similarly to the single delay case, taking into account that the delay-parameter space is not Euclidian, but a function space. In order to simplify the presentation, this case is not explicitly treated here, but it will be mentioned when some proposed results also hold in this situation.

Robustness issues The "delay-independent / delay-dependent" problems defined previously could be seen as robustness problems with respect to *delays*. The problem becomes more difficult if in addition there is uncertainty in (A, A_d). In this case we shall use an uncertain system representation of the form $\Sigma_r = (\Sigma, \mathcal{D}, \Phi)$ instead of the triplet Σ form, and all the "delay-independent / delay-dependent" notions, concepts and problems can be defined by similarity. For the sake of simplicity, we do not detail all them here.

In order to distinguish between all the considered cases, we have used the terms "delay-independent / delay-dependent" for stability analysis of Σ and the "robust delay-independent / robust-delay dependent" for the case $(\Sigma, \mathcal{D}, \Phi)$.

The term "robust" is associated to the fact that the considered property holds for any admissible uncertainty in the form $(\mathcal{D}, \Phi)$. Also if we have time-varying uncertainity we should use the "uniform asymptotic" stability concept instead of the "asymptotic" one. Notice that, for the linear system with constant parameters and without uncertainty, the notions of asymptotic, uniform asymptotic and exponential stability are equivalent (see [93] and the references therein). Only the uniform asymptotic and exponential stability notions for functional differential equations (of retarded type) are given in the appendix.

For example, for an uncertain system $(\Sigma, \mathcal{D}, \Phi)$ with a *single delay* τ, we shall say that the system is ***robustly delay-independent stable*** if the trivial solution of the associated functional differential equation is ***uniformly asymptotically stable*** for all positive values of τ and all the admissible uncertainties in the $(\mathcal{D}, \Phi)$ form. Thus, we consider implicitely that the uncertain system free of delay is ***robustly stable***[4] The associated problem can be formulated as follows:

Problem 3 *Determine if a delay system of the form $(\Sigma, \mathcal{D}, \Phi)$ satisfying Assumption 1 is robustly delay-independent stable or not. If not, find an optimal (sub-optimal) bound on the delay size which still ensures the stability property.*

Stochastic perturbations Another aspect of robustness is the sensitivity of the stability against stochastic perturbations. Stochastic models are useful when the perturbations are not directly measurable (and predictable) and fluctuate with time. With additive noise, an otherwise stable equilibrium is no longer an equilibrium solution. However there may be interest in determining the existence of a stable *distribution*. If the noise is state dependent (one often refers to such models as *bilinear* models), the deterministic equilibrium may also be a stochastic equilibrium state. In this case one may want to investigate the *stochastic* stability of this equilibrium. There are very many details and since we shall not consider stochastic issues in this book, we limit the citation to a few references. A starting point to study aspects of stochastic delay systems is the monograph by Mohammed [121] and the volumes by Mao [113, 115]. The stochastic stability of the equilibrium solution of stochastic delay equations has been studied,

[4] Due to the framework presented here, we have not considered the cases in which the delay term may induce robust stability if the system free of delay does not satisfy this property.

for various notions of stablity, by Mao [114] (asymptotic stability in probability, moment stability and exponential stability), Mohammed [122] (moment stability) and Nechayava and Khusainov [130] (exponential stability). An approach extending the Riccati equation criterion discussed in this book can be found in [52, 183, 184, 187].

1.4 Purpose of the chapter

In this chapter, some frequency and time-domain methods for analyzing stability of linear delay systems are presented and their advantages and disadvantages illustrated. The intention of the authors is twofold.

Firstly, we want to ***introduce*** the stability analysis for time-delay systems and help the reader to understand some of the issues treated in the next chapters dedicated to similar topics. In this sense, we intend to present a specific problem arising in the stability analysis of such systems: how ***robust*** is the system stability with respect to the ***delay*** term, and thus, to give a characterization of stability properties in terms of ***delays***. We have also considered the robustness analysis with respect to the other parameters of the system, i.e. uncertainty in the "present" and "delayed" matrices (A respectively A_d).

Secondly, we want to ***give*** a (non-exhaustive) guided tour of the existing results in the literature and ***present*** some interesting results obtained in the control literature on the proposed topics.

In order to simplify the presentation, we considered it necessary to include some simple examples for which the stability results are well known. In some cases, they illustrate the conservativeness of the existing results with respect to necessary and sufficient conditions. Note however that in general, such conclusions are difficult to draw. Also, we do not give complete proofs, but only some ideas of the presented results.

1.5 Outline

The chapter is organized as follows: two examples from the control literature are given in Section 2 in order to motivate the considered "delay-independent - delay-dependent" framework. In section 3, we introduce appropriate stability sets related to the considered concepts. Section 4 is dedicated to the frequency domain approach and related techniques. The time-domain approach including the Lyapunov methods and the comparison principles is presented in Section 5. Other stability results using different frameworks are given in Section 6. The robustness issues and related analysis techniques are given in Section 7. The examples given in Section 2 are reconsidered in Section 8 in order to present some applications of the techniques considered in this chapter. Some concluding remarks end the chapter. A short appendix includes some basic notions on functional differential equations stability concepts and results (the Lyapunov-Krasovskii and Razumikhin theorems).

2 Examples

In this section, some examples from the control literature are developed in order to show where one may encounter *delay-independent* and *delay-dependent* stability properties respectively. The detailed examples (chemical industry, neural network) are reconsidered later, when some of the techniques presented in this chapter are used to prove the corresponding stability results.

2.1 Chemical Industry

Consider a first order, exothermic, irreversible reaction: $A \mapsto B$ [102] [103]. Since, in practice, the conversion from A to B is not complete, one classical technique uses a recycle stream (which increases overall conversion, reduces costs of the reaction, etc.). In order to recycle, the output must be separated from the input and must flow through some length of pipe. This process does not take place "instantaneously"; it requires some "transport" time from the output to the input, and thus one may consider a system model involving a *transport delay*.

Suppose now that the unreacted A has a recycle flow rate $(1-\lambda)q$ and τ is the transport delay. Then the material and energy balances are described by a dynamical system including delayed states of the form:

$$\begin{cases} \dfrac{dA(t)}{dt} = \dfrac{q}{V}\left[\lambda A_0 + (1-\lambda)A(t-\tau) - A(t)\right] - K_0 e^{-\frac{Q}{T}} A(t) \\ \dfrac{dT(t)}{dt} = \dfrac{1}{V}\left[\lambda T_0 + (1-\lambda)T(t-\tau) - T(t)\right]\dfrac{\Delta H}{C\rho} - K_0 e^{-\frac{Q}{T}} A(t) \\ -\dfrac{1}{VC\rho}U(T(t) - T_w) \end{cases} \tag{2.1}$$

where $A(t)$ is the concentration of the component A, $T(t)$ is the temperature and λ is the recycle coefficient, which satisfies the conditions: $\lambda \in [0,1]$. The limits 1 and 0 correspond to no recycle stream and to a complete recycle, respectively. The case when we have no delay in the recycle stream ($\tau = 0$) has been completely treated in [152] (see also [15]).

For such a system, it is proved that if the steady states of the system without delay is locally asymptotically stable, then this property holds also for the system (2.1) and furthermore the local stability is of *delay-independent* type [102], [103].

We shall see later that this property holds also if we suppose to have a time-varying delay instead of a constant one.

2.2 Neural Networks

In a continuous (so called analog) neural Hopfield network [79], the state of each 'unit' is described by a voltage u_i on the input of the ith neuron, and each neuron is characterized by an input *capacitance* C_i and a *transfer function* f_i. For describing the connections between the neurons, one uses the *connection matrix* T, whose elements are of the form $-\dfrac{1}{R_{ij}}$ (or $\dfrac{1}{R_{ij}}$) when the noninverting

(inverting) output of the unit j is connected to the input of the unit i through a ***resistance*** R_{ij}.

Suppose that all the units are identical: $C_i = C$, $f_i = f$, $R_i = R$ and that there exists a ***delay*** (due, for example, to the implementation the network using VLSI), then the following model (see also [9, 116, 200]):

$$\dot{x}_i(t) = -x_i(t) + \sum_{j=1}^{n} a_{ij} f[x_j(t-\tau)], \quad 1 \le i \le n. \tag{2.2}$$

gives a good description of the system's behaviour.

The ***associative memory***, one of the oldest applications of neural networks, consists in the capacity of the system to stock ("register") information which could be recovered not via an address as in a classical memory, but giving data (not necessarily a complete information) with respect to the informations registered (see also [92]). This notion is related to the stability property of the associated dynamical system.

In this context, it is important to know what is the effect of delay on the system's stability property. In general, it is supposed that the linearized system without delay is locally asymptotically stable.

For this model, using a frequency based approach (for the linearized equation), two different situations may occur: ***delay-independent*** stability or ***delay-induced instability*** [9, 10]. If the first notion was clearly stated in the previous section, the second one states the fact that the stability property is not satisfied for all values of the delay size, but only for an interval of the form $[0, \tau^*)$. Thus, this notion is identical to the ***delay-dependent*** notion stated before.

This example, as well as the chemical reaction presented before are completely analyzed later.

2.3 Other examples

A lot of various engineering processes (robotics, machine tool vibrations) including delays are given in [93, 110] (and the references therein).

Another representative area is in the control of ***flexible structure***. In fact, in this case a delay may occur in measuring the structural response and in applying the considered active control. The effect of such delays as well as comments and references can be found in [164, 41].

In all the examples considered before, the delay effect is a ***destabilizing*** one (see also [50] for the case of contact stability of position controllers). However, there exist some cases when the existence of a delay term ***may improve*** the stability properties [1]; Indeed, consider an oscillating system of the form:

$$\ddot{y}(t) + \omega_0^2 y(t) = u(t), \quad \omega_0 \in \mathbb{R}^*.$$

It is easy to see that there does not exist any stabilizing output feedback of the form $u(t) = ky(t)$, $k \in \mathbb{R}$, but the closed-loop system is asymptotically stable for $u(t) = ky(t-\tau)$, for some real k and $\tau > 0$.

The engineering field is not the only source of delay systems examples. Notice that the last decade has witnessed important advances on modelling physiological, ecological [59], population dynamics [96] or biomedical [109] dynamical systems using delay in their representation.

In this sense, models of the form:

$$\begin{cases} \dot{x}_1(t) = -a_1 x_1(t) - f_1(x_1(t-\tau_1), x_2(t-\tau_1)) \\ \dot{x}_2(t) = -a_2 x_2(t) - f_2(x_1(t-\tau_2), x_2(t-\tau_2)) \end{cases} \tag{2.3}$$

(similar to the neural network example) are used in the study of protein hormone regulation (see also [25] and the references therein) and of population dynamics (see [165]).

From historical point of view, the first engineering studies of such dynamical systems involving delayed states started in the 30s (see the paper of Callender and Stevenson, or the 'Editorial' paper of the review 'Engineering' mentioned in Răsvan [154] or in Kolmanovskii and Nosov [93]).

3 Stability sets in parameter space

It was mentioned in the previous section that the *commensurable delays* case could be treated in a similar manner to the *single delay* case. In order to give a unitary presentation, we directly treat the commensurable delays case (with constant delays) and we shall see that this one allows a complete recovery of the single delay case.

Thus, throughout this section, the triplet $\Sigma = (A, A_d, \tau)$ which describes the delay system (1.1)-(1.2) further satisfies the condition (via an appropriate permutation):

$$\begin{cases} \tau = [\tau_1 \dots \tau_{nd}] \\ \tau_k = k\tau_0, \quad \forall k = \overline{1, n_d} \end{cases} . \tag{3.1}$$

Introduce the set:

$$\mathcal{S}(r) \quad = \quad \{(A, A_d) \quad : \quad \Sigma \quad \text{asymptotically stable at} \quad \tau = r\} \tag{3.2}$$

It is easy to see that for $r = 0$, $\mathcal{S}(0)$ becomes:

$$\mathcal{S}(0) \quad = \quad \left\{(A, A_d) \quad : \quad A + \sum_{k=1}^{n_d} A_k \quad \text{is Hurwitz stable}\right\},$$

In conclusion, a pair (A, A_d) satisfies Assumption 1 if and only if $(A, A_d) \in \mathcal{S}(0)$.

3.1 On the continuity properties

In order to see if the problem 1 is *well-posed*, let us consider the sets $\mathcal{S}(0)$ and $\mathcal{S}(\varepsilon)$, with ε a sufficiently small positive number.

Consider the characteristic equation associated with (1.1)-(1.2):

$$det\left(sI_n - A - \sum_{k=1}^{n_d} A_k e^{-sk\tau_0}\right) = 0, \tag{3.3}$$

which is a ***transcendental equation*** [47] for $\tau_0 > 0$ and has an ***infinite*** number of solutions. Some results concerning the location of the roots of the transcendental equation (3.3) relatively to the imaginary axis can be found in [12, 44] (and the references therein). These roots have some interesting properties: the number of eigenvalues with $-\alpha$ ($\alpha > 0$, arbitrary) real part in the complex plane $\mathbb{C}$ is always ***finite*** and $-\infty$ is an ***accumulation point*** (i.e. there exists an infinite subsequence of roots $\{\lambda_i\}$, such that $\lim_{i\to\infty} Re(\lambda_i) = -\infty$). Notice however that this result is not true for general functional diffrential equations. Other remarks and comments can be found, for example [165] and the references therein.

Introduce now the "quantities":

$$\begin{cases} u_h = max\left\{Re(\lambda) \le 0 \quad : \quad det\left(\lambda I_n - A - \sum_{k=1}^{n_d} A_d e^{-\lambda kh}\right) = 0\right\} \\ l_h = min\left\{Re(\lambda) \ge 0 \quad : \quad det\left(\lambda I_n - A - \sum_{k=1}^{n_d} A_d e^{-\lambda kh}\right) = 0\right\} \end{cases}, \tag{3.4}$$

with $u_h = -\infty$ and $l_h = +\infty$ if the corresponding sets are empty ("u " for upper and "l" for lower). It is clear that these quantities give the real parts of the corresponding eigenvalues (if there exists any for l_h) "closest" to the imaginary axis $j\mathbb{R}$. Using a Datko's type argument [39], the numbers u_h and l_h continuously depend on h, on all the entries of the matrices A and A_k ($k = \overline{1, n_d}$), it follows that:

Proposition 1 [133] *Consider the system (1.1)-(1.2) satisfying Assumption 1. Then the following properties hold:*

1. *If $(A, A_d) \in \mathcal{S}(0)$, then there exists an $\varepsilon > 0$ sufficiently small such that $(A, A_d) \in \mathcal{S}(h)$ for all $h \in [0, \varepsilon]$.*
2. *If $(A, A_d) \in \mathcal{S}(0)$, and if there exists a τ_1 for which the triple (A, A_d, τ_1) is not stable, then there exists an ε, $0 < \varepsilon < \tau_1$, such that $(A, A_d) \in \mathcal{S}(h)$ for all $h \in [0, \varepsilon)$ and for $h = \varepsilon$ the characteristic equation (3.3) has roots on the imaginary axis.*

This proposition can be seen as the ***continuity stability property*** for system "free" or "not" of delay. Notice that the part 1 of this proposition is more general and holds also for time-varying delays or for several delays [133]. In this sense one could use a Lyapunov second method type argument to prove it. We shall see these cases later.

Remark 2. Instead of the stability property, we could use the ***hyperbolicity*** one (i.e. no eigenvalues on the imaginary axis) with associated "delay-independent / delay-dependent" notions [143]. Furthermore, the proposition still holds in a "delay-interval" setup, i.e. if some property (stability or hyperbolicity) holds for a given value $\tau_d > 0$, then there exists a delay-interval including this value such that the considered property is still satisfied and at for (at least one of) the delay margins, we have some eigenvalues on the imaginary axis.

3.2 Definitions and related remarks

We have the following definitions:

Definition 1 Delay-independent set. [133] The set $\mathcal{S}_\infty$ defined by

$$\mathcal{S}_\infty = \{(A, A_d) \ : \ \Sigma \ \text{asymptotically stable} \ \ \forall \tau_0 \geq 0\} \tag{3.5}$$

is called the ***delay-independent stability set*** in the parameter space (A, A_d).

If a triplet Σ satisfies the condition $(A, A_d) \in \mathcal{S}_\infty$, we shall say that the triplet is $\mathcal{S}_\infty$ stable..

Definition 2 Delay-dependent set. [133] The set $\mathcal{S}_\tau$ defined by

$$\begin{aligned} \mathcal{S}_\tau = \{(A, A_d) \ : \ \Sigma \ & \text{asymptotically stable} \ \ \forall \tau_0 \in [0, \tau^*) \\ & \text{and unstable for} \ \ \tau_0 = \tau^*\} \end{aligned} \tag{3.6}$$

is called the ***delay-dependent stability set*** in the parameter space (A, A_d).

If a triplet Σ satisfies the condition $(A, A_d) \in \mathcal{S}_\tau$, we shall say that the triplet is $\mathcal{S}_\tau$ stable.

Using the stability set $\mathcal{S}(r)$ definition, one has the following natural result:

Proposition 2 [133] *The following assertions hold:*

1. $\mathcal{S}_\infty = \cap_{r \in \mathbb{R}^+} \mathcal{S}(r)$.
2. $\mathcal{S}_\tau = \mathcal{S}(0) - \mathcal{S}_\infty$.

Using the definitions given here, we have a simple algebraic equivalent formulation of Problem 1 for the commensurable (constant) delays case:

Problem 4 *Determine the maximal cone included in all $\mathcal{S}(r)$ where r is real and positive. Furtheremore, if a triplet (A, A_d, τ) satisfying (3.1) is $\mathcal{S}_\tau$ stable, then find an optimal (sub-optimal) bound τ^*.*

Suboptimality indicates that the considered method guarantees the stability for all $\tau \in [0, \tau^*]$, but there is no information on the behaviour of the system for $\tau > \tau^*$. Although the "suboptimal" notion seems to be quite conservative, we emphasize that the proposed bounds are the "maximal allowable" ones within the corresponding framework. Several comments are given later for each case analyzed.

If the delay is time-varying, the stability notions are similar with $\mathcal{S}_{v,\infty}$ (or $\mathcal{S}_{v,\beta,\infty}$) and $\mathcal{S}_{v,\tau}$ (or $\mathcal{S}_{v,\beta,\tau}$) respectively. Thus:

$$\mathcal{S}_v(r) = \{(A, A_d) \ : \ \Sigma \ \text{uniformly asymptotically stable at} \ \tau \in \mathcal{V}(r)\} \tag{3.7}$$

and

$$\mathcal{S}_{v,\infty} = \{(A, A_d) \ : \ \Sigma \ \text{uniformly asymptotically stable} \ \forall r \geq 0 \text{ and } \forall \tau \in \mathcal{V}(r)\}, \tag{3.8}$$

where

$$\mathcal{V}(r) = \left\{\tau \in \mathcal{C}^0 \ : \ 0 \leq \tau(t) \leq r, \quad \forall t \in \mathbb{R}^+\right\}.$$

It is clear that $\mathcal{S}_v(0) = \mathcal{S}(0)$. The other cases can be defined by similarity. The following result is an extension of Proposition 2:

Proposition 3 [133] *The following assertions hold:*

1. $\mathcal{S}_{v,\infty} = \cap_{r \in \mathbb{R}^+} \mathcal{S}_v(r)$.

2. $\mathcal{S}_{v,\tau} = \mathcal{S}(0) - \mathcal{S}_{v,\infty}$.

For example, consider now only the *robust* "delay-independent / delay-dependent" notions for a system with a *single delay* τ including uncertainty in the matrices A and A_d. The definitions 1 and 2 become:

Definition 3 Robust delay-independent set. The set $\mathcal{S}_{robust,\infty}$ defined by

$$\mathcal{S}_{robust,\infty} = \{(A, A_d) \ : \ (\Sigma, \mathcal{D}, \Phi) \ \text{uniformly asymptotically stable} \ \forall \tau \geq 0 \text{ and all admissible uncertainty } (\mathcal{D}, \Phi)\} \tag{3.9}$$

is called the *robust delay-independent stability set* in the parameter space (A, A_d).

If a triplet $(\Sigma, \mathcal{D}, \Phi)$ satisfies the condition $(A, A_d) \in \mathcal{S}_{robust,\infty}$, we shall say that the associated uncertain delay system is $\mathcal{S}_{robust,\infty}$ stable.

Definition 4 Robust delay-dependent set. The set $\mathcal{S}_{robust,\tau}$ defined by

$$\mathcal{S}_{robust,\tau} = \{(A, A_d) \ : \ (\Sigma, \mathcal{D}, \Phi) \ \text{uniformly asymptotically stable} \ \forall \tau \in [0, \tau^*) \ \text{and for all admissible uncertainty } (\mathcal{D}, \Phi) \ \text{and unstable for} \ \tau = \tau^* \ \text{and some admissible } (\mathcal{D}, \Phi)\} \tag{3.10}$$

is called the *robust delay-dependent stability set* in the parameter space (A, A_d).

If a triplet $(\Sigma, \mathcal{D}, \Phi)$ satisfies the condition $(A, A_d) \in \mathcal{S}_{robust,\tau}$, we shall say that the associated uncertain delay system is $\mathcal{S}_{robust,\tau}$ stable.

These definitions recover the case when we have no parameter uncertainty in the matrices (A, A_d). To the best of the authors' knowledge, a complete *characterizations* of these sets does *not exist*. We shall present later some sufficient and some necessary conditions ensuring that a pair (A, A_d) belongs to such sets for a given representation of the uncertainty $(\mathcal{D}, \Phi)$.

3.3 Scalar single delay case

Let us consider now the following simple example:

$$\begin{cases} \dot{x}(t) = -ax(t) - bx(t-\tau) \\ (a, b, \tau) \in \mathbb{R} \times \mathbb{R} \times \mathbb{R}^+ \end{cases}, \tag{3.11}$$

under appropriate initial conditions (1.2), which will be used throughout this chapter to illustrate the various analysis techniques.

Frequency-domain approach The characteristic equation associated to (3.11) is:

$$s + a + be^{-s\tau} = 0. \tag{3.12}$$

This is a *transcendental equation* having an *infinite* number of solutions. As specified before, the analysis of such a system is done in the parameter space (a, b).

Use of the *$\mathcal{D}$–decomposition method* [93] gives a parametrization of the space Oab in several regions, each region being characterized by the same number of roots with positive real parts (see also [47]). Furthermore, each region is bounded by a "hypersurface" (here a first order one), which has the property that at least one root of the characteristic equation lies on the imaginary axis for the corresponding parameters a, b and τ.

The "methodology" to be used is as follows: first, we find the "hypersurfaces" by taking $s = j\omega$ in (3.12), and second, for each region we consider one point for which the analysis of the corresponding characteristic equation is more simple.

In our case, we have two "hypersurfaces":

$$a + b = 0, \tag{3.13}$$

which corresponds to the solution $s = 0$, and:

$$\begin{cases} a + b\cos(\omega\tau) = 0 \\ \omega - b\sin(\omega\tau) = 0 \end{cases}, \quad \omega \neq 0. \tag{3.14}$$

Thus, $\mathcal{S}(r)$ is the *Oab* region, whose boundaries are parametrized by (3.13)-(3.14), for $\tau = r$. (Indeed, we can consider $b = 0$, and the system $\dot{x} = -ax$, $a > 0$ is stable, etc.).

The *delay-independent* stability region problem, which corresponds to the intersection of all $\mathcal{S}(r)$, $r > 0$ (constant, but finite), can be seen as *finding all a and b*, for which:

$$\begin{cases} a + b > 0 \\ \text{the system} \begin{cases} a + b\cos(\omega\tau) = 0 \\ \omega - b\sin(\omega\tau) = 0 \end{cases}, \quad \text{has no solution for } \omega \neq 0. \end{cases}$$

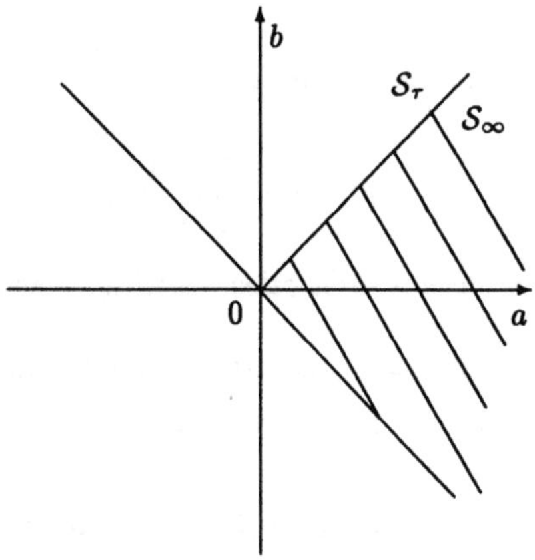

Fig. 1. Scalar single delay case. Stability regions in the parameter space Oab

Simple computations prove that the corresponding set (depicted in Fig. 1) is given by:

$$\mathcal{S}_\infty = \{ (a,b) \quad : \quad a+b>0, \; a \geq |\, b\,| \}. \tag{3.15}$$

Since $\mathcal{S}_\tau$ and $\mathcal{S}_\infty$ are complementary with respect to $\mathcal{S}(0)$, it follows that:

$$\mathcal{S}_\tau = \{ (a,b) \quad : \quad b > |\, a\,| \}. \tag{3.16}$$

The only problem here is to find the *optimal* corresponding ***bound*** τ^*. One way to compute it, is to consider the characteristic equation (3.12) as an equation in *two variables*: one real ($\omega \in \mathbb{R}^*$), and the other on the unit circle ($z \in \mathcal{C}(0,1)$) and to find:

$$min \left\{ \frac{\alpha}{\omega} \quad : \quad j\omega + a + bz = 0, \quad z = e^{-j\alpha} \right\}.$$

Notice that the above set is always ***nonempty*** if $b > |\, a\,|$.

The idea of using *two variables* for studying such stability problems is not new, and it has been used for the *delay-independent* stability problem in [87, 88, 71, 74, 75] (and the references therein). We shall talk later about the corresponding techniques.

Remark 3. Another method to compute this bound has been given in [47] using Rouché's theorem for complex functions. Another idea has been given in [17] (see also [16]), where the parameter space regions are bounded by "hypersurfaces" defined by the ***discontinuities*** of the function $\tau(a,b)$.

In conclusion, we have the following result:

Proposition 4 *The following assertions hold:*

1) The triplet $(-a,-b,\tau)$ *is* $\mathcal{S}_\infty$ *stable if and only if* $a+b>0$ *and* $a \geq |\, b\,|$.

2) The triplet $(-a, -b, \tau)$ is S_τ stable if and only if $b > | a |$. The corresponding optimal bound is given by:

$$\tau^* = \frac{\arccos\left(-\frac{a}{b}\right)}{\sqrt{b^2 - a^2}}. \tag{3.17}$$

Furthermore, there do not exist other stability regions.

It is easy to see that if $a = 0$, then the *optimal* bound τ^* becomes $\tau^* = \frac{\pi}{2b}$. This case ($a = 0$) is analyzed in the next section using a time-domain approach.

Time-domain Approach and Razumikhin Theorem This part is devoted to the introduction (via an example) of two notions largely used in the literature to develop stability results: *Lyapunov-Krasovskii* functional and *Lyapunov-Razumikhin* function.

Whereas the notion of a Lyapunov functional may seem like an obvious choice to extend the "classical" stability analysis in the sense of Lyapunov for ordinary differential equation to the infinite dimensional case, the notion of *Lyapunov-Razumikhin* is not so clear. In the latter one uses a "finite-dimensional" tool for an "infinite-dimensional" problem. It was mentioned in the Introduction that one can also interpret a functional differential equation as an evolution in an *Euclidian space*. Now the Lyapunov-Razumikhin function can be seen as the "result" of this interpretation. The main *idea* of the corresponding stability result (see the appendix for the exact formulation) can be summarized as follows: In the case of a Lyapunov-Krasovskii functional, V, a sufficient condition for stability is that the derivative, $\dot{V}$, of the candidate functional be negative *along* all the system's trajectories. In the Razumikhin based approach [155, 120] the negativity of the derivative of the Lyapunov-Razumikhin *function* $V : \mathbb{R}^n \mapsto \mathbb{R}$ is only required for the trajectories which leave at t^+ a certain set, defined by the system evolution on the interval $[t - \tau, t]$ (see also the appendix for the formulation). Other remarks on such an approach can be found in [95, 70]. To the best authors' knowledge, one of the first applications of Razumikhin theory in control is due to Thowsen [173].

Consider now the scalar case, with $a = 0$ and with a continuous, but bounded time-varying delay $\tau(t)$.

Recalling that $\mathcal{V}(\tau) = \left\{\tau \in \mathcal{C}^0 : 0 \leq \tau(t) \leq r, \forall t \in \mathbb{R}^+\right\}$, the following holds:

Proposition 5 *The triplet* $(0, -b, \tau(t))$ *(see (3.11) with* $a = 0$*) where* $\tau \in \mathcal{V}(r)$*, is delay-dependent uniformly asymptotically stable if for all t*

$$\tau(t) \leq r < \frac{1}{b}. \tag{3.18}$$

Furthermore, the result holds if b is a continuous time-dependent function $b : \mathbb{R} \mapsto \mathbb{R}^+$ *and* $\frac{1}{b}$ *is replaced by* $\frac{1}{\sup_{t \in \mathbb{R}} b(t)}$.

Sketch of the proof: Use the Lyapunov-Razumikhin function:

$$V(x(t)) = \frac{x(t)^2}{2}, \tag{3.19}$$

for the functional differential equation:

$$\dot{x}(t) = -bx(t) + b^2 \int_{-2\tau}^{-\tau} x(t+\theta)d\theta, \tag{3.20}$$

(given here for b scalar) obtained from the original system by using the Leibniz-Newton formula. For simplicity of the presentation, let us focus on the case when all terms are constant. Other comments and remarks can be found in [133, 70].

The derivative of the function (3.19) along the trajectories (3.20) is:

$$\dot{V}(x(t)) = -bx(t)^2 + b^2 x(t) \int_{-2\tau}^{-\tau} x(t+\theta)d\theta.$$

Consider now $V(x(\xi)) < q^2 V(x(t))$, $t - 2\tau \leq \xi \leq t$. It follows that the derivative $\dot{V}(x(t))$ is bounded by:

$$\dot{V}(x(t)) < -b(1 - bq\tau)x(t)^2. \tag{3.21}$$

Thus, $b\tau < 1$, implies the existence of a $q > 1$ (sufficiently small) such that $bq\tau < 1$, and the stability result follows from Razumikhin theorem.

Remark 4. We have seen that for a constant delay, the optimal bound on the delay size is given by: $\tau^* = \frac{\pi}{2b}$. However, this bound is not the optimal one if the delay is time-varying. In fact, if $\tau(t) \leq \frac{3}{2b}$ there exist oscillating solutions (see [70, 201] and the references therein).

Using a different time-domain approach, Barnea [7] (see also [70]) has improved the stability bound in the case of constant τ, to $\tau < \frac{3}{2b}$.

Consider now the general case,

$$\dot{x}(t) = -ax(t) - bx(t - \tau(t)), \quad a + b > 0, \quad \tau \in \mathcal{V}(r)\ (r > 0).$$

Using the same ideas, we have:

Proposition 6 *The following assertions hold:*

1. *The triplet* $(-a, -b, \tau(t))$, *with* $\tau \in \mathcal{V}(r)$ *is* $\mathcal{S}_{v,\infty}$ *stable if* $a > |b|$.
2. *If* $b > |a|$, *the triplet* $(-a, -b, \tau(t))$ *is* $\mathcal{S}_{v,\tau}$ *stable, and the stability is guaranteed for any* $\tau \in \mathcal{V}(r^*)$, *where:*

$$r^* = \frac{a+b}{b^2 + |ab|}. \tag{3.22}$$

Remark 5. The assertions in Proposition 6 still hold if b is a continuous time-varying bounded function $b(t)$ instead of a *constant* one. For example, the $\mathcal{S}_{v,\infty}$ stability is then reduced to test if $a > \sup_{t\in\mathbb{R}} | b(t) |$ instead of $a >| b |$ and r^* is given by:

$$r^* = \frac{a+\underline{b}}{\bar{b}^2+ | a | \bar{b}}, \quad \begin{cases} \bar{b} = \sup_{t\in\mathbb{R}} b(t) \\ \underline{b} = \inf_{t\in\mathbb{R}} b(t) \end{cases}, \quad \text{if} \quad b(t) > 0,\ a + b(t) > 0, \quad \forall t \in \mathbb{R}.$$

Other remarks and comments on the set $\mathcal{S}_{v,\infty}$ are given in [5].

Let us consider now the *delay-independent* case when the delay is *constant.* Another way to obtain the *sufficient* condition $a >| b |$ close to a *necessary and sufficient* one (see Proposition 4), is to consider the Lyapunov-Krasovskii functional:

$$V(x_t) = \frac{1}{2}x(t)^2 + \frac{b^2}{2}\int_{-\tau}^{0} x(t+\theta)^2 d\theta. \tag{3.23}$$

The advantage of such functionals is the "decoupling" between the "present" state $x(t)$ and the "previous" ones $x_t(\theta)$, $\theta \in [-\tau, 0)$. Notice that this functional can also be interpreted as a function on the space product $\mathbb{R}^n \times \mathcal{C}^v_{n,\tau}$.

For the stability problem, simple computations prove that:

$$\dot{V}(x_t) \leq -(a- | b |)x(t)^2,$$

which is *stricly negative* for $x \neq 0$ if $a >| b |$. In this case, the *delay-independent* stability condition follows from the Lyapunov-Krasovskii stability theorem (see also the Appendix).

For $a = b > 0$ the considered Lyapunov-Krasovskii functional ensures only *uniform stability* and not *uniform asymptotic* stability. A method to prove this last property is to use an extension of *Lasalle's principle* (see [70] and the references therein). Also, there exists an *Invariance principle* for the Razumikhin-based stability results (see [66]). Although this kind of approach is not used in the chapter, the corresponding *main idea* is illustrated below.

Consider the functional:

$$V_a(x_t) = \frac{1}{a}x(t)^2 + \int_{-\tau}^{0} x(t+\theta)^2 d\theta, \tag{3.24}$$

which has the derivative:

$$\dot{V}_a(x_t) = -(x(t) + x(t-\tau))^2.$$

The set for which this derivative is zero is given by:

$$S = \{\phi \in \mathcal{C}_{n,\tau} \quad : \quad \phi(0) = -\phi(-\tau)\}.$$

The largest invariant set M, which is included in S is defined by all the initial conditions for which $x(t) = -x(t-\tau)$ for all $t \in \mathbb{R}$, i.e. $\dot{x}(t) = 0$ and thus

$x(t) = c$, where c is a real constant. In conclusion, $c = -c$, and thus $c = 0$, that is the ***asymptotic stability*** property of the trivial solution.

For $\mathcal{S}_\infty$ stability, the necessary and sufficient condition is completely recovered, however the delay bound in the $\mathcal{S}_\tau$ case is not the best possible one. The difference is ***due*** to the technique that was adopted, deliberately chosen for its ease in coping with the general time-varying system uncertainty. Comparisons between this bound and the optimal one were discussed in [133]. Other details and comments about this approach are given later.

The α-stability In the previous parts, we have not considered particular constraints on the roots of the characteristic equation. In control theory, there is interest in knowing their location. If for linear systems without delays, we may consider various bounded or unbounded regions in the complex plane, for the delay case, one may consider only the ***α—stability case***, i.e. such that the solutions have a ***decay rate α***. Olbrot [148] has pointed out that increasing the delay for a given decay rate may induces ***instability*** in the system.

A simple way to analyze such a property is to use the system transformation: $y(t) = e^{\alpha t}x(t)$ and study the stability of the delay system expressed in "y". Let us illustrate this for the scalar single delay case. The system transforms to:

$$\dot{y}(t) \;=\; -(a-\alpha)y(t) - be^{\alpha\tau}y(t-\tau). \tag{3.25}$$

Due to the appearance of the delay τ in the "new" system parameters $(\bar{a}, \bar{b})$, $\bar{b} = be^{\alpha\tau}$ it follows that one cannot have ***delay-independent*** stability results, independently on the "initial" (a, b) type property. Notice however that α cannot be larger than $a + b$.

Using similar ideas to previous handled cases, we have the following result:

Proposition 7 *1) If $(-a+\alpha, -b, \tau)$ is $\mathcal{S}_\infty$ asymptotically stable, then for any τ, $0 \le \tau \le \tau_{\alpha,i}$,*

$$\tau_{\alpha,i} \;=\; \frac{1}{\alpha} log\left(\frac{a-\alpha}{|\,b\,|}\right), \tag{3.26}$$

the system is α-stable.

2) If $(-a+\alpha, -b, \tau)$ is $\mathcal{S}_\tau$ asymptotically stable, then the α-stability is guaranteed for any τ, $0 \le \tau \le \tau_{\alpha,d}$, where $\tau_{\alpha,d}$ is the unique positive solution of the transcendental equation:

$$x \;=\; e^{-\frac{arccos\left(-\frac{a-\alpha}{bx}\right)}{\sqrt{(bx)^2-(a-\alpha)^2}}}, \quad x \in \left(\frac{a-\alpha}{b}, \infty\right).$$

It is easy to see that for $\alpha \to 0$, $\tau_{\alpha,i} \to \infty$, and $\tau_{\alpha,d} \to \tau^*$ (given previously) after some simple computations. In conclusion, this result recovers the conditions given in Proposition 4. Notice that condition *1)* can be obtained using also comparison principle type techniques (see [125]).

Some remarks on the scalar multiple delays case A complete study of the stability regions in the parameter space (in the "bounded / unbounded" sense) for systems involving *two delays* has been given in [72] using the properties of some complex functions associated to the characteristic equation; the commensurable case $\{\tau, 2\tau\}$ has been considered in [160] via the $\mathcal{D}$-decomposition method.

The general *delay-independent* case has been considered in [63] (see also [28]) using $\mathcal{H}_\infty$ properties of some appropriate transfer function associated to the system.

Other comments and remarks on related techniques follow later.

Remarks and comments on the stability of second order delay systems In the previous paragraphs, we have remarked that the increasing of the delay size may have a *destabilizing* effect. Furthermore, in the single delay (constant) scalar case, there exist only *two stability regions* in the parameter space: $\mathcal{S}_\infty$ and $\mathcal{S}_\tau$, respectively, i.e. if the system becomes unstable at τ^*, it will be unstable for any τ, $\tau \in (\tau^*, \infty)$.

In this context, a natural question arises: *Does this property hold for non-scalar linear systems?* The answer is negative, and there exist some second order examples[5], which prove that the *delay* may have a *stabilizing effect*, i.e. if the system is unstable for $\tau = \tau_1$, there exists at least one delay value τ_2 for which the system becomes *stable*. Such examples are given in [1, 68]. A complete study of the stability regions in the parameter space for the second order case is given in [81] using a Pontryagin based technique. For other remarks, see [14]. Necessary and sufficient conditions for *delays-independent* asymptotic stability for systems involving several (non-commensurable) delays are given in [18].

4 Frequency Domain Approach

This section is devoted to the *frequency-domain* approach and related techniques used for the *stability analysis* of linear delay systems involving *constant* (*commensurable* or *not*) delays. We caution that these results are very difficult to generalize to the time-varying delays case. Furthermore, some of these are restricted to the *single delay* case.

Special attention has been paid to the *matrix pencil* techniques, which allow to obtain some neat results which are readily interpretable. Although we do not aim to emphasize their numerical *tractability*, we shall briefly comment on this aspect.

4.1 Analytical and Graphical Tests

Analytical tests In this class of methods, one collects all the criteria that generalize the Hurwitz method to delay systems. In fact, if we have a linear delay

[5] The term "second order" means that the dimension of the vector x in the corresponding euclidian space is 2.

system with a single or with commensurable delays, we may write the characteristic function associated to the characteristic equation in the *quasipolynomial* form:

$$P(\lambda, e^{\lambda}) = \sum_{i=0}^{p}\sum_{k=0}^{q} a_{ik}\lambda^{i}e^{k\lambda}. \tag{4.1}$$

We have considered here:

Pontryagin criteria. The main *idea* behind the Pontryagin criterion can be summarized as follows: Suppose that $P(\lambda, e^{\lambda})$ given in (4.1) has principal term (i.e. $a_{pq} \neq 0$). Let $F(\omega)$ and $G(\omega)$ denote the real and the imaginary part, respectively of the quasipolynomial $P(\cdot,\cdot)$. Then:

1. If all the roots of P are in $\mathbb{C}^-$, then the roots of $F(\omega)$ and $G(\omega)$ are real, simple, alternate, and

$$F'(\omega)G(\omega) - F(\omega)G'(\omega) > 0, \quad \forall\omega \in \mathbb{R}. \tag{4.2}$$

2. Conversely, all the roots of P are in $\mathbb{C}^-$ if one of the next conditions is satisfied:
   a) All the roots of $F(\omega)$ and $G(\omega)$ are real, simple, alternate, and the inequality (4.2) is satisfied for at least one $\omega \in \mathbb{R}$.
   b) All the roots of $F(\omega)$ (or $G(\omega)$) are real, simple and for each root the inequality (4.2) holds.

To illustrate the application of this criterion, consider the scalar case, with $a = 0$. Then $G(\omega) = \omega sin(\omega\tau) + b$ and $F(\omega) = \omega cos(\omega\tau)$. The roots of F are real and simple: $\omega_0 = 0$, $\omega_k = \dfrac{k - 1/2}{\pi}\tau$, $k = 1, 2, \ldots$. Simple computations show that the asymptotic stability property holds for all τ, satisfying: $0 \leq \tau < \dfrac{\pi}{2b}$, which recovers the necessary and sufficient condition given in the previous section. The case $a \neq 0$ can be treated similarly.

This method becomes more complicated in non scalar cases. The second order case has been considered in [14, 81]. Using similar ideas, *sufficient* conditions for *instability* have been proposed in [24].

Chebotarev criteria. This criterion can be seen as the "direct" generalization of the Routh-Hurwitz criterion to the quasipolynomial case. The application of such a criterion is not very practical since it implies the computation of a "large" number of determinants.

We shall point out here also the *existence* of *other* criteria, such as *Yesupovisch-Svirskii* criterion (see [165]), which is applicable for a restricted class of delay systems.

Notice that all these *analytical* methods seem difficult to apply to the "delay-independent / delay-dependent" stability problems considered here. However, for the stability study of particular systems, it is important to see what method is suitable with respect to the system "structure".

Root locus methods Consider the case of a delay system of the form Σ involving a *single* delay. As specified before, the roots of the characteristic equation associated to Σ continuously depend on the system parameters. The *root locus method* idea is to determine the values of the parameters for which the characteristic equation has roots on the imaginary axis. One may see these 'limit' cases as the situations for which the system behaviour changes.

In order to find "delay-independent / delay-dependent" regions in the parameter space, the methods presented here are generally combined with other algebraic or analytical methods (for example, of Routh-Hurwitz or Pontryagin type). In this class, we can include:

$\mathcal{D}$–decomposition method [131]: This method consists in obtaining a "decomposition" of the parameter space in several regions, such that each region is bounded by a *hypersurface* which corresponds to the case when at least one root lies on the imaginary axis. Furthermore, for all the parameters lying in a given region, the corresponding characteristic equation has the same number of roots with positive real part.

In this case, the stability study is reduced to the analysis of the regions without unstable roots. It is clear that each *stable region* depends on the parameters of the system Σ, i.e. the "entries" A, A_d and τ, and thus each "hypersurface" can be seen as a function of τ. If τ is considered as a parameter, the "evolution" of this hypersurface as function of τ allows to detect the particular "delay-independent / delay-dependent" regions. The scalar case can be easily analyzed (see previous sections)

Notice that the method can be extended for delay systems involving commensurable delays. The parameter regions for the linear scalar case including two commensurable delays of the form $\{\tau, 2\tau\}$ has been considered in [160]. Other examples and applications can be found in [93, 154]. However, this method is very difficult to apply in the general case.

τ–decomposition method: This method is applied only for delay systems with a single delay and requires the transformation of the characteristic equation into a form:

$$e^{\tau s} = D_0(s) \left(= \frac{P(s)}{Q(s)} \right),$$

where D_0 is a ratio of two given polynomials. The idea of the method is to analyze the behaviour of the contour $D_0(j\omega)$ ($\omega \in \mathbb{R}^+$) with the unit circle in the complex plane (since for $s = j\omega$, $e^{j\omega\tau}$ is on the unit circle). In particular, if there is no intersection with the unit circle, it is easy to conclude that the stability for the case $\tau = 0$ is preserved for all positive values of delay, that is a *delay-independent* type result. However this condition is *only* sufficient and not *necessary and sufficient.* For the scalar single delay case, it corresponds to $a > | b |$.

Thus, to conclude on the type of stability one needs a second step, which consists in analyzing the root locus behaviour in a neighborhood of the "critical" values on the unit circle thus obtained.

Let us apply this idea to the scalar case. We have:

$$D_0(s) = \frac{n(s)}{d(s)} = -\frac{b}{s+a},$$

and introduce now the polynomial W [191] defined as:

$$W(\omega^2) = n(j\omega)n(-j\omega) - d(j\omega)d(-j\omega) = -\omega^2 - a^2 + b^2.$$

Thus if $j\omega_0$ is a root of the characteristic equation associated to Σ, then ω_0^2 is a root of W.

Consider now the case $a = b > 0$. The only root of W is $\omega = 0$, but in this case, the root locus has no intersection with the imaginary axis, and thus we may conclude asymptotic stability. Furthermore, since we do not need information on the delay size, it follows that this case enters in the "delay-independent class."

Other comments and remarks can be found in [80, 101, 191, 38]. Furthermore, all these results can be extended to the more general case when P and Q are analytic functions (see Cooke and van den Driessche [34]).

Other methods: Other methods have been developed in [172], but they are relatively difficult to apply. We mention the simplification given in [147], based on a pivoting algorithm.

Argument principle methods The argument principle is one of the basic geometrical method used in the control theory of linear finite-dimensional systems, see, e.g. the Nyquist criterion, the Satche diagram or the Michailov criterion. These principles can be applied to linear systems with delayed states since the number of the unstable roots in the complex plane is *finite*. However, these criteria may become difficult (complex form of the corresponding hodographs, see also [93] for basic results and examples).

Another criterion has been proposed by [165]. Due to its wide applicability for general class systems described by functional differential equations, we have preferred to present it. For the sake of simplicity, consider a linear system involving a single delay and let Σ the corresponding triplet.

Proposition 8 [165] *The triplet Σ is asymptotically stable if and only if*

i) $\mathcal{F}(j\omega) \neq 0$, *for* $\omega \in \mathbb{R}^+$.
ii) The condition:

$$\lim_{H\to\infty} \int_{(g)} \frac{1}{\mathcal{F}(\lambda)} \cdot \frac{d\mathcal{F}(\lambda)}{d\lambda} d\lambda = 0,$$

where (g) *is the "so-called" Bromwich contour defined by* $g = \cup_{i=1}^3 g_i$, *and:*

- (g_1): $\lambda = He^{j\theta}$, *where* θ *ranges from* $-\frac{\pi}{2}$ *to* $\frac{\pi}{2}$.
- (g_2): $\lambda = j\omega$, *where* ω *ranges from* H *to* 0.
- (g_3): $\lambda = j\omega$, *where* ω *ranges from* 0 *to* $-H$.

Furthermore, if the conditions (i) *and* (ii) *hold for any* τ, *the triplet* Σ *is* $\mathcal{S}_\infty$ *asymptotically stable.*

Remark 6. Condition (i) means that the characteristic equation has no roots on the imaginary axis and condition (ii) says that there are no roots in $\mathbb{C}^+$. Some applications and further comments can be found in [165].

4.2 Special criteria

Maximum Principle Based Criteria . In this class, we have included the *Small gain theorem* based criteria and the so-called *Mori and Kokame* criterion [126].

Small gain theorem criteria. As an illustration of the *idea*, consider the $\mathcal{S}_\infty$ asymptotic stability for a single delay case $n_d = 1$.

As specified before, it requires that the matrix A is *Hurwitz stable* (it is understood that $(A, A_d) \in \mathcal{S}(0)$). Introduce now the following *finite-dimensional* dynamical system:

$$\dot{x}(t) = Ax(t) + A_d u(t), \tag{4.3}$$

which has the transfer function

$$H_{xu}(s) = (sI_n - A)^{-1} A_d.$$

Suppose now that this transfer function satisfies the condition (see also [40]):

$$\sup_{\omega \in \mathbb{R}} \|H_{xu}(j\omega)\| < 1,$$

then it follows *via* the maximum principle type argument [3], that

$$\sup_{s \in \mathbb{C}^-} \|H_{xu}(s)e^{-s\tau}\| < 1, \quad \forall \tau \in \mathbb{R}^+.$$

This condition leads to:

$$det\left(I_n - (sI_n - A)^{-1} A_d e^{-s\tau}\right) \neq 0, \quad \forall \tau \in \mathbb{R}^+, \quad s \in \mathbb{C}^+,$$

which allows to conclude the *delay-independent* stability property. Related criteria can be found in [181] (resulting from the "Strict bounded real lemma" and the delay Riccati equation [182]) or in [27] (a singular value decomposition technique).

Due to its facility to treat the $\mathcal{S}_\infty$ type problems for the non commensurable delays case, we present this case here. Thus, we have delay-independent stability (with respect to each delay) if the matrix A is Hurwitz stable and

$$\mu_{\mathcal{X}_{n_d}}(\mathcal{M}(j\omega)) < 1, \quad \omega > 0,$$

where $\mathcal{M}(s)$ is defined as follows:

$$\mathcal{M}(s) = [I_n \dots I_n]^T (sI_n - A)^{-1} [A_{d1} \dots A_{dn_d}].$$

Here, $\mathcal{X}_{n_d}$[6] is the corresponding family of block diagonal matrices [27, 150], defined by:

$$\mathcal{X}_{n_d} = \left\{ diag\left(\delta_1 I_{k_1}, \ldots \delta_{n_d} I_{k_{n_d}}\right) \;:\; \delta_k \in \mathbb{C}, \;\; |\, \delta_k \,| \leq 1 \right\},$$

and $\sum_{i=1}^{n_d} k_i = nn_d$.

Using similar ideas combined with *matrix measures* properties (see Desoer and Vidyasagar [42]), one can obtain various delay-independent [29] or delay-dependent conditions. Thus, for the single delay case, the triplet Σ is $\mathcal{S}_\infty$ asymptotically stable if:

$$\|PA_d\| < 1,$$

where P is the unique symmetric and positive-definite solution of the Lyapunov equation:

$$A^T P + PA = -2I_n.$$

Notice also that this technique allows to recover the result due to Mori *et al.* [124] (using a comparison principle technique), which can be summarized as follows: the triplet Σ is $\mathcal{S}_\infty$ asymptotically stable if:

$$\mu(A) + \|A_d\| < 0,$$

where $\mu(A)$ is the corresponding matrix measure of A (see notations).

For the single delay scalar case, *delay-independent* asymptotic stability is implied by the following conditions:

$$a > 0, \quad \|\frac{b}{j\omega + a}\| < 1, \quad \forall \omega \in \mathbb{R},$$

condition which is equivalent to $a > |\, b \,|$.

Remark 7. We have developed here *only* sufficient conditions, which are close to the *necessary and sufficient* ones, which require $\omega \in \mathbb{R}^*$, instead of $\omega \in \mathbb{R}$, etc. (see [27] and the references therein). Furthermore, notice that relationships between the scalar case with single or several non commensurable delays and some $\mathcal{H}_\infty$ norm of some associated transfer function have been considered in [63, 28].

---

[6] The general form is $\mathcal{X}_{n_d}(\gamma)$ defined by:

$$\mathcal{X}_{n_d}(\gamma) = \left\{ diag\left(\delta_1 I_{k_1}, \ldots \delta_{n_d} I_{k_{n_d}}\right) \;:\; \delta_k \in \mathbb{C}, \;\; |\, \delta_k \,| \leq \gamma \right\},$$

where 1 is replaced by γ.

Mori and Kokame criterion [126]. The *idea* of this criterion is based on the maximum principle of an harmonic or subharmonic[7] function [3] combined with the following root property of the characteristic equation associated to Σ: if there exist unstable roots of the characteristic equation, then they are located in a *compact* domain in $\mathbb{C}^+$. Thus, the stability problem is reduced to the computation of a given function on the boundary of a compact domain. Several methods to restrict the compact domain have been proposed in [192] or in [194]. Other comments can be found in [169, 168].

Polynomial criteria In this class of criteria, we have included the following cases:

- *One variable* polynomial: the Tsypkin criterion [93] and the *matrix pencil* techniques [133, 26, 27].
- *Several variables* polynomials: two variables (see [87, 88]) or several variables (see [71]).

Tsypkin criterion. This criterion is one of the first results of *delay-independent* closed-loop *stability* type. Let us consider a transfer function of the form:

$$H_0(s) = \frac{P(s)}{Q(s)}e^{-s\tau},$$

where $P(s)$ and $Q(s)$ are real polynomials of degrees $(n-1)$ and n, respectively. Then we have the following result (see also [93] and the references therein):

Proposition 9 (Tsypkin criterion) *If $Q(s)$ is a stable polynomial, then the closed-loop system:*

$$H_b(s) = \frac{P(s)e^{-s\tau}}{Q(s) + P(s)e^{-s\tau}}$$

is $\mathcal{S}_\infty$ asymptotically stable if and only if the following condition

$$\mid Q(j\omega) \mid \; > \; \mid P(j\omega) \mid,$$

holds for all $\omega \in \mathbb{R}$.

Remark 8. A generalisation of this result in the *multiple delays* case has been given in [47]. For the sake of simplicity we do not consider it here.

Two variables polynomial criteria To the best authors knowledge, the connection between *delay-independent* stability of linear system with commensurable delays and the roots distribution of an associated *two variables* polynomial have been firstly considered by Kamen [87, 88], and has been largely treated in the literature. The basic *idea* of such an approach can be summarized as follows:

[7] See also [20] for the applications of such functions in control.

First, the corresponding characteristic equation (associated to Σ) with respect to imaginary axis,

$$det\left(j\omega I_n - A - \sum_{k=1}^{n_d} A_{dk}e^{-j\omega k\tau}\right) = 0, \quad \omega \in \mathbb{R}$$

can be interpreted as a *two* independent *variables* equation:

- one on the imaginary axis "$j\omega$" and
- the other one on the unit circle "$z = e^{-j\omega\tau}$", since τ is a "free" parameter.

Second, due to the continuity properties presented in the previous section, the *delay-independent* stability can be reduced to check:

- the Hurwitz stability of the matrix $A + \sum_{k=1}^{n_d} A_{dk}$, and
- if the corresponding characteristic equation has *no roots* on the imaginary axis, i.e. if there are no roots which cross the imaginary axis, and thus the corresponding upper and lower bounds u_h and l_h, respectively satisfy the conditions $u_h < 0$ and $l_h = +\infty$.

Thus, we have the following result [71]:

Proposition 10 *The following assertions are equivalent:*

i) The triple Σ is $\mathcal{S}_\infty$ asymptotically stable
ii) $(A, A_d) \in \mathcal{S}(0)$ and

$$det\left(j\omega I_n - A - \sum_{i=1}^{n_d} A_{di}z^i\right) \neq 0, \quad \omega \in \mathbb{R}^*, \quad z \in \mathcal{C}(0,1) \qquad (4.4)$$

Remark 9. Similar results for more general differential equations including delayed states have been developed in [33]. To the best authors konwledge, there is no general way to reduce the computation difficulty of such problems, excepting the case of *commensurable* delays.

Remark 10. We have preferred to present only the delay-independent stability case here. It is clear that the delay-dependent case corresponds to the situation when the equation (4.4) has some roots on the imaginary axis and unit circle, respectively. Furthermore the condition *(ii)* seems difficult to be verified by direct computation.

Remark 11. For the sake of simplicity, we have not considered the *multiple delays* criteria with non-comensurate delays. In this case, we should handle a polynomial with $n_d + 1$ *variables*, where n_d is the number of *non-commensurable* delays (we may have also the "mixed" case: some commensurable delays combined with non-commensurable ones). Thus, for a triplet Σ, the corresponding polynomial (in the hypothesis when all the n_d delays are not commensurable) becomes:

$$\mathcal{P}(\omega, z_1, \ldots z_{n_d}) = det\left(j\omega I_n - A - \sum_{k=1}^{n_d} A_{dk} z_k\right).$$

More details are given in [74, 75] and the references therein.

Other two variables type methods can be found in [31] (delay-dependent type results using two polynomials) or in [156] (delay-independent type results for the single delay case only). The Repin's idea was to use a different form for the condition *(ii)* in Proposition 10. Thus, for the single delay triplet $\Sigma = (A, A_d, \tau)$, one has $\mathcal{S}_\infty$ asymptotic stability if:

- A is a Hurwitz stable matrix, and
- for every $\omega \in \mathbb{R}^*$, the solutions of the equation

$$det\begin{bmatrix} \lambda A + A_d & -\lambda\omega I_n \\ \lambda\omega I_n & \lambda A + A_d \end{bmatrix} = 0,$$

  satisfy the condition $|\lambda| < 1$ (see also [71]).

Notice here that for ***delay-independent*** stability one needs implicitly the Hurwitz stability of the matrices A and $A + A_d$ (or $A + \sum_{k=1}^{n_d} A_{dk}$). These aspects will be considered later, when some simple ***sufficient*** tests for $\mathcal{S}_\tau$ are given.

Matrix pencils techniques A different way to handle such "delay-independent / delay-dependent" criteria is to use the ***matrix pencil*** techniques. We have briefly presented before the ***two variables*** polynomial approach. One of the major inconvenient of the corresponding results consists in the difficulty to check the condition on numerical example. Thus, we need to simplify it, and one of the way is to ***reduce*** the variables number from ***two*** to ***one***; the "reduced" one can be, for example, the ***imaginary axis*** variable ($j\omega$, $\omega \in \mathbb{R}^*$).

This fruitfull idea has been exploited by Chen *et al.* [28] (matrix pencil framework) and by Su [169] (eigenvalues computation of an appropriate complex matrix with a larger size than the system's matrices). The approach in [28] consists in computing the generalized eigenvalue distribution with respect to the unit circle for an associated ***constant***, and ***finite dimensional*** matrix pencil. Notice that this matrix pencil is obtained using a linearization technique for matrix polynomials. In the framework presented here, this matrix pencil is the so-called matrix pencil associated to ***finite delays***. In order to give a complete characterization, one needs to use the generalized eigenvalue distribution of a second matrix pencil, the so-called matrix pencil associated to ***infinite delays***.

Other matrix pencils techniques have been considered in [26], where the stability properties are reduced to the generalized eigenvalues distribution for some associated frequency-dependent matrix pencils. A different approach has been considered in [141], where a sufficient condition for ***delay-independent*** stability is given in terms of some appropriate algebraic properties of a matrix pencil, possibly singular. The construction of such matrix pencils is similar to the one encountered in the optimal control theory for linear systems without delay.

Before giving the main results, we introduce the following notion:

Definition 5. Let us consider two real matrices: $M, N \in \mathbb{R}^{h\times h}$.

1) The matrix pencil $\Lambda = zM + N$, $z \in \mathbb{C}$ is called simply dichotomic relatively to the unit circle if it has no eigenvalue on the unit circle.
2) The matrix pencil $\Lambda = zM + N$, $z \in \mathbb{C}$ is called dichotomically separable relatively to the unit circle if there exist r eigenvalues λ_i, $i = \overline{1,r}$, $1 \leq r < h$ such that $|\lambda_i| > 1 > |\lambda_j|$, for all $i = \overline{1,r}$, for all $j = \overline{r+1,h}$ (i.e. r eigenvalues outside the unit circle and all the others inside the unit circle). Furthermore, if $h = 2r$, then the matrix pencil is called symmetrically dichotomically separable relatively to the unit circle.

Consider the matrix pencils:

$$\Lambda_i(z) \;=\; zM_i + N_i, \quad i = 1, 2. \tag{4.5}$$

associated to the triplet Σ, where M_1, $N_1 \in \mathbb{R}^{(2n_dn^2)\times(2n_dn^2)}$ M_2, $N_2 \in \mathbb{R}^{(n_dn)\times(n_dn)}$ are given by:

$$M_1 = \begin{bmatrix} I_{n^2} & 0 & \dots & 0 & 0 \\ 0 & I_{n^2} & \dots & 0 & 0 \\ & & \ddots & & \\ 0 & 0 & \dots & I_{n^2} & 0 \\ 0 & 0 & \dots & 0 & A_{n_d}\otimes I_n \end{bmatrix}, N_1 = \begin{bmatrix} 0 & -I_{n^2} & 0 & \dots & 0 \\ 0 & 0 & -I_{n^2} & \dots & 0 \\ & & & \ddots & \\ 0 & 0 & 0 & \dots & -I_{n^2} \\ B_{-n_d} & B_{-n_d+1} & B_{-n_d+2} & \dots & B_{n_d-1} \end{bmatrix}, \tag{4.6}$$

$$M_2 = \begin{bmatrix} I_n & 0 & \dots & 0 & 0 \\ 0 & I_n & \dots & 0 & 0 \\ & & \ddots & & \\ 0 & 0 & \dots & I_n & 0 \\ 0 & 0 & \dots & O & A_{n_d} \end{bmatrix}, \quad N_2 = \begin{bmatrix} 0 & -I_n & 0 & \dots & 0 & 0 \\ 0 & 0 & -I_n & \dots & 0 & 0 \\ & & & \ddots & & \\ 0 & 0 & 0 & \dots & 0 & -I_n \\ A & A_1 & A_2 & \dots & A_{n_d-2} & A_{n_d-1} \end{bmatrix}, \tag{4.7}$$

with B_{-k} $(k = \overline{1,n_d})$, B_i $(i = \overline{1,n_d-1})$ given by:

$$B_{-k} \;=\; I_n \otimes A_k^T, \quad B_i = A_i \otimes I_n,$$
$$B_0 = A \oplus A^T,$$

where $\otimes$, $\oplus$ are the product and the sum of Kronecker (see Lancaster and Tismenetsky [99]). Following [133], the matrix pencil $\Lambda_1(z)$ is associated to the case of ***finite delays*** and $\Lambda_2(z)$ to the case of ***infinite delay***.

Denote by $\sigma(\Lambda)$ the set of eigenvalues of the matrix pencil Λ and let $\sigma_a = \sigma(\Lambda_1) - \sigma(\Lambda_2)$ (i.e. the generalized eigenvalues of the matrix pencil Λ_1, which are not eigenvalues of Λ_2).

With these notations and definitions, the following results:

Theorem 1 (Delay-independent stability) [133, 143] *The following statements are equivalent:*

(i) The triplet Σ is $\mathcal{S}_\infty$ asymptotically stable.
(ii) The pair $(A, A_d) \in \mathcal{S}(0)$ and the matrix polynomial

$$\mathcal{P}_1(z) = (A_{n_d} \otimes I_n)z^{2n_d} + B_0 z^{n_d} + B_{-n_d} + \sum_{i=1}^{n_d-1} \left(B_k z^{n_d+i} + B_{-k} z^{n_d-i}\right) \tag{4.8}$$

has either no roots on the unit circle; or if it does, all the roots z_0 of $\mathcal{P}_1(z)$ on the unit circle are roots of the matrix polynomial $\mathcal{P}_2(z)$:

$$\mathcal{P}_2(z) = A + \sum_{k=1}^{n_d} A_k z^k. \tag{4.9}$$

(iii) The pair $(A, A_d) \in \mathcal{S}(0)$ and the matrix pencil Λ_1 is either dichotomically separable relatively to the unit circle or if not, all the generalized eigenvalues z_0 of Λ_1 on the unit circle are eigenvalues of Λ_2.

Remark 12. It is easy to see that the *matrix pencils* techniques can be included into the *one variable polynomial* type *criteria*, but it seemed better to treat it separately due to its applicability for numerical treatments.

Following [57], the matrix pencils Λ_1 and Λ_2 are the *linearizations* of the matrix polynomials $\mathcal{P}_1$ and $\mathcal{P}_2$, respectively. In conclusion, all the results obtained in the matrix pencil framework can be easily converted into a polynomial framework and conversely. For the numerical implementation, we prefer to present only the matrix pencil formulation. Methods and algorithms for the computation of the corresponding generalized eigenvalues can be found, for example, in [58], etc.

Remark 13. It is easy to see that the triplet Σ is $\mathcal{S}_\infty$ asymptotically stable if the corresponding system free of delay is stable and the matrix pencil Λ_1 is *dichotomically separable* with respect to the unit circle. This condition is *only* a sufficient condition, but close to a "necessary and sufficient" one (for other comments and comparisons, see also [133]).

Remark 14. Since we are in the *delay-independent* stability case, the condition $(A, A_d) \in \mathcal{S}(0)$ in (ii) or (iii) can be replaced by $(A, A_d) \in \mathcal{S}(r)$, for some $r \geq 0$. We prefer the first variant due to its simplicity in verifying such stability conditions.

Remark 15. Suppose that Σ is associated to a single delay system. From Theorem 1, simple computations prove that the following conditions:

- $A + A_d$ is Hurwitz stable, and
- *det* $\left[(A + A_d\bar{z})^T \oplus (A + zA_d)\right] \neq 0$, for all $z \in \mathcal{C}(0,1)$

are only *sufficient* to guarantee that Σ is $\mathcal{S}_\infty$ asymptotically stable (see also [169, 133]).

We have the following:

Theorem 2 (Delay-dependent stability) [133, 143] *The following statements are equivalent:*

(i) The triplet Σ is $\mathcal{S}_\tau$ asymptotically stable.
(ii) The pair $(A, A_d) \in \mathcal{S}(0)$ and the matrix pencil Λ_1 has at least one generalized eigenvalue z_0 on the unit circle which is not eigenvalue of the matrix pencil Λ_2. Furthermore, the optimal bound on the delay size is given by:

$$\tau^* = \min_{1\le k\le 2n_d n^2} \min_{1\le i\le n} \frac{\alpha_k}{\omega_{ki}}, \tag{4.10}$$

where $\alpha_k \in [0, 2\pi]$, $e^{-j\alpha_k} \in \sigma_a$ and $j\omega_{ki}$ is an eigenvalue of the complex matrix $A + \sum_{i=1}^{n_d} A_i e^{-j\alpha_k i}$.

Remark 16. One of the "delay-independent / delay-dependent" stability problems is the computation of the generalized eigenvalues of *two* finite-dimensional *matrix pencils* of *large* dimension. For a reduction of this complexity, based on a *tensor product* approach, see [145] (basic notions on such an approach can be found in [117]).

For example, consider the single delay scalar case. The corresponding Λ_1 and Λ_2 are given by:

$$\begin{aligned} \Lambda_1 &= z\begin{bmatrix} 1 & 0 \\ 0 & -b \end{bmatrix} + \begin{bmatrix} 0 & -1 \\ -b & -2a \end{bmatrix}, \\ \Lambda_2 &= -zb - a. \end{aligned}$$

Suppose now that $b \neq 0$. In this case, the corresponding generalized eigenvalues are $z_{1,2}$ (Λ_1) and z' (Λ_2):

$$z_1 = -\frac{a}{b} + \sqrt{\left(\frac{a}{b}\right)^2 - 1}, \quad z_2 = -\frac{a}{b} - \sqrt{\left(\frac{a}{b}\right)^2 - 1}, \quad z' = -\frac{a}{b}.$$

Thus, if Λ_1 is dichotomically separable with respect to the unit circle then $a > |b|$; Λ_1 and Λ_2 have the same eigenvalues on the unit circle if $a = b > 0$ (we have the hypothesis: $a + b > 0$); Λ_1 has eigenvalues on the unit circle, that are not eigenvalues of Λ_2 if $b > |a|$, etc.

This technique can be easily extended to the following type problems:

delay-interval stability analysis, i.e. for a given delay value τ_0 for which the stability property is guaranteed, to find the *optimal* bounds $\underline{\tau} \le \tau_0$, $\overline{\tau} \ge \tau_0$, such that the corresponding triplet Σ is asymptotically stable (or hyperbolic) for all $\tau \in (\underline{\tau}, \overline{\tau})$ (see, for example, [144]).

delay-independent - *delay-dependent* or *delay-interval hyperbolicity*, i.e. to find if there exists or not roots of the characteristic equation on the imaginary axis under relaxed assumptions on the linear system free of delay (see, for example, [143, 144]).

For the briefty of the presentation, we consider here only the *delay-independent hyperbolicity* with a given number of roots in $\mathbb{C}^+$ and the *delay-interval stability* type problems.

We have the following results:

Proposition 11 [143] *Consider the triplet Σ satisfying the eigenvalue distribution $In\left(A+\sum_{k=1}^{n_d} A_{dk}\right) = (n_\pi, n_\nu, 0)$ for $\tau = 0$. Then the following statements are equivalent:*

(i) Σ is delay-independent hyperbolic with n_ν roots with positive real part (in $\mathbb{C}^+$) of the characteristic equation.
(ii) The matrix pencil Λ_1 is dichotomically separable relatively to the unit circle or if not, all the generalized eigenvalues z_0 of Λ_1 on the unit circle are either eigenvalues of Λ_2, either they satisfy:

$$In\left(A+\sum_{k=1}^{n_d} A_{dk}\right) = In\left(A+\sum_{k=1}^{n_d} A_{dk} z_0^k\right),$$

for all $z_0 \in \mathcal{C}(0,1) \cap \sigma_a$.

For a given real $r > 0$, introduce now the sets:

$$\begin{aligned}
\sigma_{r,+} &= \Bigg\{(\tau_{k_i}, \alpha_k) \;:\; \tau_{k_i} = \frac{\alpha_k}{\omega_{ki}} > r \;:\; e^{-j\alpha_k} \in \sigma_a, \\
&\quad j\omega_{ki} \in \Lambda\left(A+\sum_{h=1}^{n_d} e^{-jh\alpha_k} A_h\right) - \{0\}, \quad 1 \le k \le 2n^2, \quad 1 \le i \le n\Bigg\} \\
\sigma_{r,-} &= \Bigg\{(\tau_{k_i}, \alpha_k) \;:\; \tau_{k_i} = \frac{\alpha_k}{\omega_{ki}} < r \;:\; e^{-j\alpha_k} \in \Lambda_d, \\
&\quad j\omega_{ki} \in \Lambda\left(A+\sum_{h=1}^{n_d} e^{-jh\alpha_k} A_h\right) - \{0\}, \quad 1 \le k \le 2n^2, \quad 1 \le i \le n\Bigg\}
\end{aligned}$$

Proposition 12 [144] *Consider the triplet Σ. The following statements are equivalent:*

(i) The triplet Σ is delay-interval asymptotically stable.
(ii) There exists a $r > 0$ such that $(A, A_d) \in \mathcal{S}(r)$ and such that the sets $\sigma_{r,+}$ and $\sigma_{r,-}$ are not empty. The exact bounds on the delay interval including r are

$$\begin{aligned}
\overline{\tau} &= \min \{\tau \;:\; (\alpha, \tau) \in \sigma_{r,+}\} \\
\underline{\tau} &= \max \{\tau \;:\; (\alpha, \tau) \in \sigma_{r,-}\}
\end{aligned}$$

Furthermore, if $\tau \in \{\underline{\tau}, \overline{\tau}\}$, the corresponding characteristic equation has at least two complex conjugate eigenvalues on the imaginary axis.

Remark 17. It is clear that if $\sigma_{r,+}$ is an empty set, then we have the stability guaranteed for all delays in $(\underline{\tau}, +\infty)$; if $\sigma_{r,-}$ is empty, then we have the "delay-dependent" stability, with the corresponding $[0, \overline{\tau})$; and, if both sets $\sigma_{r,+}$ and $\sigma_{r,-}$ are empty, then we have "delay-independent" stability. The results can be developed similarly for the hyperbolic case.

5 Time-Domain Approach

5.1 Lyapunov's Second Method

It was mentioned earlier that there are *two* ways to develop the second method of Lyapunov for time-delay systems,

- One is based on the theory of *Lyapunov-Krasovskii functionals.*
- The other is based on the theory of *Lyapunov-Razumikhin functions.*

These approaches were briefly sketched for the scalar single delay case (The corresponding stability theorems can be found in the Appendix). One of the ways to extend these results to the general case is given below.

Lyapunov-Krasovskii functional method We limit the analysis to the single delay and time invariant case. In the cited references, various extensions are given: from the single delay case to multiple delays, time-varying delays, etc. Basically the complexity increases somewhat but the main ideas remain the same.

Delay-independent type criteria. Consider first a triplet $\Sigma = (A, A_d, \tau)$, including a *single delay* and introduce the following Lyapunov-Krasovskii functional:

$$\begin{cases} V(x_t) = x(t)^T P x(t) + \int_{-\tau}^{0} x(t+\theta)^T S x(t+\theta) d\theta \\ P > 0, \quad S > 0 \end{cases} . \tag{5.1}$$

Following the same idea as in the scalar case, this functional yields:

Proposition 13 *The triplet Σ is S_∞ asymptotically stable if there exists a triple of symmetric positive definite matrices $P > 0$, $S > 0$ and $R > 0$ satisfying the (delay) Riccati equation:*

$$A^T P + P A + P A_d S^{-1} A_d^T P + S + R \;=\; 0. \tag{5.2}$$

For details, see [179, 182]. Using the Schur complement property, the inequality (5.2) can be transformed into an LMI ("linear matrix inequality", see [21] and the references therein) form:

$$\left\{ \begin{array}{l} \begin{bmatrix} A^T P + PA + S & PA_d \\ A_d^T P & -S \end{bmatrix} < 0 \\ P > 0, \quad S > 0 \end{array} \right. \tag{5.3}$$

and thus the $\mathcal{S}_\infty$ problem is reduced to the *feasibility* of an LMI, i.e. *finding* if there exist matrices P and S which satisfy simultaneously the set of constraints (5.3).

Remark 18. A *necessary* condition for the existence of a triple of positive definite matrices solving the Riccati equation (5.2) is the Hurwitz stability of the matrix A. This condition was also seen to be *necessary* for *delay-independent* stability in the previous section.

Remark 19. Using the Bounded Real Lemma, one can relax the conditions in (5.2). In fact, we may look for solutions satisfying $P \geq 0$ for the Riccati equation [137]:

$$A^T P + PA + PA_d S^{-1} A_d^T P + S \; = \; 0.$$

Another relaxation ($P \geq 0$ and $S \geq 0$) has been considered in [141], where the stability problem is reduced to the existence of positive-semidefinite solutions for some appropriate Lur'e system, via some appropriate (possibly singular) matrix pencils.

Remark 20. An equivalent frequency domain interpretation is given in [181].

Furthermore, if there exists a decomposition of the matrix A_d given by:

$$A_d \; = \; BD, \quad B \in \mathbb{R}^{n\times m}, \quad D \in \mathbb{R}^{m\times n}, \tag{5.4}$$

where $m \leq n$ and $\mathbf{rank}(D) = m$, then it is possible to modify Proposition 13 to:

Proposition 14 *The triplet Σ satisfying (5.4) is $\mathcal{S}_\infty$ asymptotically stable if there exists a symmetric and positive definite solution $P > 0$ to the following Riccati inequality:*

$$A^T P + PA + PBS^{-1}B^T P + D^T SD \; < \; 0, \tag{5.5}$$

where $S \in \mathbb{R}^{m\times m}$ is a symmetric and positive definite matrix.

An advantage of a such decomposition of A_d lies in the fact that the associated LMI problem has smaller dimension: $(m+n)\times(m+n)$, instead of $2n \times 2n$. The corresponding Lyapunov-Krasovskii functional is:

$$\left\{ \begin{array}{l} V(x_t) = x(t)^T P x(t) + \int_{-\tau}^{0} x(t+\theta)^T D^T SDx(t+\theta)d\theta \\ P > 0, \quad S > 0 \end{array} \right.$$

This Lyapunov-Krasovskii based Riccati equation formalism was extended to the time-varying case (including time varying delays) in [180]. See also [107]. For example, in the case when in the triplet Σ, the delay τ is *time-varying* with bounded derivatives ($\dot{\tau}(t) \leq \beta < 1$), choose a Lyapunov-Krasovskii functional (5.1) is ([137]):

$$\begin{cases} V(x_t) = x(t)^T Px(t) + \dfrac{1}{1-\beta} \displaystyle\int_{-\tau}^{0} x(t+\theta)^T Sx(t+\theta)d\theta \\ P > 0, \quad S > 0 \end{cases}, \tag{5.6}$$

and Proposition 13 becomes:

Proposition 15 *The triplet Σ is S_∞ stable if the there exists a symmetric and positive definite solution $P > 0$ to the following Riccati inequality:*

$$A^T P + PA + PA_d S^{-1} A_d^T P + \frac{1}{1-\beta} S \;<\; 0, \tag{5.7}$$

where $S \in \mathbb{R}^{n \times n}$ is a symmetric and positive definite matrix.

The case with the decomposition on $A_d = BD$ runs similarly.

Consider now the α-stability case. Using the same ideas as in the scalar case, we have the following:

Proposition 16 *The triplet Σ satisfying (5.4) is α-stable for all $\tau \geq 0$ for which there exists a symmetric and positive definite solution $P > 0$ to the following Riccati inequality:*

$$A^T P + PA + e^{2\alpha\tau} PBS^{-1}B^T P + D^T SD + 2\alpha P \;<\; 0, \tag{5.8}$$

where $S \in \mathbb{R}^{m \times m}$ is a symmetric and positive definite matrix.

By the Schur complement property (5.8) is equivalent to the following LMIs:

$$\begin{cases} \begin{bmatrix} A^T P + PA + D^T SD + 2\alpha P & e^{\alpha\tau} PB \\ e^{\alpha\tau} B^T P & -S \end{bmatrix} < 0 \\ P > 0, \quad S > 0 \end{cases}. \tag{5.9}$$

Notice that for $\alpha = 0$, these LMIs completely recover the case of S_∞ stability. Due to the form of (5.9), we may see that the computation of the *suboptimal* (in the sense "maximal allowable") τ^* is a *generalized eigenvalue optimization type* problem [21] in the variables P and S. A "convex / quasi-convex" alternating procedure has been proposed in [146].

Consider now the case with *multiple delays*, which are not *commensurable.* For the simplicity of the presentation, we take $n_d = 2$. Ways to extend the Lyapunov-Krasovskii functional (5.1) to this case are:

$$\begin{cases} V(x_t) = x(t)^T Px(t) + \displaystyle\sum_{i=1}^{2} \int_{-\tau_i}^{0} x(t+\theta)^T S_i x(t+\theta)d\theta \\ P > 0, \quad S_1 > 0, \quad S_2 > 0 \end{cases}. \tag{5.10}$$

or:

$$\begin{cases} V(x_t) = x(t)^T P x(t) + \sum_{i=1}^{2} \int_{-\tau_i}^{-\tau_{i-1}} x(t+\theta)^T S_i x(t+\theta) d\theta \quad (\tau_0 = 0) \\ P > 0, \quad S_1 > 0, \quad S_2 - S_1 > 0 \end{cases} . \tag{5.11}$$

if we assume $\tau_1 \leq \tau_2$ (this restriction turns out to be in fact immaterial for the resulting criterion, see e.g. ([179, 182]).

Since the corresponding delays-independent conditions are similar, we develop here only the results obtained via the Lyapunov-Krasovskii functional of the form (5.10).

We have the following result expressed in the LMI form:

Proposition 17 *Suppose that $A_i = B_i D_i$, $i = \overline{1,2}$, with D_i of maximal rank. Then Σ is S_∞ stable if the following LMIs hold:*

$$\begin{cases} \begin{bmatrix} A^T P + PA + D_1^T S_1 D_1 + D_2^T S_2 D_2 & PB_1 & PB_2 \\ B_1^T P & -S_1 & 0 \\ B_2^T P & 0 & -S_2 \end{bmatrix} < 0 \\ P > 0, \quad S_1 > 0, \quad S_2 > 0 \end{cases} \tag{5.12}$$

Using similar ideas, these results can be easily extended for the time-varying delays case. Other remarks on this method can be found in [21, 48, 133, 180].

The Riccati-equation technique (or LMI) was also adapted to systems with distributed delays [185, 187]. Analogous results have also been found in the discrete case (i.e., delay-difference systems). Such equations are already finite dimensional (for commensurable delays), but treating them as 'delay'-systems allows a considerable reduction in dimension. For details, see [84, 186]. For delay systems with *stochastic* perturbations, it can be shown that the considered Lyapunov-Krasovskii functionals have the supermartingale property if again a Riccati type equation holds for some positive definite matrices. By a straightforward extension of Khasminskii's theory, stochastic stability can then be concluded. Details can be found in [52, 184]. Finally, some preliminary ideas for applications of the approach to nonlinear delay systems are presented in [188, 189].

Delay-dependent type criteria Consider first the triplet Σ, when one has a single and constant delay. As in the scalar case, the analysis will be performed on the $[t - 2\tau, t]$ system:

$$\dot{x}(t) = (A + A_d)x(t) + A_d \int_{-\tau}^{0} [Ax(t+\theta) + A_d x(t+\theta-\tau)] \, d\theta, \tag{5.13}$$

obtained using the Leibniz-Newton formula for the "original" system.

The associated Lyapunov-Krasovskii functional is:

$$\begin{cases} V(x_t) = \sup_{\theta \in [-2\tau, 0]} e^{\delta\theta} x(t+\theta) P x(t+\theta) \\ P > 0, \quad \delta = \frac{1}{\tau} log(1+\gamma) : \quad \gamma \in \mathbb{R}^+ \end{cases} . \tag{5.14}$$

Although the form of this functional is quite complex, we see that it depends only on *one positive definite* matrix P. Connections between this functional and the Lyapunov-Razumikhin function $V(x(t)) = x(t)^T Px(t)$ have been considered in [89] via a comparison principle.

Perhaps a general view is that it is preferable to use Lyapunov-Krasovskii functionals for delay-independent criteria and Lyapunov-Razumikhin functions for delay-dependent type results.

Lyapunov-Razumikhin function approach. As in the Lyapunov-Krasovskii functional approach, we shall present the following cases:

Delay-independent type results. Consider first the single delay case with a time-varying delay $\tau(t) \in \mathcal{V}(r)$, with r positive, but arbitrary. The corresponding Lyapunov-Razumikhin function is:

$$\begin{cases} V(x(t)) = x(t)^T Px(t) \\ P > 0 \end{cases} . \tag{5.15}$$

The same methods as in the scalar case lead to:

Proposition 18 *The triplet $\Sigma = (A, A_d, \tau(t))$ is $\mathcal{S}_{v,\infty}$ stable if one of the following (equivalent) conditions holds:*

(i) there exists a symmetric and positive-definite solution P to the following Riccati inequality:

$$A^T P + PA + \beta^{-1} PA_d P^{-1} A_d^T P + \beta P \ < \ 0, \tag{5.16}$$

(ii) there exists a symmetric and positive-definite solution Q to the following matrix inequality:

$$QA^T + AQ + \beta^{-1} A_d Q A_d^T + \beta Q \ < \ 0, \tag{5.17}$$

where β is a positive scaling.

Remark 21. The condition (5.17) is obtained from (5.16) by pre and postmultiplying the last one by P^{-1} and taking $Q = P^{-1}$.

A way to relax (5.16) is to consider the Riccati inequality:

$$A^T P + PA + \beta^{-1} PA_d S^{-1} A_d^T P + \beta P \ < \ 0, \tag{5.18}$$

with the constraint $P \geq S$. Notice that this technique will be applied to uncertain systems. Thus, if $\beta = 1$, the stability problem is reduced to the *feasibility* problem of the following set of LMIs:

$$\begin{cases} \begin{bmatrix} A^T P + PA + P & PA_d \\ A_d^T P & -S \end{bmatrix} < 0 \\ P \geq S, \quad S > 0 \end{cases} .$$

All the results proposed in the Lyapunov-Krasovskii functional framework may be developed along the same lines as for Proposition 18. For the sake of brevity, we do not develop them here.

Delay-dependent type results. As shown earlier, *delay-dependent* conditions are obtained from the Razumikhin approach on the associated $[t-2\tau, t]$ system:

$$\dot{x}(t) = (A + A_d)x(t) + A_d \int_{-\tau(t)}^{0} [Ax(t+\theta) + A_d x(t+\theta-\tau)]\, d\theta. \quad (5.19)$$

Thus, we have the following result:

Proposition 19 *The triplet $\Sigma = (A, A_d, \tau(t))$ is uniformly asymptotically stable for any $\tau(t) \in \mathcal{V}(\tau^*)$ if one of the following (equivalent) conditions holds:*

(i) there exists a symmetric and positive-definite solution P to the following Riccati inequality:

$$\begin{aligned} &(A + A_d)^T P + P(A + A_d) + \\ &+\tau^* \left[\beta_1^{-1} P A_d A P^{-1} A_d^T A^T P + \beta_1^{-1} P (A_d)^2 P^{-1} (A_d^T)^2 P + (\beta_1 + \beta_2) P\right] < 0 \end{aligned} \quad (5.20)$$

(ii) there exists a symmetric and positive-definite solution Q to the following matrix inequality:

$$\begin{aligned} &Q(A + A_d)^T + (A + A_d)Q \\ &+\tau^* \left[\beta_1^{-1} A_d A Q A^T A_d^T + \beta_1^{-1} (A_d)^2 Q (A_d^T)^2 + (\beta_1 + \beta_2) Q\right] < 0, \end{aligned} \quad (5.21)$$

where β_1 and β_2 are positive scalings.

Remark 22. Proposition 19 becomes very restrictive if the system is $\mathcal{S}_{v,\infty}$. In conclusion, we consider that the *time-domain* techniques should be applied as follows:

- First, check if the system is *delay-independent* stable using a Lyapunov-Krasovskii functional approach;
- Second, if we have failed, we give a *suboptimal* bound (in the sense "maximal allowable") on the delay size using a Lyapunov-Razumikhin function approach.

However, there are some simple cases for which we may conclude that the stability is of *delay-dependent* type. We have seen before, that *delay-independent* asymptotic stability *implies* that the matrix A is *Hurwitz stable.* Thus, if $(A, A_d) \in \mathcal{S}(0)$, but A is an *unstable matrix,* then the corresponding triplet Σ is $\mathcal{S}_\tau$ stable. Using similar arguments, we have the following result:

Proposition 20 *Assume that $(A, A_d) \in \mathcal{S}(0)$. Then the triplet Σ is $\mathcal{S}_\tau$ stable if one of the following assertions holds:*

a) If the delays are commensurable,
i) the matrix A is unstable.
ii) the matrix $A+\sum_{i=1}^{n_d}(-1)^i A_{di}$, *is strictly unstable (i.e. there exist at least one eigenvalue with strictly positive real part).*
b) If the delays are not commensurable, the previous condition i) and
ii') the matrices $A+\sum_{i=1}^{n_d}(-1)^{j_i} A_{di}$, *(where* j_i *is any value* $j_i \in \{-1,1\}$, *excepting the case* $j_i = 1$ *for all i) are strictly unstable.*

Remark 23. For the sake of simplicity, consider now the single delay case $\Sigma = (A, A_d, \tau)$, with $A + A_d$, A Hurwitz stable matrices, but with ***no strictly unstable*** eigenvalues for the matrix $A - A_d$ (i.e. in $\mathbb{C}^-$ or on $j\mathbb{R}$).

If the only eigenvalues on the imaginary axis are in 0, it is still possible to have *delay-independent* stability. This is the case of a single delay system of the form $\Sigma = (-a, -a, \tau)$, with $a > 0$. Other examples can be found in [133]. Some remarks on the limit case, i.e. $A_d = \alpha A$, with $\alpha \in (-1, 1]$ can be found in cite2hale1 (real eigenvalues for the matrix A) or in [27] (general case).

Remark 24. For the single delay scalar case, the above conditions completely recover the ***necessary and sufficient*** condition for ***delay-dependent*** stability. However, for the general case, these conditions are not ***necessary.***

Remark 25. Proposition 20 can be extended to time-varying delays using similar arguments [133].

Combining Proposition 19 and 20, it follows:

Proposition 21 *Assume that the pair* (A, A_d) *satisfies the conditions of the Proposition 20. Then the triplet* $\Sigma = (A, A_d, \tau(t))$ *is* $S_{v,\tau}$ *uniformly asymptotically stable for any* $\tau(t) \in \mathcal{V}(\tau^*)$ *if one of the following (equivalent) conditions holds:*

(i) there exists a symmetric and positive-definite solution P to the following Riccati inequality:

$$(A + A_d)^T P + P(A + A_d) + \\ +\tau^* \left[\beta_1^{-1} P A_d A P^{-1} A_d^T A^T P + \beta_1^{-1} P A_d A_d P^{-1} A_d^T A_d^T P + (\beta_1 + \beta_2) P\right] < 0 \quad (5.22)$$

(ii) there exists a symmetric and positive-definite solution Q to the following matrix inequality:

$$Q(A + A_d)^T + (A + A_d)Q + \\ +\tau^* \left[\beta_1^{-1} A_d A Q A^T A_d^T + \beta_1^{-1} (A_d)^2 Q (A_d^T)^2 + (\beta_1 + \beta_2) Q\right] < 0, (5.23)$$

where β_1 *and* β_2 *are positive scalings.*

The *computation* of a *suboptimal* bound (in the sense *maximal allowable*) on the delay size can be reduced to an LMI optimization problem. Thus, for example, if the scalings $\beta_i = 1$ ($i \in \{1,2\}$), then we have the following:

$$\begin{cases} \max_{Q=Q^T>0} \tau^* \quad \text{such that} \\ (5.23) \quad \text{holds,} \end{cases} \tag{5.24}$$

which is a standard *LMI generalized eigenvalue problem*, which is a quasiconvex one (see [21]). An alternative "convex / quasiconvex" (using the scalings β_i as parameters) algorithm has been proposed in [138].

For the sake of simplicity, we do not consider here the α-stability case. The "methodology" is similar, but makes use of a different associated functional differential equation.

Consider now the *multiple delays* case, and suppose that the delays are *not commensurable*. For the sake of simplicity, we present only the *two delays case*, but the results can be easily extended to the general case. Using similar arguments, we have the following result:

Proposition 22 *Asssume that the pair* (A, A_d), *with* $A_d = [A_{d1} \ A_{d2}]$ *satisfies Proposition 20. The triplet* Σ *with* $\tau = (\tau_1(t), \tau_2(t))$, *where* $\tau_i \in \mathcal{V}(\tau_i^*)$ $(i \in \{1,2\})$ *is* $\mathcal{S}_{v,\tau}$ *uniformly asymptotically stable, if there exist a symmetric and positive-definite matrix* P *and scalars* β_j $(j = \overline{1,6})$ *such that the following matrix inequality holds:*

$$\begin{bmatrix} \begin{pmatrix} (A + A_{d1} + A_{d2})^T P + P(A + A_{d1} + A_{d2}) \\ + \sum_{j=1}^{6} \beta_j P \end{pmatrix} & \tau_1^* P A_{d1} M & \tau_2^* P A_{d2} M \\ \tau_1^* M^T A_{d1}^T P & -\tau_1^* R_1 & 0 \\ \tau_2^* M^T A_{d2}^T P & 0 & -\tau_2^* R_2 \end{bmatrix} < 0, \tag{5.25}$$

where:

$$\begin{aligned} M &= \begin{bmatrix} A & A_{d1} & A_{d2} \end{bmatrix}, \\ R_1 &= diag\,(\beta_1 P, \beta_2 P, \beta_3 P), \\ R_2 &= diag\,(\beta_4 P, \beta_5 P, \beta_6 P). \end{aligned}$$

The matrix inequality (5.25) is not an LMI in the set of all variables. An alternative "convex / quasi-convex" feasibility algorithm for given delay values τ_1^* and τ_2^* can be given [133].

If we consider the delays as *constant* parameters, "sub-optimal" ellipsoïds (or other convex regions) in the delay space can be computed [139]. Other remarks can be found in [133].

Mixed delay-independent / delay-dependent stability As specified in the previous section, in the multiple delays systems case, we can consider the "mixed" *delay-independent / delay-dependent* stability with respect to a given "partition" of the delay set $\tau = [\tau_1, \dots \tau_{nd}]$.

For a simplified presentation, we shall consider the case of *two* delays: τ_1 and τ_2, respectively, and without loss of generality, we shall impose that the corresponding triple Σ is *delay-independent* "in" τ_2 and *delay-dependent* "in" τ_1. To obtain such results, the analysis will be performed on $[t - 2\tau_1, t]$ for the system obtained by integrating the "original" functional differential equation. Then, we have the following result:

Proposition 23 *The triplet Σ with $\tau = (\tau_1(t), \tau_2(t))$, where $\tau_i \in \mathcal{V}(\tau_i^*)$ ($i \in \{1,2\}$) is $S_{v,\tau}/S_{v,\infty}$ uniformly asymptotically stable, if there exist a symmetric and positive-definite matrix P and scalars β_j ($j = \overline{1,3}$) such that the following matrix inequalities hold:*

$$\begin{cases} \begin{bmatrix} \begin{pmatrix} (A + A_{d1}^T P + P(A + A_{d1}) \\ + \sum_{j=1}^{3} \beta_j P \end{pmatrix} & \tau_1^* P A_{d1} M & P A_{d2} \\ \tau_1^* M^T A_{d1}^T P & -\tau_1^* R_1 & 0 \\ A_{d2}^T P & 0 & -R_2 \end{bmatrix} < 0, \\ P \geq R_2, \quad P > 0 \end{cases} \tag{5.26}$$

where:

$$\begin{aligned} M &= \begin{bmatrix} A & A_{d1} & A_{d2} \end{bmatrix}, \\ R_1 &= diag\,(\beta_1 P, \beta_2 P, \beta_3 P), \end{aligned}$$

Remark 26. It is easy to see that by setting $A_{d2} \equiv 0$ and $\beta_3 \equiv 0$, we have a single delay system and this result completely recovers the *delay-dependent* type result via a Lyapunov-Razumikhin function approach.

Similarly, the corresponding *delay-independent* type result is obtained if we set $A_{d1} \equiv 0$ and $\tau_1 \equiv 0$, etc.

5.2 Comparison Principle

The idea is to find an ordinary differential equation, or a functional differential equation, called (B), with known asymptotic behaviour such that its (asymptotic) stability implies the (asymptotic) stability for the initial time-delay system, called (A). In this case, we say that the system (B) is a *comparison system* for the system (A).

The first comparison principles have been estabilished by Halanay [67], Lakshmikhantam and Leela [97] and Driver [45]. Notice that such approaches allow to give different proofs for classical stability theorems relaxing some of the conditions, or to propose new analyzing techniques for general functional differential equations (see, e.g. [54]). The tool which seems best adapted to such an approach is the *vector Lyapunov functions*, see [11, 119].

Guided tours of the techniques used to develop such criteria is given in [38] (delay-independent type results) and in [62] (delay-dependent results).

Using the *comparison principle* ideas combined with some *matrix techniques*, the following approaches seem the most interesting:

Matrix measures One of the first results in this framework is due to Mori *et al.* [124], where the delay-independent stability of a single delay-system described by the triplet Σ is reduced to check the condition:

$$\mu(A) + \|A_d\| \; < \; 0. \tag{5.27}$$

Using a different approach, Hmamed [76] has proved that the triplet Σ is $\mathcal{S}_\infty$ stable if:

$$\mu(A) + \mu(zA_d) \; < \; 0, \quad z \in \mathbb{C}, \quad |\, z \,| = 1.$$

If one uses a Lyapunov vector function approach, one needs to use the Lyapunov function: $V(x) = \|x\|$, combined with a technique of Tokumaru *et al.* [175] (see also [38] and the references therein).

Remark 27. This condition (5.27) is only sufficient, and special interest has been paid to reduce its conservativness (see the paragraph dedicated to *Mori and Kokame* criterion).

Extension for the α-stability case can be found in [125], and the corresponding stability condition is:

$$\mu(A) + \|A_d\| e^{\alpha\tau} + \alpha \; < \; 0.$$

Some improvements of this result using different ideas can be found in [19, 76].

Notice that all these results can be easily extended to the *multiple* delays case; using the technique described in the previous paragraphs for obtaining *delay-dependent* type results, this idea can be used also in this framework. For the sake of brevity, we detail only the *time-varying delay* case, which allows to recover completely the constant delay case.

Time-varying delays For the sake of simplicity, we consider here only the single delay case. The multiple delays case is treated in [136]. A different technique from the one presented here can be found in [104] (delay-independent type results). If the delay is constant, one recovers similar results from the literature (see [193] and the references therein).

Let us consider the triplet $\Sigma = (A, A_d, \tau(t))$, where $\tau(t)$ is a continuous function with bounded derivative. Introduce now the following *system*:

$$\dot{y}(t) \; = \; -\eta_A y(t) + q(t) y(t - \tau(t)), \tag{5.28}$$

where

$$q(t) \; = \; \left(\eta_A - \frac{\sigma}{\tau(t)}\right) exp\left(-\sigma \int_{t-\tau(t)}^{t} \frac{d\theta}{\tau(\theta)}\right).$$

A direct verification shows that

$$y(t) = C_0\, exp\left(-\sigma \int_0^t \frac{d\theta}{\tau(\theta)}\right),$$

where C_0 is a real constant, is a solution of (5.28).

Using this scalar system in order to compare its solution with the solution of the triplet Σ, we have the following result:

Proposition 24 [157, 138] *Consider the triplet Σ and assume that A is a Hurwitz stable matrix satisfying*

$$\| exp(At) \| \leq k_A \cdot exp(-\eta_A t) \tag{5.29}$$

for some real numbers $k_A \geq 1$ and $\eta_A > 0$. If the following inequality

$$\frac{k_A}{\eta_A} \| A_d \| < 1 \tag{5.30}$$

holds, then the transient response of $x(t)$ satisfies

$$\| x(t) \| < M \sup_{\theta \in \bar{\mathcal{E}}_0} \{\| \phi(\theta) \|\}\, exp\left(-\sigma \int_0^t \frac{d\theta}{\tau(\theta)}\right), \quad \forall\, t \geq 0 \quad M \geq 1,$$

where $\sigma > 0$ is the unique positive solution of the transcendental equation

$$1 - \frac{\sigma}{\eta_A \tau(0)} = \frac{k_A}{\eta_A} \| A_d \|\, exp\left(\frac{\sigma}{1-\alpha}\right). \tag{5.31}$$

Furthermore, the triplet Σ is exponentially stable with a decay rate $\frac{\sigma}{\bar{\tau}}$.

Using some basic results on matrix measures [42], we have the following result:

Corollary 1 *Consider the triplet Σ and assume that A is a Hurwitz stable matrix. If the inequality*

$$\mu(A) + \| A_d \| < 0$$

holds, then the triplet Σ is exponentially stable with a decay rate $\frac{\sigma}{\bar{\tau}}$, where $\sigma > 0$ is the unique positive solution of the transcendental equation

$$\mu(A) + \frac{\sigma}{\tau(0)} + \| A_d \|\, exp\left(\frac{\sigma}{1-\alpha}\right) = 0.$$

M-matrices The basic ideas can be summarized as follows:

- First, to introduce a ***comparison system*** which can be, for example (single delay case, with a constant delay), of the form:

$$\dot{y}(t) = My(t) + Ny(t-\tau),$$

with some appropriate matrices M and N computed starting from the original triplet Σ.

- Second, to test if the matrix $M+N$ is the ***"opposed"*** form of an M-***matrix***[8].

In terms of Lyapunov vector functions, one uses (see also [37, 38, 62, 61] and the references therein):

$$V(x) = \begin{bmatrix} \mid x_1 \mid \\ \vdots \\ \mid x_n \mid \end{bmatrix}, \quad x \in \mathbb{R}^n.$$

Other references, and further remarks on this technique are given in the chapter 10 due to Richard *et al.*

6 Other Stability Results and Remarks

Although our interest is focused on stability criteria when the delay system is interpreted as a ***functional differential equation***, we do not want to end this part without mentioning how ***other interpretations*** of delay systems can cope with the "delay-independent / delay-dependent" stability presented before.

Since "delay-independent / delay-dependent" stability is not the only problem treated in the delay system control literature, we briefly mention also other ***problems***, ***techniques***, and ***remarks*** on different topics.

A special attention has been paid to the ***complexity*** of stability problems involving multiple delays which are not commensurable.

6.1 Various interpretations of delay systems

Without loss of generality, we consider the following cases:

Differential Equations over Rings The basic idea of this approach is to rewrite the system (1.1) using the translation operator, and to interpret (1.1) as a differential equation on the ring $\mathbb{R}[z]$. For exemplification, let us consider the single delay case:

$$\dot{x}(t) = Ax(t) + A_d x(t-\tau). \tag{6.1}$$

[8] A matrix D is called an M-*matrix* if the elements on the diagonal are non-positive, the matrix D is not singular, and furthermore, all the elements of D^{-1} are non-negative.

The translation operator $\mathcal{D}_\tau$, defined by: $\mathcal{D}_\tau f(t) = f(t-\tau)$, allows to rewrite the considered system as:

$$\dot{x}(t) = F(\mathcal{D}_\tau)x(t), \tag{6.2}$$

where $F(\mathcal{D}_\tau) = A + \mathcal{D}_\tau$ is an operator acting on the evolution "$x(\theta)$" for $\theta \in [-\tau, 0]$ of the system.

To equation (6.2), we can associate the differential equation on the ring $\mathbb{R}[z]$ (see also [86]) given by:

$$\dot{x}(t) = F(z)x(t). \tag{6.3}$$

Some connections between the characteristic equation associated to (6.3) and (6.1) are given in [86, 87, 88]. The *delay-independent* criterion given in [71] can be seen also in this framework.

Without discussing all the possible interpretations, we cite the *delay-independent* stability condition given in [22], where the *delay-independent* stability property is reduced to the existence of a *hermitian* and *positive-definite* solution to a complex Lyapunov equation.

$2-D$ Equations The basic idea of the approach is to rewrite the differential equations associated to the system as a *$2-D$ equation*. Thus, for the scalar case, the corresponding functional differential equation

$$\dot{x}(t) = -ax(t) - bx(t-\tau),$$

can be rewritten as:

$$\begin{bmatrix} x_1(t+\tau) \\ x_2(t) \end{bmatrix} = \begin{bmatrix} 0 & 1 \\ -b & -a \end{bmatrix} \cdot \begin{bmatrix} x_1(t) \\ x_2(t) \end{bmatrix}, \tag{6.4}$$

which *combines* an "ordinary" differential equation and a "functional equation". Sufficient *delay-independent* stability conditions expressed in terms of *2-D Lyapunov equations* have been given in [2]. Other criteria have been considered in [32], etc.

Matrix Characteristic Equation Approach The basic idea of this approach is to transform the functional differential equation associated to the system (1.1) into an *ordinary* differential equation via an appropriate linear transformation. Thus, for the single delay system case (6.1), the transformation is:

$$z(t) = T_{x_t}x_t = x(t) + \int_{-\tau}^{0} e^{-A_{mc}(\theta+t)}A_d x(t+\theta)d\theta, \tag{6.5}$$

where the matrix A_{mc} satisfies the *characteristic matrix equation*:

$$A_{mc} = A + e^{-A_{mc}\tau}A_d. \tag{6.6}$$

The corresponding linear system is:

$$\dot{z}(t) = A_{mc}z(t). \tag{6.7}$$

Notice that the equation (6.6) is a *transcendental* matrix equation and it is hard to use it for "delay-independent / delay-dependent" type results. However, we can mention the algorithms for computing the "corresponding" A_{mc} matrix given in [49, 205].

6.2 On the complexity of multiple delays stability problems

We have shown in the previous paragraphs that it was possible to solve the "delay-independent / delay-dependent" stability problems in polynomial time for a class of linear systems with commensurable delays.

The following natural question arises: *Does the same property hold for the multiple delays case?* Unfortunately, the answer is negative. Recently, a paper of Toker and Ozbay [174] has proved that such problems are *$\mathcal{NP}$-hard* using the *$\mathcal{NP}$-hardness* of complex bilinear programming over the unit polydisk. Definitions for *$\mathcal{NP}$-hardness* can be found in [55] and the references therein; other *$\mathcal{NP}$-hard* problems arising in robust control theory are presented in [132].

In conclusion, it is rather unlikely to find efficient procedures (of polynomial-time type) for such problems in the general case. However, we should point out that better approximation schemes can be thought of to improve the "sufficient" (relatively simple) "delay-independent / delay-dependent" conditions presented above.

6.3 Other stability problems

We have considered in this section only stability results expressed in terms of their *robustness* with respect to the "delay", viewed as a *free parameter*, in several different cases: time-varying or constant delays, commensurable (similar to single case) or non commensurable delays. The analysis has been given under the hypothesis of the asymptotic stability of the system free of delay.

This situation is not the only one which can be considered. Thus, we have mentioned the *delay-interval* stability analysis (matrix pencil framework) for linear systems with commensurable delays. Another problem consists in analyzing the relationships between the commensurable delays and the multiple delays (constant, but independent) stability cases (see [108]). Generally, it is clear that, if the system involving the delays as "independent" parameter is asymptotically stable, then also the commensurable delays stability case problem presented before is guaranteed. The question is if this condition is also *necessary*. The answer is positive in some cases, and we do not consider them here.

Another problem is to analyze if, in the *delay-dependent* stability case (always for commensurable delays), the stability region thus obtained is the *only existing one*. The examples given in [1] or in [71] prove that one can have a sequence *stability / instability / stability* for non scalar systems. Some remarks on this kind of problems have been considered in [140].

Other remarks In the previous paragraphs, we have considered *stability* issues for some classes of linear systems with delayed state, using particular forms for the Lyapunov-Krasovskii functionals or Lyapunov-Razumikhin functions.

Other constructing methods for Lyapunov-Krasovskii functionals have been proposed in [83] (constant delays), or in [107] (extension to time-varying delays). The main advantage of such methods is that the corresponding conditions are closed to "necessary and sufficient" conditions, but they do not allow to easily handle numerical examples. We should point out also the constructing method due to Barnea [7], not only for the stability test, but also for instability.

In this sense, one can mention also the extension of the classical Lyapunov equation theory to the case of linear differential equations of delayed type [82] (see also [70]).

Other interesting *stability criteria* consist in computing the optimal size of the delay which ensures the stability of the closed-loop system:

$$H(s) = H_0(s)e^{-s\tau}, \quad \tau > 0,$$

where $H_0(s)$ corresponds to the nominal transfer function ("non-delayed"), and the delayed term can be seen as an *uncertainty* one. For this problem, we want to mention the approach proposed in [43] (a sub-optimal bound on the delay size using a modified form of the classical Nevalinna-Pick interpolation theory basic results). A partial solution to the problem can be also found in [46].

And last, *another* stability necessary and sufficient condition to compute the *optimal* bound on the delay-size in the S_τ case is given in [169] in terms of the eigenvalues distribution of some appropriate "large" matrices.

We did deliberately not consider all these approaches here to guarantee the unity and simplicity of the presented materials.

7 Robust Stability

This section is devoted to the *stability analysis* results for *uncertain delay systems* in $(\Sigma, \mathcal{D}, \Phi)$ representation. Some of the results presented here can be obtained as simple extensions of the "nominal" cases treated before. In some cases, we have not detailed the existing results, since some of the next chapters handle such problems. The structure of the section follows the nominal case description, but with the particularity that we have tried to present various uncertainty representations. In this sense, each paragraph uses a different $(\mathcal{D}, \Phi)$ uncertainty description.

7.1 Frequency-Domain Approach

Quasipolynomials Robustness We shall start by considering a quasipolynomial family with n_d commensurable delays described by (see also Kogan [91]):

$$\mathcal{Q} = \left\{ q(s,\delta,\tau) = \sum_{k,l=0}^{m,n_d} t_{k,l}(\delta) s^k e^{-sl\tau} \quad \delta \in \mathcal{D}, \tau \in [\underline{\tau}, \overline{\tau}] \right\}, \tag{7.1}$$

where $\mathcal{D}$ is a compact and convex subset of $\mathbb{C}^{(m+1)\times(m+1)}$ and $t_{k,l}(\delta)$ are continuous (complex or real valued) functions of δ.

A basic result for such a quasipolynomial family is the following *zero exclusion criterion*, which can be summarized as follows:

Proposition 25 (Zero exclusion criterion) *If there exist two couples (d_0, τ_0) and (d_1, τ_1) such that $q(s, d_0, \tau_0)$ and $q(s, d_1, \tau_1)$ are stable and respectively unstable quasipolynomials, then there exist a couple (d, τ) and an $\omega \in \mathbb{R}$, such that the quasipolynomial $q(s, d, \tau)$ satisfies the condition:*

$$q(j\omega, d, \tau) = 0. \tag{7.2}$$

Other remarks. Fu *et al.* [53] have extended the Edge theorem (see also the paper of Bartlett, Hollot and Lin [8]) to quasipolynomials family with constant delays and coefficients depending affinely on parameters. A different result has been proposed in [6] for a quasipolynomial family with *interval* delays. Tsypkin and Fu [178] have proposed a graphical test for quasipolynomial family with an *interval* delay. A different approach based on *convex directions* has been considered by Kharitonov and Zabko [90]. Algorithms as well as further results on convex approach for stability analysis of such kinds of quasipolynomials can be found in [91].

Delay-independent stability results for *interval-quasipolynomial* (the coefficients are inside some specified intervals, etc.) with constant delays are given in [149]. *Delay-independent* and *delay-dependent* stability conditions for the commensurable as well as for non-commensurable delays cases are given in [78]. Boese [17] derived necessary and sufficient stability conditions for quasipolynomials with interval coefficients and *one* interval delay. Other robust stability results are given in [77].

Special criteria In this class, we shall consider only the maximum principle type extensions (in all the parameter sets).

Maximum Principle Based Approach In this class, one considers the *Mori and Kokame criterion* extensions and the *structured singular value* techniques:

- *Mori and Kokame criterion* [126] extensions. Consider now the triplet Σ_r associated to the system (single delay case):

$$\dot{x}(t) = (A + \Delta A)x(t) + (A_d + \Delta A_d)x(t - \tau), \tag{7.3}$$

with ΔA and ΔA_d time-invariant uncertainty satisfying the following boundness condition:

$$\|\Delta A\| < \beta, \quad \|\Delta A_d\| < \beta_d. \tag{7.4}$$

The basic idea is to verify the inequality:

$$\mu(A + A_d e^{-s\tau}) + \beta + \beta_d < 0$$

on the boundary of a given compact in $\mathbb{C}^+$. If this inequality holds, then the ***robust stability*** property holds (see, for example, [171]).

- ***Structured singular value*** techniques. Consider the triplet Σ_r described by (7.3) and suppose that:

$$\overline{\sigma}(\Delta A) < \gamma_a, \quad \overline{\sigma}(\Delta A_d) < \gamma_d$$

Introduce now the following sets (see also [27]):

$$\begin{aligned} \mathcal{Y}_p(\gamma) &= \{diag(\Delta_1, \ldots, \Delta_p) \;:\; \Delta_k \in \mathbb{R}^{n\times n}, \overline{\sigma}(\Delta_k) \le \gamma\} \\ \mathcal{Z}_p(\gamma) &= \{diag(\Delta_1, \Delta_2) \;:\; \Delta_1 \in \mathcal{X}_1(\gamma), \Delta_2 \in \mathcal{Y}_p(\gamma)\} \end{aligned}$$

Then we have ***robust delay-independent*** stability of the triplet Σ_r if:

$$\mu_{\mathcal{Y}_1}\left(\gamma_a (j\omega I_n - A)^{-1}\right) < 1, \quad \omega \in \mathbb{R}^+,$$

and

$$\mu_{\mathcal{Z}_2}(\mathcal{M}(j\omega)) < 1, \quad \omega \in \mathbb{R}^+,$$

where

$$\mathcal{M}(s) = \begin{bmatrix} (sI_n - A)^{-1}A_d & (sI_n - A)^{-1} & (sI_n - A)^{-1} \\ -\gamma_d I_n & 0 & 0 \\ \gamma_a (sI_n - A)^{-1} & \gamma_a (sI_n - A)^{-1} & \gamma_a (sI_n - A)^{-1} \end{bmatrix}.$$

For further comments, see [27].

7.2 Time-Domain Approach

Similarly to the nominal case, we consider here the ***Lyapunov's second method*** via Lyapunov-Krasovskii and Lyapunov-Razumikhin techniques and the ***comparison principle*** methods (only in the matrix measure framework), respectively.

Lyapunov's Second Method Consider now the triplet $\Sigma_r = (\Sigma, \mathcal{D}, \Phi)$ described by the following functional differential equation:

$$\dot{x}(t) = [A + \Delta A(t)]x(t) + \sum_{i=1}^{n_d}[A_{di} + \Delta A_{di}(t)]x(t - \tau_i), \tag{7.5}$$

where $\Delta A(t)$ and $\Delta A_{di}(t)$ $(i = \overline{1, n_d})$ are *time-varying* and *norm-bounded* uncertainties satisfying:

$$\begin{cases} \Delta A(t) = D_a F_a(t) E_a, & D_a \in \mathbb{R}^{n \times m_a}, \quad E_a \in \mathbb{R}^{n_a \times n}, \\ & F_a(t) \in \mathbb{R}^{m_a \times n_a} \\ \Delta A_{di}(t) = D_{di} F_{di}(t) E_{di}, & D_{di} \in \mathbb{R}^{n \times m_{di}}, \quad E_{di} \in \mathbb{R}^{n_{di} \times n}, \\ & F_{di}(t) \in \mathbb{R}^{m_{di} \times n_{di}} \quad (i = \overline{1, n_d}). \end{cases} \tag{7.6}$$

where $F_a(t)$ and $F_{di}(t)$ are the uncertain matrices for the *'actual'* state $x(t)$ and for the *'delayed'* state $x(t - \tau_i)$ respectively and D_a, D_{di}, E_a, E_{di} $i = \overline{1, n_d}$ are known real matrices which characterize how the unknown parameters in $F_a(t)$ and $F_{di}(t)$ enter the nominal matrices A and A_{di} respectively.

$$F_a(t), F_{di}(t) \in \mathcal{F} = \{F(t) \quad : \quad F(t)^T F(t) \leq I\}.$$

(where the elements of $F(t)$ are supposed Lebesgue measurable).

For the sake of simplicity, the following results correspond to the case when we have *no uncertainty* on the matrices A_{di}, $i = \overline{1, n_d}$, i.e. $\Delta A_{di} = 0$.

Since the ideas are completely similar to the nominal case, via the same Lyapunov-Krasovskii and Lyapunov-Razumikhin forms, we shall present only two simple extensions: *robust delay-independent* stability using a Lyapunov-Krasovskii approach and *robust delay-dependent* stability via a Lyapunov-Razumikhin approach, respectively, for the single delay case. The results can be summarized as follows:

Proposition 26 *Assume that the pair (A, A_d) satisfies Assumption 1 ($A + A_d$ Hurwitz stable). Then the triplet $\Sigma_r = (\Sigma, \mathcal{D}, \Phi)$ (7.5)-(7.6) (with $D_d = 0$, $E_d = 0$) is delay-independent robustly stable if one the following assertions holds:*

(i) there exists a couple of symmetric and positive definite matrices $P > 0$ and $S > 0$ satisfying the following Riccati inequality:

$$A^T P + PA + P[A_d S^{-1} A_d^T + D_a D_a^T]P + E_a^T E_a + S < 0.$$

(ii) there exists a couple of symmetric and positive definite matrices $P > 0$ and $S > 0$ satisfying the following LMI:

$$\begin{bmatrix} A^T P + PA + E_a^T E_a + S & PA_d & PD_a \\ A_d^T P & -S & 0 \\ D_a^T P & 0 & -I_{m_a} \end{bmatrix} < 0.$$

The proof makes use of the same Lyapunov-Krasovskii functional:

$$V(x_t) = x(t)^T P x(t) + \int_{-\tau}^{0} x(t+\theta)^T S x(t+\theta) d\theta,$$

The general *delay-independent* cases can be found in [198] (Riccati equation approach using a similar Lyapunov-Krasovskii functional with $P > 0$ and $S = I_n$) or in [137] (Riccati equation approach for time-varying delays).

Proposition 27 *Assume that the pair (A, A_d) satisfies Assumption 1. Then the triplet $\Sigma_r = (\Sigma, \mathcal{D}, \Phi)$ (7.5)-(7.6) (with $D_d = 0$, $E_d = 0$) with $\tau(t) \in \mathcal{V}(\tau^*)$ is uniformly asymptotically stable if there exist a symmetric and positive definite matrix $P > 0$ and scalars $\varepsilon > 0$, $\beta_1 > 0$ and $\beta_2 > 0$ satisfying the following LMIs:*

$$\begin{bmatrix} \begin{pmatrix} (A+A_d)^T P + P(A+A_d) + \\ +E_a^T E_a + \frac{\tau^*}{\varepsilon} D_a D_a^T + \\ +\tau^*(\beta_1+\beta_2)P \end{pmatrix} & PD_a & \tau^* P A_d \mathcal{Q} \\ D_a^T P & -I_{m_a} & 0 \\ \tau^* \mathcal{Q}^T A_d^T P & 0 & -\tau^* \mathcal{R} \end{bmatrix} < 0, \quad (7.7)$$

$$P - \varepsilon E_a^T E_a > 0,$$

where

$$\mathcal{Q} = \begin{bmatrix} A & A_d & D_a A \end{bmatrix},$$

$$\mathcal{R} = \begin{bmatrix} \beta_1 P & 0 & 0 \\ 0 & \beta_2 P & 0 \\ 0 & 0 & P - \varepsilon E_a^T E_a \end{bmatrix}. \quad (7.8)$$

Remark 28. The computation of the suboptimal bound τ^* on the delay size can be reduced to a standard LMI optimization problem (see the stability results section).

The general *delay-dependent* cases can be found in [134] (a Riccati equation approach using an appropriate Lyapunov-Krasovskii functional; single delay case), [135] (the same approach for multiple delays case, i.e. bounded sets) [196] (an LMI approach based on a Lyapunov-Krasovskii functional) or [197] (an LMI approach via a Razumikhin type technique). Delay-dependent results (single delay case) can be also found in [170] (see also [199]) or in [167] (Razumikhin type technique, different uncertainty representations).

Comparison Principle Using the "classical" scheme for this section, we shall start by introducing a "new" class of time-varying uncertainty. Consider now the following Σ_r triplet described by the functional differential equation:

$$\dot{x}(t) = Ax(t) + \sum_{i=1}^{n_d} A_{di} x(t-\tau_i) + f(x(t),t) + \sum_{i=1}^{n_d} f_{di}(x(t-\tau_i),t), \quad (7.9)$$

where $f(x(t),t)$ and $f_{di}(x(t-\tau_i),t)$ $(i=\overline{1,n_d})$ are *non-linear, continuous* and *time-varying uncertainty*, satisfying the following boundedness condition: there exist non-negative numbers β and β_{di}, $i=\overline{1,n_d}$, such that for all $x \in \mathbb{R}^n$ and for all t:

$$\begin{cases} \| f(x,t) \| \leq \beta \| x \| \\ \| f_{d_i}(x,t) \| \leq \beta_{di} \| x \|, \quad i=\overline{1,n_d}. \end{cases} \tag{7.10}$$

Simple computations allow to define the corresponding $\mathcal{D}$ and Φ respectively in order to define the triplet $\Sigma_r = (\Sigma, \mathcal{D}, \Phi)$. It is necessary to assume that the corresponding *time-varying* functional differential equation is well-defined, etc.

In the sequel we shall present results concerning the robust exponential stability of Σ_r, using a *matrix measure* based *comparison principle* method. In fact, for the brevity of the paper, we consider only the *time-varying single delay* case. The results can be summarized as follows:

Proposition 28 *Consider the triplet Σ_r, and assume that A is a Hurwitz stable matrix satisfying*

$$\| exp(At) \| \;\leq\; k_A \cdot exp(-\eta_A t)$$

for some real numbers $k_A \geq 1$ and $\eta_A > 0$. If the inequality

$$\frac{k_A}{\eta_A}(\| A_d \| + \beta + \beta_d) \;<\; 1 \tag{7.11}$$

holds then, the transient response of $x(t)$ satisfies

$$\| x(t) \| \;\leq\; M \sup_{\theta \in \mathcal{E}_0} \{\| \phi(\theta) \|\} exp\left(-\sigma \int_0^t \frac{d\theta}{\tau(\theta)}\right), \qquad \forall t \geq 0, \quad M \geq 1, \tag{7.12}$$

where $\sigma > 0$ is the unique positive solution of the transcendental equation

$$1 - \frac{k_A}{\eta_A}\beta - \frac{\sigma}{\eta_A \tau(0)} = \frac{k_A}{\eta_A}(\| A_d \| + \beta_d) exp\left(\frac{\sigma}{1-\alpha}\right). \tag{7.13}$$

Remark 29. When there are no uncertainties, i.e. $\beta = 0$ and $\beta_d = 0$, we recover Proposition 24. Using the measure of the matrix A by letting $k_A = 1$ and $\eta_A = -\mu(A)$, we can easily rewrite Proposition 28, similarly to Corollary 1.

Remark 30. The technique used to prove this result is similar to the one described in the previous section dedicated to stability result, but it uses a different system for comparison:

$$\dot{y}(t) \;=\; -(\eta_A - k_A\beta)y(t) + q(t)y(t-\tau(t)), \tag{7.14}$$

where

$$q(t) = \left(\eta_A - k_A\beta - \frac{\sigma}{\tau(t)}\right) exp\left(-\sigma \int_{t-\tau(t)}^{t} \frac{d\theta}{\tau(\theta)}\right).$$

Simple verifications show that the corresponding solution is

$$y(t) = C_0\, exp\left(-\sigma \int_0^t \frac{d\theta}{\tau(\theta)}\right),$$

where C_0 is a real constant. If $\beta \equiv 0$, one obtains the same comparison system form as in the case without uncertainty (5.28).

For the *delay-dependent* case, one has the following result:

Proposition 29 *Consider the triplet Σ_τ and assume that $A + A_d$ is a Hurwitz stable matrix satisfying*

$$\| exp((A + A_d)t) \| \leq k\, exp(-\eta t) \tag{7.15}$$

for some real numbers $k \geq 1$ and $\eta > 0$. If the inequality

$$\frac{k}{\eta}[\bar{\tau}(\| A_dA \| + \| A_d{}^2 \| + \| A_d \| \beta + \| A_d \| \beta_d) + \beta + \beta_d] < 1 \tag{7.16}$$

holds then, the transient response of $x(t)$ satisfies

$$\begin{aligned} \| x(t) \| &\leq M \sup_{\theta\in\mathcal{E}_0} \{\| \phi(\theta) \|\} exp\left(-\sigma \int_0^t \frac{d\theta}{\tau(\theta)}\right), \\ & \forall\, t \geq 0, \quad M \geq 1, \end{aligned} \tag{7.17}$$

where $\sigma > 0$ is the unique positive solution of the transcendental equation

$$\begin{aligned} 1 - \frac{k}{\eta}\beta - \frac{\sigma}{\eta\tau(0)} &= \frac{k}{\eta} exp\left(\frac{\sigma}{1-\alpha}\right) [\bar{\tau} \| A_dA \| \\ &+ \| A_d \| \beta) + \beta_d + \bar{\tau}(\| A_d^2 \| + \| A_d \| \beta_d)\, exp\left(\frac{\sigma}{1-\alpha}\right)]. \end{aligned} \tag{7.18}$$

7.3 Other Remarks

In this section, we have considered the *delay-independent / delay-dependent* stability results in the case when some restrictions have been imposed on the uncertainties. A different problem of interest consists in computing some bounds (in the sense "maximal allowable") for the uncertainty such that the corresponding property (robust delay-independent or delay-dependent) still holds.

For example, consider now the following system:

$$\dot{x}(t) = Ax(t) + f_d(x(t-\tau), t), \tag{7.19}$$

where f_d is a continuous time-varying and nonlinear function satisfying:

$$\|f_d(x,t)\| \leq \beta_d\|x\|, \quad x \in \mathbb{R}^n. \tag{7.20}$$

The robust stability problem which can be considered consists in computing the *maximal bound* on β_d, such that the system (7.19) is robustly stable. Such problems have been considered in [195] (comparison principle techniques), [30] (Razumikhin based approach), and in [177] (Lyapunov-Krasovskii functionals).

For a particular form of uncertainty, real and complex stability radii have been proposed in [176] using an infinite-dimensional representation of the considered system. Other results and comments on robust stability problems can be found in [133].

8 The Examples Revisited

In the previous sections we have presented several analyzing techniques for *delay-independent / delay-dependent* stability for systems including delayed state. Here we are interested to apply some of them to the study of *local* asymptotic stability properties for the examples considered in Section 2.

8.1 Chemical Example

Consider the nonlinear delay system (2.1), whose linearization around the *stationnary point* $x_s = [A_s\ T_s]^T$ is given by:

$$\dot{x}(t) = Ax(t) + A_d x(t-\tau), \tag{8.1}$$

where $x(t) = [A(t)\ T(t)]^T$ and the matrices A and A_d are:

$$\begin{cases} A = \begin{bmatrix} -\frac{q}{V} - K_0 e^{-\frac{Q}{T_s}} & -\frac{K_0 Q A_s}{T_s^2} e^{-\frac{Q}{T_s}} \\ -\frac{(-\Delta H) Q K_0 A_s}{C\rho} e^{-\frac{Q}{T_s}} & -\frac{q}{V} - \frac{(-\Delta H) Q K_0 A_s}{T_s^2 C\rho} e^{-\frac{Q}{T_s}} - \frac{U}{VC\rho} \end{bmatrix}, \\ A_d = \begin{bmatrix} \frac{q(1-\lambda)}{V} & 0 \\ 0 & \frac{q(1-\lambda)}{V} \end{bmatrix}. \end{cases} \tag{8.2}$$

We have the following result:

Proposition 30 [142] *If the linearized system (8.1) without delay is asymptotically stable, then it is $\mathcal{S}_\infty$ asymptotically stable. Furthermore if the delay $\tau(t)$ is a time-varying function in the $\mathcal{V}(r)$ class, then the linearized system is $\mathcal{S}_{v,\tau}$ uniformly asymptotically stable.*

In conclusion, the system (2.1) is delay-independent locally asymptotically stable.

Proof: Since the system without delay is asymptotically stable, it follows that there exists a symmetric and positive definite matrix P such that:

$$(A + A_d)^T P + P(A + A_d) \; < \; 0. \tag{8.3}$$

Since $A_d = 2\beta I_2$, with $\beta = \dfrac{q(1-\lambda)}{2V}$, the condition (8.3) can be rewritten as:

$$\begin{bmatrix} A^T P + PA + \beta P & PA_d \\ A_d^T P & -\beta P \end{bmatrix} \; < \; 0,$$

which is the LMI form of the Riccati inequality (5.16), and thus the property follows via Proposition 18.

Remark 31. Another method to prove the result is based on the *matrix measure* property presented in Section 4. Indeed, due to the particular structure of the matrices A and A_d, the stability property of the system without delay implies that:

$$\mu(A) + ||A_d|| \; < \; 0,$$

and thus the stability property follows.

Remark 32. The constant delay case has been proved in [103] via a frequency-based technique. The same result can be obtained using the matrix pencil technique presented in Section 4 (see also [133]).

8.2 Neural Network Example

Bélair [9] considers a particular structure for the delay neural network (2.2), i.e. of the form:

$$\dot{x}_i(t) \;=\; -x_i(t) + \sum_{j=1}^{n} a_{ij} tanh\left[x_j(t-\tau)\right], \quad 1 \leq i \leq n. \tag{8.4}$$

In order to analyze local stability properties for such systems, consider its *linearization* around 0, i.e. of the form:

$$\dot{x}(t) \;=\; -x(t) + A_d x(t-\tau), \tag{8.5}$$

where the matrix A_d is given by:

$$A_d \;=\; \beta\left[a_{ij}\right]_{1\leq i,j\leq n}, \quad \beta = \frac{d(tanh(s))}{ds}(0).$$

Only for simplification, suppose now that the matrix A_d has real eigenvalues d_j, $j = \overline{1,n}$. Then we have the following result:

Proposition 31 [143] *Consider the system (8.5) satisfying the hypothesis given above. Let Σ be the associated triple. Then the following assertions hold:*

i) Σ is $\mathcal{S}_\infty$ asymptotically stable if and only if the eigenvalues $d_j \in [-1, 1)$ for all $j = \overline{1, n}$.

ii) Σ is $\mathcal{S}_\tau$ asymptotically stable if and only the eigenvalues $d_j < 1$ for all $j = \overline{1, n}$, but there exists at least one eigenvalue d_{j1}, $1 \leq j1 \leq n$, such that $d_{j1} < -1$.

In this case, the optimal bound on the delay size is given by:

$$\tau^* = \min_{1 \leq j \leq n} \frac{arccos(\frac{1}{d_j})}{\sqrt{d_j^2 - 1}}, \tag{8.6}$$

where we consider only the eigenvalues d_j satisfying the condition $d_j < -1$.

The *proof* can be given using the matrix pencil techniques presented in Section 4, after some algebraic manipulations (particular structure for $A = -I_n$, and real eigenvalues for A_d, etc.).

Remark 33. The results are similar to the one given in [9] using a different frequency-based technique, but without taking into account the "limit" $\mathcal{S}_\infty$ case (i.e. corresponding to $d_j = -1$). Other comments are given in [133]. Notice that $d_j \in (-1, 1)$ is equivalent to the *strong* delay-independent stability result.

9 Concluding Remarks

In this chapter, some topics on time-delay systems stability and robust stability have been considered. The delay systems are described by linear differential equations with delayed state including a single or multiple delays, constant or time-varying. Furthermore, state uncertainty may be present. A specific problem has been considered throughout the chapter: The ***influence*** of the delay size on the ***asymptotic stability*** (***robust stability***) property, i.e. delay-independent or delay-dependent. Some algebraic tools have been considered in detail, other tools have been only mentioned in order to reduce the "overlap" with other chapters of this monography. The intention of the authors was not only to classify existing results (and the corresponding methods), but also to present some trends in this field.

Using similar ideas we can analyze the ***stabilization*** problem in terms of the closed-loop system "delay-independent / delay-dependent" stability. Such studies have been considered in [133], where *memoryless* feedback laws have been used. Some *prescriptive* stabilization methods were presented in [52] and [183], for the more general deterministic and stochastic case. See also Lehman et al. [105] for a different technique. Other remarks and comments on the stabilization problem and related topics can be found in the next chapters.

Acknowledgement

Special thanks to Huaizhong Li for several discussions that the authors have with him during his stay at Grenoble. We also thank Michel Dambrine for the fruitfull discussions on the "comparison principle" techniques.

A Stability theory

This appendix recalls the basic notions and definitions used in the Lyapunov second method for functional differential equations.

A.1 Basic definitions

Consider the functional differential equation of retarded type

$$\begin{cases} \dot{x}(t) = f(t, x_t), & t \geq t_0 \\ x_{t_0}(\theta) = \phi(\theta), & \forall\, \theta \in [-\tau, 0] \end{cases} \tag{A.1}$$

where $x_t(\cdot)$, for a given $t \geq t_0$, denotes the restriction of $x(\cdot)$ to the interval $[t-\tau, t]$ translated to $[-\tau, 0]$, i.e.

$$x_t(\theta) = x(t+\theta), \quad \forall\, \theta \in [-\tau, 0].$$

It is assumed that $\phi \in C^v_{n,\tau}$ and the map $f(t,\phi) : \mathbb{R}^+ \times C^v_{n,\tau} \mapsto \mathbb{R}^n$ is continuous and Lipschitzian in ϕ and $f(t,0) = 0$.

Let us denote by $x(t_0, \phi)$ the solution of the functional differential equation (A.1) with the initial condition $(t_0, \phi) \in \mathbb{R}^+ \times C^v_{n,\tau}$.

Definition 1 *The trivial solution $x(t) \equiv 0$ of (A.1) is said to be 'uniformly asymptotically stable' if:*

(a) for every $\kappa > 0$ and for every $t_0 \geq 0$ there exists a $\delta = \delta(\kappa)$ independent of t_0 such that for any $\phi \in C^\delta_{n,\tau}$ the solution $x(t_0, \phi)$ of (A.1) satisfies $x_t(t_0, \phi) \in C^\kappa_{n,\tau}$ for all $t \geq t_0$;

(b) for every $\eta > 0$ and for every $t_0 \geq 0$ there exist a $T(\eta)$ independent of t_0 and a $v_0 > 0$ independent of η and t_0 such that for any $\phi \in C_{n,\tau}$, $\| \phi \|_c < v_0$ implies that $\| x_t(t_0, \phi) \|_c < \eta$, $\forall\, t \geq t_0 + T(\eta)$.

(2) The trivial solution $x(t) \equiv 0$ of (A.1) is said to be 'exponentially stable' if there exist a $B > 0$ and an $\alpha > 0$ such that for all initial conditions $\phi \in C^v_{n,\tau}$, $\|\phi\|_c \leq v_0 \leq v$, the solution satisfies the inequality:

$$\|x(t_0, \phi)(t)\| \;\leq\; B e^{-\alpha(t - t_0)} \|\phi\|_c.$$

We recall that condition (a) implies uniform stability. Furthermore, if the system is linear, the 'uniform asymptotic stability' property is equivalent to the 'asymptotic stability' or to the 'exponential stability' property [93].

Consider now the case of a linear autonomous and homogeneous equations:

$$\dot{x}(t) = L(x_t), \tag{A.2}$$

where the functional $L : C_{n,\tau} \mapsto \mathbb{R}^n$ is continuous. In order to simplify the presentation, we shall focus on the linear systems (with finite point delays) of the form:

$$\dot{x}(t) = Ax(t) + \sum_{i=1}^{n_d} A_i x(t - \tau_i). \tag{A.3}$$

We have the following definitions:

Definition 6. [165] The function $\mathcal{F} : \mathbb{C} \mapsto \mathbb{C}$ given by:

$$\mathcal{F}(\lambda) = det\left(\lambda I_n - A - \sum_{i=1}^{n_d} A_i e^{-\lambda\tau_i}\right), \tag{A.4}$$

is called the ***characteristic function*** corresponding to the linear system (A.3).

Definition 7. [165] The characteristic function $\mathcal{F}$ given in Definition (6) is called ***stable*** if the following condition holds:

$$\{\lambda \in \mathbb{C} \ : \ Re(\lambda) \geq 0, \ \ \mathcal{F}(\lambda) = 0\} = 0. \tag{A.5}$$

In the case of ordinary differential equations the stability of the characteristic function is equivalent to the exponential stability of the trivial solution. The same property holds for the case considered here, but it is not true for general functional differential equations (one needs supplimentary assumptions if the system has infinite delays, etc.; see [165] and the references therein).

A.2 Lyapunov's second method

As we have mentioned before, there are two different ways to develop Lyapunov's second method type results, function on the way of interpreting the solution of the considered functional differential equation, as an ***evolution in a function space*** (Lyapunov-Krasovskii functional) or as an ***evolution in an Euclidian space*** (Lyapunov-Razumikhin function).

We have the following results:

Theorem 3 (Krasovskii Stability Theorem) [70] *Suppose that the function $f : \mathbb{R} \times C_{n,\tau} \mapsto \mathbb{R}^n$ takes bounded sets of $C_{n,\tau}$ in bounded sets of $\mathbb{R}^n$ and suppose that $u(s)$, $v(s)$ and $w(s)$ are continuous, nonnegative and nondecreasing functions with $u(s)$, $v(s) > 0$ for $s \neq 0$ and $u(0) = v(0) = 0$.*

If there is a continuous function $V : \mathbb{R} \times C_{n,\tau} \mapsto \mathbb{R}$ such that

(i) $u(\|\phi(0)\|) \leq V(t,\phi) \leq v(\|\phi\|_c)$,
(ii) $\dot{V}(t,\phi) \leq -w(\|\phi(0)\|)$

then the solution $x = 0$ of the equation (A.1) is uniformly stable.

If $u(s) \to \infty$ as $s \to \infty$ the solutions are uniformly bounded.

If $w(s) > 0$ for $s > 0$, then the solution $x = 0$ is uniformly asymptotically stable.

Theorem 4 (Razumikhin Stability Theorem) [70] *Consider the functional differential equation (A.1). Suppose $u, v, w, p : \mathbb{R}^+ \mapsto \mathbb{R}^+$ are continuous, nondecreasing functions, $u(s)$, $v(s)$, $w(s)$ positive for $s > 0$, $u(0) = v(0) = 0$ and $p(s) > s$ for $s > 0$. If there is a continuous function $V : \mathbb{R} \times \mathbb{R}^n :\mapsto \mathbb{R}$ such that*

(a) $u(\| x \|) \leq V(t,x) \leq v(\| x \|), \quad t \in \mathbb{R}, x \in \mathbb{R}^n$
(b) $\dot{V}(t,x(t)) \leq -w(\| x(t) \|)$ *if* $V(t+\theta,x(t+\theta)) < p(V(t,x(t))), \quad \forall\theta \in [-\tau,0]$

Then, the trivial solution of (A.1) is uniformly asymptotically stable.

References

1. Abdallah, G., Dorato, P., Benitez-Read, J. and Byrne, R.: Delayed positive feedback can stabilize oscillatory systems. *Proc. American Contr. Conf.* (1993) 3106-3107.
2. Agathoklis, P. and Foda, S.: Stability and matrix Lyapunov equation for delay differential systems. *Int. J. Contr.*, **49** (1989) 417-432.
3. Ahlfors, L. V.: *Complex Analysis*, 3rd Ed., McGraw-Hill Book Company, New York, 1979.
4. Amemyia, T: Delay-independent stability of higher-order systems. *Int. J. Contr.*, **50** (1989) 139-149.
5. Amemyia, T.: On the delay-independent stability of a delayed differential equation of a 1st order, *J. Math. Anal. Appl.*, **142** (1989) 13-25.
6. Barmish, B. R. and Shi, Z.: Robust stability of perturbed systems with time-delays. *Automatica*, **25** (1989) 371-381.
7. Barnea, D. I.: A method and new results for stability and instability of autonomous functional differential equations. *SIAM J. Appl. Math.*, **17** (1969) 681-697.
8. Bartlett, A. C., Hollot, C. V. and Lin, H.: Root locations of an entire polytope of polynomials: it suffices to check the edges. *Math. Contr., Sign. & Syst.*, **1** (1988) 61-71.
9. Bélair, J.: Stability in delayed neural networks. in *Ordinary and delay differential equations*, J. Wiener, J. K. Hale (Editors), Pitman Research Notes Math. Series, **272**, John Wiley & Sons, (1992) 6-9.
10. Bélair, J., Campbell, S. A. and van den Driessche, P.: Frustration, stability and delay-induced oscillations in a neural network model *SIAM J. Appl. Math.*, **56** (1996) 245-255.
11. Bellman, R. E.: Vector Lyapunov functions. *SIAM J. Contr.*, Ser. A, **1** (1962) 33-34.

12. Bellman, R. E. and Cooke, K. L.: *Differential-Difference Equations*, Academic Press, New York, 1963.
13. Bensoussan, A., Da Prato, G., Defour, M. C. and Mitter, S. K.: *Representation and control of infinite dimensional systems.* Systems & Control: Foundation & Applications, 2 volumes, Birkhäuser, Boston, 1993.
14. Bhatt, S. J. and Hsu, C. S.: Stability criteria for second-order dynamical systems with time lag. *J. Applied Mechanics* (1966) 113-118.
15. Bilous, O. and Admundson, N.: Chemical reactor stability and sensitivity. *AI ChE Journal*, **1** (1955) 513-521.
16. Boese, F. G.: Stability conditions for the general linear difference-differential equation with constant coefficients and one constant delay. *J. Math. Anal. Appl.*, **140** (1989) pp. 136-176.
17. Boese, F. G.: Stability in a special class of retarded difference-differential equations with interval-valued parameters. *J. Math. Anal. Appl.*, **181** (1994) 227-247.
18. Boese, F. G.: Stability criteria for second-order dynamical systems involving several time delays *SIAM J. Math. Anal.* **5** (1995) 1306-1330.
19. Bourlès, H: α-stability of systems governed by a functional differential equation - extension of results concerning linear delay systems. *Int. J. Contr.*, **45** (1987) 2233-2234.
20. Boyd, S. and Desoer, C. A.: Subharmonic functions and performance bounds in linear time-invariant feedback systems. *IMA J. Math. Contr. Information*, **2** (1985) 153-170.
21. Boyd, S., El Ghaoui, L., Feron, E. and Balakrishnan, V.: *Linear matrix inequalities in system and control theory,* SIAM Studies in Applied Mathematics, **15**, 1994.
22. Brierley, S. D., Chiasson, J. N., Lee, E. B. and Zak, S. H.: On stability independent of delay for linear systems. IEEE Trans. Automat. Control **AC-27** (1982) 252-254.
23. Burton, T. A.: *Stability and periodic solutions of ordinary and functional differential equations.* Academic Press, Orlando, **178**, 1985.
24. Buslowicz, M.: Sufficient conditions for instability of delay differential systems. *Int. J. Contr.* **37** (1983) 1311-1321.
25. Campbell, S. A. and Bélair, J.: Multiple-delayed differential equations as models for biological control systems. *Proc. World Math. Conf.* (1993) 3110-3117.
26. Chen, J.: On computing the maximal delay intervals for stability of linear delay systems. *IEEE Trans. Automat. Contr.* **40** (1995) 1087-1093.
27. Chen, J. and Latchman, H. A.: Frequency sweeping tests for stability independent of delay. *IEEE Trans. Automat. Contr.*, **40** (1995) 1640-1645.
28. Chen, J. Gu, G. and Nett, C. N.: A new method for computing delay margins for stability of linear delay systems. *Proc. 33rd IEEE CDC, Lake Buena Vista, Florida, U.S.A.*, (1994) 433-437.
29. Chen, J., Xu, D. and Shafai, B.: On sufficient conditions for stability independent of delay. *Proc. 1994 American Contr. Conf.*, Baltimore, Maryland (1994) 1929-1933.
30. Cheres, E., Gutman, S. and Palmor, Z. J.: Quantitative measures of robustness for systems including delayed perturbations. *IEEE Trans. Automat. Contr.*, **34** (1989) 1203-1204.
31. Chiasson, J.: A method for computing the interval of delay values for which a differential-delay system is stable. IEEE Trans. Automat. Contr. **33** (1988) 1176-1178.
32. Chiasson, J. N., Brierley, S. D. and Lee, E. B.: A simplified derivation of the Zeheb-Walach 2-D stability test with applications to time-delay systems. *IEEE Trans.*

Automat. Contr., **AC-30** (1985) 411-414; corrections in *IEEE Trans. Automat. Contr.*, **AC-31** (1986) 91-92.

33. Cooke, K. L. and Ferreira, J. M.: Stability conditions for linear retarded functional differential equations. *J. Math. Annal. Appl.*, **96** (1983) 480-504.
34. Cooke, K. L. and van den Driessche, P.: On zeroes of some transcendental equations *Funkcialaj Ekvacioj* **29** (1986) 77-90.
35. Curtain, R. F.: A synthesis of time and frequency domain methods for the control of infinite-dimensional systems: A system theoretic approach. In H. T. BANKS (Ed.), *Control and estimation in distributed parameter system* (1992) 171-224.
36. Curtain, R. F. and Pritchard, A. J.: *Infinite-dimensional linear systems theory.* Lecture Notes in Contr. and Inf. Sciences, **8**, Springer-Verlag, Berlin, 1978.
37. Dambrine, M. and Richard, J. P.: Stability analysis on time-delay systems. *Dynamic Syst. Appl.*, **2** (1993) 405-414.
38. Dambrine, M.: *Contributions à l'étude de la stabilité des systèmes à retard.* Ph.D. Thesis, LAIL URA CNRS D1440, Ecole Centrale de Lille, 1994.
39. Datko, R.: A procedure for determination of the exponential stability of certain differential-difference equations. *Quart. Appl. Math.* **36** (1978) 279-292.
40. Datko, R.: Remarks concerning the asymptotic stability and stabilization of linear delay differential equations. *J. Math. Anal. Appl.*, **111** (1985) 571-584.
41. Datko, R.: Not all feedback stabilized hyperbolic systems are robust with respect to small time delays in their feedbacks. *SIAM J. Contr. Optimization*, **26** (1988) 697-713.
42. Desoer, C. A. and Vidyasagar, M.: *Feedback System: Input-Output Properties.* Academic Press, New York, 1975.
43. Devanathan, R.: A lower bound for limiting time delay for closed-loop stability of an arbitrary SISO plant. *IEEE Trans. Automat. Contr.*, **40** (1995) 717-721.
44. Diekmann, O., von Gils, S. A., Verduyn Lunel, S. M. and Walther, H. -O.: *Delay equations, Functional-, Complex- and Nonlinear Analysis.* Appl. Math. Sciences Series, **110**, Springer-Verlag, New York, 1995.
45. Driver, R. D.: Existence and stability of a delay-differential system. *Arch. Rational Mech. Anal.*, **10** (1962) 401-426.
46. El Sakkary: Estimating robust dead time for closed loop stability. *IEEE Trans. Automat. Contr.*, **35** (1990) 209-210.
47. El'sgol'ts, L. E. and Norkin, S. B.: *Introduction to the theory and applications of differential equations with deviating arguments.* Mathematics in Science and Eng., **105**, Academic Press, New York, 1973.
48. Feron, E., Balakrishnan, V. and Boyd, S.: A design of stabilizing state feedback for delay systems via convex optimization. *Proc. 31st IEEE Conf. Dec. Contr.*, Tuscon, Arizona, USA, (1992) 147-148.
49. Fiagbedzi, Y. A. and Pearson, A. E.: Feedback stabilization of linear autonomous time lag systems. *IEEE Trans. Automat. Contr.*, **AC-31** (1986) 847-855.
50. Fiala, J. and Lumia, R.: The effect of time delay and discrete control on the contact stability of simple position controllers. *IEEE Trans. Automat. Contr.*, **39** (1994) 870-873.
51. Fliess, M. and Mounier, H.: Quelques propriétés structurelles des systèmes linéaires à retards constants. *C.R. Acad. Sci. Paris*, **I-319** (1994) 289-294.
52. Florchinger, P. and Verriest, E. I.: Stabilization of Nonlinear Stochastic Systems with Delay Feedback. *Proc. 32nd IEEE Conf. Decision and Control* San Antonio TX (1993) 859-860.

53. Fu, M., Olbrot, A. W. and Polis, M. P.: Robust stability for time-delay systems: The edge theorem and graphical tests. *IEEE Trans. Automat. Contr.*, **34** (1989) 813-820.
54. Furumochi, T.: Stability and boundedness in functional differential equations. *J. Math. Anal. Appl.*, **113** (1986) 473-489.
55. Garey, M. and Johnoson, D.: *Computers and intractability: A guide to the theory of $\mathcal{NP}$-completeness.* Freeman, San Francisco, 1979.
56. Glader, C., Hognas, G., Makila, P. and Toivonen, H. T.: Approximation of delay systems - a case study. *Int. J. Contr.*, **53** (1991) 369-390.
57. Gohberg, I., Lancaster, P. and Rodman, L.: *Matrix Polynomials.* Computer Science & Appl. Math., Academic Press, New York, 1982.
58. Golub, G. H. and van Loan, Ch. F.: *Matrix computations.* John Hopkins Univ. Press, Baltimore & London, 1989.
59. Gopalsamy, K.: *Stability and oscillations in delay differential equations of population dynamics.* Kluwer Academic Publishers, Math. Its Appl. Series, **74**, 1992.
60. Górecki, H., Fuksa, S., Gabrowski, P. and Korytowski, A.: *Analysis and Synthesis of Time Delay Systems.* John Wiley & Sons (PWN), Warszawa Poland 1989.
61. Goubet, A., Dambrine, M. and Richard, J. P.: An extension of stability criteria for linear and nonlinear time-delay systems, *Proc. IFAC Syst. Struct. Contr.*, Nantes, France (1995) 278-283.
62. Goubet-Bartholomeüs, A.: *Sur la stabilité et la stabilisation des systèmes retardés: Conditions en fonction du retard* (in French), Ph.D. Thesis, Univ. des Sciences et Technologies de Lille, 1996.
63. Gu, G. and Lee, E. B.: Stability testing of time-delay systems. *Automatica*, **25** (1989) 777-780.
64. Gu, G., Khargonekar, P. P., Lee, E. B. and Misra, P.: Finite dimensional approximations of unstable infinite-dimensional systems. *SIAM J. Contr. Opt.*, **30** (1992) 704-716.
65. Habets, L.: *Algebraic and computational aspects of time-delay systems.* Ph. Thesis, Eindhoven Univ. Technology, 1994.
66. Haddock, J. R. and Terjeki, J.: Liapunov-Razumikhin functions and an invariance principle for functional differential equations. *J. Diff. Eq.*, **48** (1983) 95-122.
67. Halanay, A.: *Differential Equations: Stability, Oscillations, Time Lags.* Academic Press, New York, 1966.
68. Hale, J. K.: Dynamics and delays. In S. Busenberg, M. Martelli (Editors) *Delay Differential Equations and Dynamical Systems*, Lecture Notes in Math., **1475** (1991) 16-30, Springer Verlag, Berlin..
69. Hale, J. K., Magalhaes, L. T. and Oliva, W. M.: *An introduction to infinite dynamical systems - Geometric theory.* Applied Math. Sciences, **47**, Springer Verlag, New York, 1985.
70. Hale, J. K. and Verduyn Lunel, S. M.: *Introduction to Functional Differential Equations* . Applied Math. Sciences, **99**, Springer-Verlag, New York, 1991.
71. Hale, J. K., Infante, E. F. and Tsen, F. S. P.: Stability in linear delay equations. *J. Math. Anal. Appl.*, **105** (1985) 533-555.
72. Hale, J. K. and Huang, W.: Global geometry of the stable regions for two delay differential equations. *J. Math. Anal. Appl.*, **178** (1993) 344-362.
73. Hale, J.K., Effects of delays on stability and control. Report CDSN97-270, Center for Dynamical Systems and Nonlinear Studies, Georgia Institute of Technology, 1997.

74. Hertz, D., Jury, E. I. and Zeheb, E.: Stability independent and dependent of delay for delay differential systems. *J. Franklin Inst.*, **318** (1984) 143-150.
75. Hertz, D., Jury, E. I. and Zeheb, E.: Root exclusion from complex polydomains and some of its applications. *Automatica*, **23** (1987) 399-404.
76. Hmamed, A.: On the stability of time-delay systems: New results. *Int. J. Contr.*, **43** (1986) 321-324.
77. Hocherman, J. and Zeheb, E.: Robust stability of time delay systems under uncertainty conditions. *ECCTD'93-Circuit Theory and Design* (1993) 409-414.
78. Hocherman, J., Kogan, J. and Zeheb, E.: On exponential stability of linear systems and Hurwitz stability of characteristic quasipolynomials. *Syst. & Contr. Lett.* **25** (1995) 1-7.
79. Hopfield, J. J.: Neural networks and physical systems with emergent collective computation abilities. *Proc. National Acad. Science U.S.A.*, **79** (1982) 2554-2558.
80. Hsu, C. S.: Application of the τ-decomposition method to dynamical systems subjected to retarded follower forces. *J. Appl. Mechanics*, **37** (1970) 258-266.
81. Hsu, C. S. and Bhatt, S. J.: Stability charts for second-order dynamical systems with time lag. *J. Appl. Mechanics* (1966) 119-124.
82. Huang, W.: Generalization of Lyapunov's theorem in a linear delay system. *J. Math. Anal. Appl.*, **142** (1989) 83-94.
83. Infante, E. F. and Castelan, W. B.: A Lyapunov functional for a matrix difference-differential equation. *J. Diff. Eq.*, **29** (1978) 439-451.
84. Ivanov, A. F. and Verriest, E. I.: Robust Stability of Delay-Difference Equations. In *Systems and Networks: Mathematical Theory and Applications*, (U. Helmke, R. Mennicken, and J. Saurer, eds.) University of Regensburg (1994) 725-726.
85. Jacobson, C. A. and Nett, C. N.: Linear state-space systems in infinite-dimensional space: The role and characterization of joint stabilizability/detectability. *IEEE Trans. Automat. Contr.*, **33** (1988) 541-549.
86. Kamen, E.W.: *Lectures on Algebraic System Theory: Linear Systems over Rings.* NASA Contractor Report 3016 1978.
87. Kamen, E. W.: On the relationship between zero criteria for two-variable polynomials and asymptotic stability of delay differential equations. *IEEE Trans. Automat. Contr.*, **AC-25** (1980) 983-984.
88. Kamen, E. W.: Linear systems with commensurate time delays: Stability and stabilization independent of delay. *IEEE Trans. Automat. Contr.*, **AC-27** (1982) 367-375; corrections in *IEEE Trans. Automat. Contr.*, **AC-28** (1983) 248-249.
89. Kato, J.: Liapunov's second method in functional differential equations. *Tôhoku Math. Journ.*, **332** (1980) 487-492.
90. Kharitonov, V. L. and Zhabko, A. P.: Robust stability of time-delay systems. *IEEE Trans. Automat. Contr.*, **39** (1994) 2388-2397.
91. Kogan, J.: *Robust stability and convexity* LNCIS, vol. 201, Springer-Verlag, Berlin, 1995.
92. Kohonen, T.: *Self organization and Associative Memory.* Springer Verlag, Berlin, 1984.
93. Kolmanovskii, V. B. and Nosov, V. R.: *Stability of Functional Differential Equations.* Mathematics in Science and Eng., **180**, Academic Press, New York, 1986.
94. Kolmanovskii, V. and Myshkis, A.: *Applied Theory of Functional Differential Equations.* Kluwer, Dordrecht the Netherlands 1992.
95. Krasovskii, N. N.: *Stability of motion.* Stanford University Press, 1963.
96. Kuang, Y.: *Delay differential equations with applications in population dynamics.* Academic Press, Boston, 1993.

97. Lakshmikantam, V. and Leela, S.: *Differential and integral inequalities*. Academic Press, New York, 1969.
98. Lam, J.: Convergence of a class of Padé approximations for delay systems. *Int. J. Contr.*, **52** (1990) 989-1008.
99. Lancaster, P. and Tismenetsky, M.: *The theory of matrices* (2nd Edition). Comp. Science Appl. Math. Series, Academic Press, Orlando, 1985.
100. Lee, E. B., Lu, W. -S. and Wu, N. E.: A Lyapunov theory for linear time-delay systems. *IEEE Trans. Automat. Contr.*, **AC-31** (1986) 259-261.
101. Lee, H. and Hsu, C.: On the τ-decomposition method of stability analysis for retarded dynamical systems. *SIAM J. Contr.*, **7** (1969) 242-259.
102. Lehman, B.: Stability of chemical reactions in a CSTR with delayed recycle stream. *Proc. 1994 Amer. Contr. Conf.*, Baltimore, Maryland, U.S.A., (1994) 3521-3522.
103. Lehman, B. and Verriest, E. I.: Stability of a continuous stirred reactor with delay in the recycle streams. *Proc. 30th IEEE Conf. Dec. Contr.*, Brighton, England, (1991) 1875-1876.
104. Lehman, B. and Shujaee, K.: Delay independent stability conditions and decay estimates for time-varying functional differential equations. *IEEE Trans. Automat. Contr.*, **39** (1994) 1673-1676.
105. Lehman, B., Bentsman, J., Verduyn-Lunel, S., and Verriest, E. I.: Vibrational control of nonlinear time-lag systems with arbitrarily large but bounded delay: Averaging theory, stabilizability, and transient behavior. *IEEE Trans. Automatic Control*, **AC-39** 5 (1994) 898-912.
106. Lewis, R. M. and Anderson, B. D. O.: Necessary and sufficient conditions for delay independent stability of linear autonomous systems. *IEEE Trans. Automat. Control*, **AC-25** (1980) 735-739.
107. Louisell, J.: A stability analysis for a class of differential-delay equations having time-varying delay. In S. BUSENBERG, M. MARTELLI (Eds.): *Delay Differential Equations and Dynamical Systems*, Lecture Notes in Math., **1475** (1991) 225-242, Springer Verlag, Berlin.
108. Louisell, J.: Absolute stability in linear delay-differential systems: Ill-possedness and robustness. *IEEE Trans. Automat. Contr.*, **40** (1995) 1288-1291.
109. MacDonald, N.: *Time lags in biological models*. Lecture Notes in Biomathematics, **27**, Springer Verlag, Berlin, 1978.
110. Malek-Zavarei, M. and Jamshidi, M.: *Time Delay Systems: Analysis, Optimization and Applications*. North-Holland Systems and Control Series, **9**, Amsterdam, 1987.
111. Manitius, A.: Necessary and sufficient conditions of approximate controllability for linear retarded systems. *SIAM J. Contr. Opt.*, **19** (1981) 516-532.
112. Manitius, A. and Triggiani, R.: Function space controllability of retarded systems: a derivation from abstract operator conditions. *SIAM J. Opt. Contr.*, **16** (1978) 599-645.
113. Mao, X.: *Stability of stochastic differential equations with respect to semimartingales*. Pitman Research Notes in Mathematics Series **251** Longman Harlow UK 1991.
114. Mao, X.: Robustness of stability of nonlinear systems with stochastic delay perturbations. *Systems & Control Lett.* **19** (1992) 391-400.
115. Mao, X.: *Exponential Stability of stochastic differential equations*. Marcel Dekker New York 1994.

116. Marcus, C. M. and Westervelt, R. M.: Stability of analog neural networks with delay. *Phys. Rev.*, **A 39** (1989) 347-359.
117. Marcus, M.: *Finite dimensional multiplinear algebra* (vol. I). Marcel Dekker, New York, 1973.
118. Marshall, J. E., Górecki, H., Walton, K. and Korytowski, A.: *Time-delay systems: Stability and performance criteria with applications*. Ellis Horwood, New York, 1992.
119. Matrosov, V. M.: Comparison principle and vector Lyapunov functions. *Diff. Urav.* **4** (1968) 1374-1386.
120. Mikoljska, Z.: Une remarque sur des notes de Razumichin et Krasovskij sur la stabilité asymptotique. *Annales Polonici Mathematici* **22** (1969) 69-72.
121. Mohammed, S.-E. A.: *Stochastic functional differential equations*. Pitman Research Notes in Mathematics **99** London UK 1984.
122. Mohammed, S.-E. A.: Stability of linear delay equations under a small noise. *Proc. Edinburgh Math. Soc.* **29** (1986) 233-254.
123. Mori, T.: Criteria for asymptotic stability of linear time-delay systems. *IEEE Trans. Automat. Contr.*, **AC-30** (1985) 158-160.
124. Mori, T., Fukuma, N. and Kuwahara, M.: Simple stability criteria for single and composite linear systems with time delay *Int. J. Contr.*, **34** (1981) 1175-1184.
125. Mori, T., Fukuma, N. and Kuwahara, M.: On an estimate of the decay rate for stable linear delay systems. *Int. J. Contr.*, **36** (1982) 95-97.
126. Mori, T. and Kokame, H.: Stability of $\dot{x}(t) = Ax(t) + Bx(t-\tau)$ *IEEE Trans. Automat. Contr.*, **AC-34** (1989) 460-462.
127. Morse, A. S.: Ring models for delay differential systems *Automatica*, **12** (1976) 529-531.
128. Mounier, H.: *Propriétés structurelles des systèmes linéaires à retards: Aspects théoriques et pratiques*. Université Paris-Sud, Orsay, 1995.
129. Myshkis, A. D.: General theory of differential equations with delay. *Uspehi, Mat. Nauk*, **4** (1949) 99-141 (*Engl. Transl. AMS*, **55**, 1-62, 1951).
130. Nechayeva, I. G. and Khusainov, D. Ya.: Exponential estimates of solutions of linear stochastic differential functional systems. *Ukrainian Math. J.* **42** (10) (1990).
131. Neimark, J.: D-subdivisions and spaces of quasi-polynomials. *Prikl. Math. Mech.*, **13** (1949) 349-380.
132. Nemirovskii, A.: Several $\mathcal{NP}$-hard problems arising in robust stability analysis. *Math. Contr. Signals, Syst*, **6** (1993) 99-105.
133. Niculescu, S. -I.: *On the stability and stabilization of linear systems with delayed-state*, (in French). Ph.D. Thesis, Laboratoire d'Automatique de Grenoble, INPG, February, 1996.
134. Niculescu, S. -I., de Souza, C. E., Dion, J. -M. and Dugard, L.: Robust stability and stabilization of uncertain linear systems with state delay: Single delay case (I). *Proc. IFAC Symp. Robust Contr. Design*, Rio de Janeiro, Brasil (1994) 469-474.
135. Niculescu, S. -I., de Souza, C. E., Dion, J. -M. and Dugard, L.: Robust stability and stabilization of uncertain linear systems with state delay: Multiple delays case (II). *Proc. IFAC Symp. Robust Contr. Design*, Rio de Janeiro, Brasil (1994) 475-480.
136. Niculescu, S. -I., de Souza, C. E., Dugard, L. and Dion, J. -M.: Robust exponential stability of uncertain linear systems with time-varying delays. *Proc. 3rd European Contr. Conf.*, Rome, Italy (1995) 1802-1808.

137. Niculescu, S. -I., de Souza, C. E., Dion, J. -M. and Dugard, L.: Robust $\mathcal{H}_\infty$ memoryless control for uncertain linear systems with time-varying delay. *Proc. 3rd European Contr. Conf.*, Rome, Italy (1995) 1814-1819.
138. Niculescu, S. -I., Trofino-Neto, A., Dion, J. -M. and Dugard, L.: Delay-dependent stability of linear systems with delayed state: An LMI approach *Proc. 34th IEEE Conf. Dec. Contr.*, New Orleans, United States (1995) 1495-1497.
139. Niculescu, S. -I., Dion, J. -M. and Dugard, L.: Delays-dependent stability for linear systems with several delays: An LMI approach *Proc. 13th IFAC World Congr.*, San Francisco **D** (1996) 165-170.
140. Niculescu, S. -I., Dion, J. -M. and Dugard, L.: A matrix pencil approach for asymptotic stability of linear systems with delayed state. *MTNS'96*, Saint Louis, 1996.
141. Niculescu, S. -I. and Ionescu, V.: On delay-independent stability criteria: A matrix pencil approach. *Internal Note LAG 95*, to appear in *IMA Journal Math. Contr. Information*, 1997.
142. Niculescu, S. -I., Dion, J. -M. and Dugard, D.: On the stability of time-delay systems (in French). In J. -M. DION, D. POPESCU (Eds.) *Commande optimale. Conception optimisée des systèmes*, Diderot, Paris (1996) 249-283.
143. Niculescu, S. -I.: Stability and hyperbolicity of linear systems with delayed state: A matrix pencil approach. To appear in *IMA J. Math. Contr. Information* (1997).
144. Niculescu, S. -I.: Delay-interval stability and hyperbolicity of linear time-delay systems: A matrix pencil approach. *4th European Contr. Conf.*, Brussels, Belgium, July 1997.
145. Niculescu, S. -I. and Collado, J.: Stability and hyperbolicity of linear time-delay systems: A matrix pencil tensor product approach. to be presented at *4th IFAC Syst. Struct. Contr.*, Bucharest, Romania, October 1997.
146. Niculescu, S. -I.: $\mathcal{H}_\infty$ memoryless control with an α-stability constraint for time delays systems: An LMI approach. accepted in *IEEE Trans. Automat. Contr.* (1997).
147. Nishioka, K., Adachi, N. and Takeuki, K.: Simple pivoting algorithm for root-locus method of linear systems with delay. *Int. J. Contr.*, **53** (1991) 951-966.
148. Olbrot, A. W.: A sufficient large time-delay in feedback loop must destroy exponential stability of any decay rate. *IEEE Trans. Automat. Contr.*, **AC-29** (1984) 367-3687.
149. Olbrot, A. W. and Igwe, C. U. T.: Necessary and sufficient conditions for robust stability independent of delays and coefficient perturbations. *34th IEEE CDC*, New Orleans, Louisiana, (1995).
150. Packard, A. and Doyle, J.: The complex structured singular value. *Automatica* **29** (1993) 71-109.
151. Partington, J. R.: Approximation of delay systems by Fourrier-Laguerre series. *Automatica* **27** (1991) 569-572.
152. Perlmutter, D.: *Stability of chemical reactors*, Prentice Hall, New Jersey, 1972.
153. Picard, P.: *Sur l'observabilité et la commande des systèmes linéaires à retards modélisés sur un anneau* (in French). Ph.D. Thesis, Ecole Centrale de Nantes, 1996.
154. Răsvan, V.: *Absolute stability of automatic control systems with delays* (in Romanian). Eds. Academiei RSR, Bucharest, Romania, 1975.
155. Razumikhin, B. S.: On the stability of systems with a delay. *Prikl. Math. Meh.*, **20** (1956) 500-512.

156. Repin, Yu. M.: On conditions for the stability of systems of differential equations for arbitrary delays. *Uchen. Zap. Ural.*, **23** (1960) 31-34.
157. Rozkhov, V. I. and Popov, A. M.: Inequalities for solutions of certain systems of differential equations with large time-lag. *Diff. Eq.* **7** (1971) 271-278.
158. Salamon, D.: Structure and stability of finite dimensional approximations for functional differential equations. *SIAM J. Contr. Opt.*, **23** (1985) 928-951.
159. Salamon, D.: On controllability and observability of time-delay systems. *IEEE Trans. Automat. Contr.*, **AC-29** (1984) 432-439.
160. Schoen, G. M. and Geering, H. P.: Stability condition for a delay differential system. *Int. J. Contr.*, **58** (1993) 247-252.
161. Sename, O.: *Sur la commandabilité et le découplage des systèmes linéaires à retards*, Thèse Université de Nantes - Ecole Centrale de Nantes, 1994.
162. Singh, T. and Vadali, S. R.: Robust time-delay control. *J. Dynamical Syst., Meas. and Contr.*, **115** (1993) 303-306.
163. Shyu, K. -K. and Yan, J. -J.: Robust stability of uncertain time-delay systems and its stabilization by variable structure control. *Int. J. Contr.*, **57** (1993) 237-246.
164. Sloss, J. M., Sadek, I. S., Bruch Jr. J. C. and Adali, S.: The effects of time delayed active displacement control of damped structures. *Control Theory Advanced Tech.* **10** (1995) 973-992.
165. Stépán, G.: *Retarded dynamical systems: stability and characterisitc function*, Research Notes in Math. Series, **210**, John Wiley & Sons, 1989.
166. Sontag, E.D., Linear systems over commutative rings: a survey. *Richerche Automat.* **7** (1976), 1-34.
167. Su, J. H: Further results on the robust stability of linear systems with a single delay. *Syst. & Contr. Lett.* **23** (1994) 375-379.
168. Su, J. H., Fong, I. K., and Tseng, C. L.: Stability analysis of linear systems with time delay. *IEEE Trans. Automat. Contr.* **39** (1994) 1341-1344.
169. Su, J. H.: The asymptotic stability of linear autonomous systems with commensurate delays. *IEEE Trans. Automat. Contr.* **40** (1995) 1114-1118.
170. Su, T. J. and Huang, C. G.: Robust stability of delay dependence for linear uncertain systems. *IEEE Trans. Automat. Control* **37** (1992) 1656-1659.
171. Su, T. J. and Liu, P. -L.: Robust stability for linear time-delay systems with delay-dependence. *Int. J. Syst. Science* **24** (1993) 1067-1080.
172. Suh, I. H. and Bien, Z.: A root-locus technique for linear systems with delay. *IEEE Trans. Automat. Contr.* **AC-27** (1982) 205-208.
173. Thowsen, A.: Uniform ultimate boundness of the solutions of uncertain dynamic delay systems with state-dependent and memoryless feedback control. *Int. J. Contr.* **37** (1983) 1153-1143.
174. Toker, O. and Ozbay, H.: Complexity issues in robust stability of linear delay-differential systems. *Math., Contr., Signals, Syst.*. **9** (1996) 386-400.
175. Tokumaru, H., Adachi, N. and Amemyian, T.: Macroscopic stability of interconnected systems. *Proc. 6th IFAC Congress*, paper ID 44.4, Academic Press, New York, 1966.
176. Townley, S. and Pritchard, A. J.: On problems of robust stability for uncertain systems with time-delay. *Proc. 1st European Contr. Conf.* Grenoble France (1991) 2078-2083.
177. Trinh, H. and Aldeen, M.: On the stability of linear systems with delayed perturbations. *IEEE Trans. Automat. Contr.* **39** (1994) 1948-1951.
178. Tsypkin, Ya. Z., and Fu, M.: Robust stability of time-delay systems with an uncertain time-delay constant. *Int. J. Contr.*, **57** (1993) 865-879.

179. Verriest, E. I. and Ivanov, A. F.: Robust stabilization of systems with delayed feedback. *Proc. Second Int'l Symposium on Implicit and Robust Systems* Warszawa Poland (1991) 190-193.
180. Verriest, E. I.: Robust stability of time varying systems with unknown bounded delays. *Proc. 33rd IEEE CDC* Lake Buena Vista FL (1994) 417-422.
181. Verriest, E. I., Fan, M. K. H. and Kullstam, J.: Frequency domain robust stability criteria for linear delay systems. *Proc. 32nd IEEE CDC* San Antonio TX (1993) 3473-3478.
182. Verriest, E. I. and Ivanov, A. F.: Robust stability of systems with delayed feedback. *Circ., Syst., Signal Proc.* **13** (1994) 213-222.
183. Verriest, E. I.: Stabilization of deterministic and stochastic systems with uncertain time delays. *Proc. 33rd IEEE CDC* Orlando FL (1994) 3829-3834.
184. Verriest, E. I. and Florchinger, P.: Stability of stochastic systems with uncertain time delays. *Systems & Control Letters* **24** 1 (1995) 41-47.
185. Verriest, E. I.: Stability of systems with distributed delays. *Preprints of the IFAC System Structure and Control* Nantes France (1995) 294-299.
186. Verriest, E. I. and Ivanov, A. F.: Robust stability of delay-difference equations. *Proc. 34th IEEE CDC* New Orleans LA (1995) 386-391.
187. Verriest, E. I.: Stability and stabilization of stochastic systems with distributed delays. *Proc. 34th IEEE CDC* New Orleans LA (1995) 2205-2210.
188. Verriest, E. I. and Aggoune, W.: Stability of nonlinear differential delay systems. *Proc. CESA-96 IMACS* (1996) Lille France 790-795.
189. Verriest, E. I. and Fan, M. K. H.: Robust stability of nonlinearly perturbed delay systems. *Proc. 35th IEEE CDC* Kobe Japan (1996) 2090-2091.
190. Verriest, E. I. and Aggoune, W.: Stability of nonlinear differential delay systems. To appear in *Mathematics and Computers in Simulation* (1997).
191. Walton, K. and Marshall, J. E.: Direct method for TDS stability analysis. *IEE Proc.* **134** part D (1987) 101-107.
192. Wang, S. S.: Further results on stability of $\dot{x}(t) = Ax(t) + Bx(t-\tau)$. *Syst. & Contr. Lett.* **19** (1992) 165-168.
193. Wang, S. S., Chen, B. S. and Lin, T. P.: Robust stability of uncertain time-delay systems. *Int. J. Control* **46** (1987) 963-976.
194. Wang, W. J. and Wang, R. J.: New stability criteria for linear time-delay systems. *Control - Theory and Advanced Tech.* **10** (1995) 1213-1222.
195. Wu, H. and Mizukami, K.: Quantitative measures of robustness for uncertain time-delay dynamical systems. *Proc. 32nd IEEE CDC* San Antonio TX (1993) 2004-2005.
196. Xi, L. and de Souza, C. E.: LMI approach to delay-dependent robust stability and stabilization of uncertain linear delay systems. *Proc. 34th IEEE Conf. Dec. Contr.* New Orleans LA (1995) 3614-3619.
197. Xi, L. and de Souza, C. E.: Criteria for robust stability of uncertain linear systems with time-varying state delays. *Proc. 13th IFAC World Congr.* San Francisco CA **H** (1996) 137-142.
198. Xie, L. and de Souza, C. E.: Robust stabilization and disturbance attenuation for uncertain delay system. *Proc. 2nd European Contr. Conf.* Groningen The Netherlands (1993) 667-672.
199. Xu, B.: Comments on "Robust Stability of Delay Dependence for Linear Uncertain Systems". *IEEE Trans. Automat. Contr.* **AC-39** (1994) 2365.

200. Ye, H. Michel, A. M. and Wang, K.: Stability of nonlinear dynamical systems with parameter uncertainties with an application to neural networks. *Proc. 1995 American Contr. Conf.* Seattle WA (1995) 2772-2776.
201. Yoneyama, T.: On the $\frac{3}{2}$ stability theorem for one-dimensional delay-differential equations. *J. Math. Anal. Appl.* **125** (1987) 161-173.
202. Yoneyama, T.: On the stability region of scalar delay-differential equations. *J. Math. Anal. Appl.* **134** (1988) 408-425.
203. Yu, W., Sobel, K. M. and Shapiro, E. Y.: A time domain approach to the robustness of time delay systems. *Proc. 31st IEEE CDC* Tucson, AZ (1992) 3726-3727.
204. Zhang, D. -N. Saeki, M. and Ando, K.: Stability margin calculation of systems with structured time-delay uncertainties. *IEEE Trans. Automat. Contr.* **37** (1992) 865-868.
205. Zheng, F., Cheng M. and Gao, W.: Feedback stabilization of linear systems with point delays in state and control variables *Proc. 12th IFAC World Congr.*, Sydney, Australia, **2** (1993) 375-378.
206. Zhou, K., Doyle, J. and Glover, K.: *Robust and optimal control* Prentice Hall, New Jersey, 1995.

Convex directions for stable polynomials and quasipolynomials: A survey of recent results

L. Atanassova[1], D. Hinrichsen[1], V.L. Kharitonov[2]

[1] Institut für Dynamische Systeme, Universität Bremen, D-28334 Bremen, Germany, Telefax: +49 421 218-4235, Phone: +49 421 212227, E-mails: atanassova@mathematik.uni-bremen.de, dh@mathematik.uni-bremen.de
[2] CINVESTAV-IPN, Control Automatico, A.P. 14-740, 07000 Mexico D.F, Mexico, Telefax: 52 5 7477089, Phone: 52 5 747000 ext. 3226, E-mail: khar@ctrl.cinvestav.mx

Abstract. The purpose of this chapter is to give a survey of recent results on convex directions for the sets of stable polynomials and quasipolynomials. It presents a number of analytic criteria ensuring Hurwitz and Schur stability of segments of polynomials. Convex directions are characterized in terms of root loci and it is shown that these root loci behave differently in the real and the complex case. The convex direction problem for sets of stable quasipolynomials is also discussed. Applying similar methods as in the polynomial case analytic stability criteria are obtained for segments of quasipolynomials of delay and of neutral type.

1 Introduction

In recent years robust stability analysis of systems with uncertain parameters has received a good deal of attention, see e.g. [3], [5] and the references therein. A time-invariant linear difference, differential or differential-difference system is exponentially stable if and only if the associated characteristic polynomial (quasipolynomial) is stable. For systems with uncertain parameters this leads to the *robust stability problem* of checking the stability of *sets* of polynomials or quasipolynomials.

Many problems of robust stability can be reduced to the problem of ascertaining the stability of a *polytope* of polynomials or quasipolynomials (i.e. the stability of all the (quasi-)polynomials contained in the polytope). By the Edge Theorem (see [10], [12]) this problem can be reduced to the problem of checking the stability of a *segment* of polynomials (or quasipolynomials). The Edge Theorem states that a polytope Π of polynomials or quasipolynomials is stable if and only if all the edges of Π are stable. A further reduction of the problem is obtained if the directions defined by the edges are *convex directions*. Intuitively speaking, a (quasi-) polynomial q is a convex direction if the set of stable (quasi-) polynomials behaves like a convex set in the direction of q. Hence if all the edges of Π are convex directions the stability of the *vertices* of Π alone ensures the stability of the whole polytope of polynomials (quasipolynomials). This explains why the study of convex directions is of importance for robust stability analysis.

However, it is also of independent theoretical interest regarding the geometry of stable polynomials and quasipolynomials.

The first characterization of convex directions was given by Rantzer in [20]. More precisely, Rantzer's condition characterizes the polynomials q having the property that for *all* (Hurwitz or Schur) stable polynomials p with $\deg p > \deg q$, the stability of $p+q$ implies the stability of the whole segment $[p, p+q]$. Some special classes of convex directions have been identified in [13] and [19]. An attempt to describe the whole set of these polynomials has been made in [14] where some algebraic conditions were derived. A new characterization of convex directions for real polynomials in terms of root loci was presented in [16]. This characterization has led to a new concept of ***convex direction for a given Hurwitz polynomial*** which can be characterized by a graphical test [16].

Some of the stability results for polynomials admit a natural extension to quasipolynomials. Convex directions for Hurwitz stable quasipolynomials of delay and of neutral type were introduced and characterized in [18]. The ***root loci approach*** of [16] was extended to the the stability analysis of segments of real quasipolynomials in [1].

Another promising area of application for the concept of convex directions is the stability theory of multivariable polynomials and quasipolynomials [7], [4], [9]. Multivariable polynomials play an important role in stability analysis of passive multidimensional systems which are used e.g. in image processing [6]. In order to extend the convex direction concept to the case of multivariable polynomials and quasipolynomials one needs characterizations of convex directions for *complex* polynomials and quasipolynomials [2].

The purpose of this chapter is to give an up to date survey of published and unpublished results concerning stable convex directions for polynomials and quasipolynomials (in one variable). For proofs and further technical details we refer to the original papers. We will only present the proofs of unpublished results.

The chapter is organized as follows. In Section 2 and 3 we formulate the convex direction problem for stable polynomials and quasipolynomials in the real and the complex cases. Moreover we describe the basic convex direction conditions given in [20] and [18]. Sections 4 and 5 deal with root loci characterizations of convex directions for stable polynomials and quasipolynomials. It is shown that in the complex case the root loci of the polynomial pencils $p_0(z) + \mu q(z)$, $\mu \geq 0$ (p_0 stable, q a convex direction) never return to the stability region once they have left it. In the real case, however, there are holes in the stability boundary through which a root may return (Theorems 10, 11 and 16). The root loci characterizations in Sections 4 and 5 provide a basis for studying a weaker concept of convex direction which is less conservative than Rantzer's concept if a stable reference (quasi-) polynomial is given. The associated modified convex direction problem is briefly discussed at the end of Sections 4 and 5.

2 Convex Directions for Stable Polynomials

For $n \geq 1$ let $\mathcal{P}_n(\mathbb{K})$ denote the $(n+1)$-dimensional vector space of polynomials of degree at most n with coefficients in the field $\mathbb{K}$, i.e.

$$\mathcal{P}_n(\mathbb{K}) := \left\{ p \, ; \; p(z) = \sum_{k=0}^{n} a_k z^k \, , \; a_k \in \mathbb{K} \right\} . \tag{2.1}$$

Throughout the chapter we suppose that $\mathbb{K} = \mathbb{R}$ or $\mathbb{K} = \mathbb{C}$.

Let $\mathbb{C} = \mathbb{C}_g \dot{\cup} \mathbb{C}_b$ be a given nontrivial partition of the complex plane $\mathbb{C}$ into a "good" and a "bad" region, where the "good" region $\mathbb{C}_g$ is assumed to be open. A non-constant polynomial is called *$\mathbb{C}_g$-stable* if all its roots belong to the "good" region $\mathbb{C}_g$. $\mathcal{S}_n(\mathbb{K}, \mathbb{C}_g)$ denotes the set of all $\mathbb{C}_g$-stable polynomials of degree n with coefficients in $\mathbb{K}$. Thus $\mathcal{S}_n(\mathbb{K}, \mathbb{C}_g) \subset \mathcal{P}_n(\mathbb{K}) \setminus \mathcal{P}_{n-1}(\mathbb{K})$. *Throughout this chapter we assume $n \geq 2$.*

The set $\mathcal{S}_n(\mathbb{K}, \mathbb{C}_g)$ is a non-convex cone. However, in some directions it behaves like a convex set. The following definition is due to Rantzer.

Definition 1. A polynomial $q \in \mathcal{P}_{n-1}(\mathbb{K})$ is called a *convex direction* for the set $\mathcal{S}_n(\mathbb{K}, \mathbb{C}_g)$ if, for all the polynomials $p \in \mathcal{S}_n(\mathbb{K}, \mathbb{C}_g)$, the $\mathbb{C}_g$-stability of $p+q$ implies the $\mathbb{C}_g$-stability of the segment $[p, p+q] = \{p + \mu q \, ; \; \mu \in [0,1]\}$, i.e. if q satisfies

$$p, p+q \in \mathcal{S}_n(\mathbb{K}, \mathbb{C}_g) \quad \Rightarrow \quad [p, p+q] \subset \mathcal{S}_n(\mathbb{K}, \mathbb{C}_g).$$

for all $p \in \mathcal{S}_n(\mathbb{K}, \mathbb{C}_g)$. The set of all these convex directions will be denoted by $\mathcal{D}_{n-1}(\mathbb{K}, \mathbb{C}_g)$.

Note that convex directions for $\mathcal{S}_n(\mathbb{K}, \mathbb{C}_g)$ are, by definition, of degree $< n$ so that all the polynomials in the segment $[p, p+q]$ are of degree n. Since $\mathcal{S}_n(\mathbb{K}, \mathbb{C}_g)$ is a cone one obtains the following simple consequence of Definition 1:
A polynomial $q \in \mathcal{P}_{n-1}(\mathbb{K})$ is a convex direction if and only if, for every stable polynomial $p \in \mathcal{S}_n(\mathbb{K}, \mathbb{C}_g)$, the set

$$M(p,q) \;\; = \;\; \{\mu \geq 0; \; p + \mu q \in \mathcal{S}_n(\mathbb{K}, \mathbb{C}_g)\} \tag{2.2}$$

is an interval in $\mathbb{R}_+ = [0, \infty)$. In particular, the set of convex directions for $\mathcal{S}_n(\mathbb{K}, \mathbb{C}_g)$ is itself a cone.

Remark 1. Another way of expressing the preceding necessary and sufficient condition is: $q \in \mathcal{P}_{n-1}(\mathbb{K})$ is a convex direction for $\mathcal{S}_n(\mathbb{K}, \mathbb{C}_g)$ if and only if the intersection of the ray $R(p,q) = \{p + \mu q; \; \mu \geq 0\}$ with the set $\mathcal{S}_n(\mathbb{K}, \mathbb{C}_g)$ is convex for *all* $p \in \mathcal{S}_n(\mathbb{K}, \mathbb{C}_g)$. Given a stable polynomial $p_0 \in \mathcal{S}_n(\mathbb{K}, \mathbb{C}_g)$ we say that $q \in \mathcal{P}_{n-1}(\mathbb{K})$ is a *convex direction for $\mathcal{S}_n(\mathbb{K}, \mathbb{C}_g)$ relative to p_0* if

$$p_0 + \mu q \in \mathcal{S}_n(\mathbb{K}, \mathbb{C}_g) \quad \Rightarrow \quad [p_0, p_0 + \mu q] \subset \mathcal{S}_n(\mathbb{K}, \mathbb{C}_g) \tag{2.3}$$

holds for all $\mu \in \mathbb{R}_+$. In other words, if a point moving away from p_0 on the ray $R(p_0, q)$ leaves the set $\mathcal{S}_n(\mathbb{K}, \mathbb{C}_g)$ it never returns to this set. The set

$\mathcal{D}_{n-1}(\mathbb{K}, \mathbb{C}_g; p_0)$ of all convex directions for $\mathcal{S}_n(\mathbb{K}, \mathbb{C}_g)$ relative to p_0 depends not only on $\mathcal{S}_n(\mathbb{K}, \mathbb{C}_g)$ but also on the polynomial p_0. The intersection of all these sets coincides with the set $\mathcal{D}_{n-1}(\mathbb{K}, \mathbb{C}_g)$ of convex directions in the sense of Definition 1. In other words, a convex direction in the sense of Definition 1 is a convex direction for the set $\mathcal{S}_n(\mathbb{K}, \mathbb{C}_g)$ relative to every point $p_0 \in \mathcal{S}_n(\mathbb{K}, \mathbb{C}_g)$. In order to emphasize this we will sometimes call them *global* convex directions for $\mathcal{S}_n(\mathbb{K}, \mathbb{C}_g)$.
Note that in order to conclude the stability of a given segment $[p_0, p_0 + \mu q]$ from the stability of the endpoints p_0 and $p_0 + \mu q$ it suffices to know that q is a convex direction for $\mathcal{S}_n(\mathbb{K}, \mathbb{C}_g)$ relative to p_0. It can be shown by examples, see [16], that the set $\mathcal{D}_{n-1}(\mathbb{K}, \mathbb{C}_g; p_0)$ of convex directions for $\mathcal{S}_n(\mathbb{K}, \mathbb{C}_g)$ relative to a given p_0 may be much larger than the set $\mathcal{D}_{n-1}(\mathbb{K}, \mathbb{C}_g)$ of global convex directions for $\mathcal{S}_n(\mathbb{K}, \mathbb{C}_g)$.

In this chapter we restrict our considerations to the classical stability domains, namely the open complex left half-plane

$$\mathbb{C}_- = \{z \in \mathbb{C}\,;\ \ \operatorname{Re} z < 0\}$$

and the open complex unit disk

$$\mathbb{C}_1 = \{z \in \mathbb{C}\,;\ \ |z| < 1\}.$$

A non-constant polynomial $p \in \mathcal{P}_n(\mathbb{K})$ is said to be *Hurwitz stable* or a *Hurwitz polynomial* if all its roots belong to $\mathbb{C}_-$; it is called *Schur stable* or a *Schur polynomial* if all its roots are contained in $\mathbb{C}_1$. Thus $\mathcal{S}_n(\mathbb{K}, \mathbb{C}_-)$ and $\mathcal{S}_n(\mathbb{K}, \mathbb{C}_1)$ are the sets of Hurwitz and of Schur polynomials, respectively, with coefficients in $\mathbb{K}$.

The following characterizations of convex directions for Hurwitz and Schur polynomials are due to Rantzer [20]. We first discuss the Hurwitz case.

2.1 Convex directions for Hurwitz polynomials: $\mathbb{C}_g = \mathbb{C}_-$

Theorem 2 [20].
(i) A polynomial $q \in \mathcal{P}_{n-1}(\mathbb{C})$ is a convex direction for the set $\mathcal{S}_n(\mathbb{C}, \mathbb{C}_-)$ if and only if the following inequalities hold:

$$\frac{\partial \arg(q(\imath\omega))}{\partial \omega} \leq 0\,, \quad \omega \in \{w \in \mathbb{R}\,;\ q(\imath w) \neq 0\}\,. \tag{2.4}$$

(ii) A polynomial $q \in \mathcal{P}_{n-1}(\mathbb{R})$ is a convex direction for the set $\mathcal{S}_n(\mathbb{R}, \mathbb{C}_-)$ if and only if the following inequalities hold:

$$\frac{\partial \arg(q(\imath\omega))}{\partial \omega} \leq \left| \frac{\sin(2\arg(q(\imath\omega)))}{2\omega} \right|\,, \quad \omega \in \{w > 0\,;\ q(\imath w) \neq 0\}\,. \tag{2.5}$$

It follows from (2.4) and (2.5) that if $q \in \mathcal{P}_{n-1}(\mathbb{K})$ is a convex direction for the set $\mathcal{S}_n(\mathbb{K}, \mathbb{C}_-)$, then cq is also a convex direction, for arbitrary $c \in \mathbb{K}$. In particular, if q is a convex direction for $\mathcal{S}_n(\mathbb{K}, \mathbb{C}_-)$ then also $-q$ is a convex direction for this set. Moreover, since the conditions (2.4) and (2.5) do not depend upon n, a convex direction q for $\mathcal{S}_n(\mathbb{K}, \mathbb{C}_-)$ is also a convex direction for $\mathcal{S}_{n'}(\mathbb{K}, \mathbb{C}_-)$, if $\deg q < n'$. Finally, every real polynomial which is a convex direction for $\mathcal{S}_n(\mathbb{C}, \mathbb{C}_-)$ is also a convex direction for $\mathcal{S}_n(\mathbb{R}, \mathbb{C}_-)$. But, the converse is not true.

Conditions (2.4) and (2.5) contrast with the following well-known *phase increasing properties of complex and real Hurwitz polynomials* which play an important role in the proof of Theorem 2, see [20]. Note that these properties again do not depend upon the degree $n \geq 2$.

Theorem 3.
(i) If $p \in \mathcal{S}_n(\mathbb{C}, \mathbb{C}_-)$, then

$$\frac{\partial \arg(p(\imath\omega))}{\partial \omega} > 0, \quad \omega \in \mathbb{R}. \tag{2.6}$$

(ii) If $p \in \mathcal{S}_n(\mathbb{R}, \mathbb{C}_-)$, then

$$\frac{\partial \arg(p(\imath\omega))}{\partial \omega} > \left| \frac{\sin(2\arg(p(\imath\omega)))}{2\omega} \right|, \quad \omega > 0. \tag{2.7}$$

(Recall that we assume $n \geq 2$ throughout the chapter. The strict inequality in (2.7) has to be replaced by an equality if p is a Hurwitz polynomial of degree 1).

2.2 Convex directions for Schur polynomials: $\mathbb{C}_g = \mathbb{C}_1$

Theorem 4 [20].
(i) A polynomial $q \in \mathcal{P}_{n-1}(\mathbb{C})$ is a convex direction for the set $\mathcal{S}_n(\mathbb{C}, \mathbb{C}_1)$ if and only if the following inequalities hold:

$$\frac{\partial \arg(q(e^{\imath\theta}))}{\partial \theta} \leq \frac{n}{2}, \quad \theta \in \{\phi \in [0, 2\pi) \, ; \, q(e^{\imath\phi}) \neq 0\}. \tag{2.8}$$

(ii) A polynomial $q \in \mathcal{P}_{n-1}(\mathbb{R})$ is a convex direction for the set $\mathcal{S}_n(\mathbb{R}, \mathbb{C}_1)$ if and only if the following inequalities hold:

$$\frac{\partial \arg(q(e^{\imath\theta}))}{\partial \theta} \leq \frac{n}{2} + \left| \frac{\sin(2\arg(q(e^{\imath\theta})) - n\theta)}{2\sin(\theta)} \right|, \quad \theta \in \{\phi \in (0, \pi) \, ; \, q(e^{\imath\phi}) \neq 0\} \tag{2.9}$$

Schur stable polynomials of degree ≥ 2 have the following *phase increasing property*, see e.g. [20], [8].

Theorem 5.
(i) If $p \in \mathcal{S}_n(\mathbb{C}, \mathbb{C}_1)$, *then*

$$\frac{\partial \arg(p(e^{\imath\theta}))}{\partial\theta} > \frac{n}{2}, \quad \theta \in [0, 2\pi). \tag{2.10}$$

(ii) If $p \in \mathcal{S}_n(\mathbb{R}, \mathbb{C}_1)$, *then*

$$\frac{\partial \arg(p(e^{\imath\theta}))}{\partial\theta} > \frac{n}{2} + \left|\frac{\sin(2\arg(p(e^{\imath\theta})) - n\theta)}{2\sin(\theta)}\right|, \quad \theta \in (0, \pi). \tag{2.11}$$

Theorem 5 shows that, in contrast with the Hurwitz case, the phase increasing property of Schur polynomials explicitly depends upon the degree of the polynomial. This explains why the characterization of convex directions given in Theorem 4 depends upon the degree n of the Schur polynomials in $\mathcal{S}_n(\mathbb{K}, \mathbb{C}_1)$. Thus, in contrast with the Hurwitz case, a convex direction for $\mathcal{S}_n(\mathbb{K}, \mathbb{C}_1)$ is not necessarily a convex direction for all $\mathcal{S}_{n'}(\mathbb{K}, \mathbb{C}_1)$ with $n' > \deg q$. However, it is a convex direction for all $\mathcal{S}_{n'}(\mathbb{K}, \mathbb{C}_1)$ with $n' > n$.

Similarly to the Hurwitz case a pair of polynomials $p_0 \in \mathcal{S}_n(\mathbb{K}, \mathbb{C}_1)$, $q \in \mathcal{P}_{n-1}(\mathbb{K})$ may not satisfy the conditions (2.8) or (2.9) although q is a convex direction for $\mathcal{S}_n(\mathbb{K}, \mathbb{C}_1)$ at p_0 in the sense of the remark in Subsection 2.1.

3 Convex Directions for Stable Quasipolynomials

In this section we extend the previous results to stable quasipolynomials. For proofs, we refer the reader to [18].

A quasipolynomial is an entire function of the form

$$f(z) = p_0(z)e^{\tau_0 z} + p_1(z)e^{\tau_1 z} + \cdots + p_m(z)e^{\tau_m z} = \sum_{j=0}^{m}\sum_{k=0}^{n} a_{kj} z^k e^{\tau_j z} \tag{3.1}$$

where $p_j(z) = \sum_{k=0}^{n} a_{kj} z^k$, $j = 0, 1, \ldots, m$ are polynomials with coefficients $a_{kj} \in \mathbb{K}$ and $\tau_0 < \tau_1 < \cdots < \tau_m$ are real numbers representing "time shifts" or "delays". The largest degree of the polynomials $p_j(z)$ is said to be the degree of $f(z)$ and is denoted by $\deg f$. If $a_{nm} \neq 0$ then $\deg f = n$ and we call $a_{nm} z^n e^{\tau_m z}$ the ***principal term*** of $f(z)$.

We begin by reviewing some basic properties of quasipolynomials, see [11]. A quasipolynomial $f(z)$ of the form (3.1) with $m > 0$ has an infinite number of roots, but within any bounded region in the complex plane it has only a finite number of roots. Far from the origin the roots belong to a finite number of asymptotic root chains. The positions of these chains are determined by a small number of terms in (3.1). A careful study of the geometry of these chains can be found in [21]. There exist three types of quasipolynomials: delay type, neutral type and advanced type quasipolynomials. The class of delay type quasipolynomials only has ***delay-type root chains***, i.e. asymptotic root chains going "deep" into the left half plane. The class of neutral type quasipolynomials has at least one asymptotic chain of roots in a vertical strip of the complex plane but no asymptotic chain of roots that goes "deep" into the right half plane. If a quasipolynomial has at least one asymptotic chain of roots that goes "deep" into the right half plane, then it belongs to the class of advanced type quasipolynomials. Since we deal in this section with Hurwitz stable quasipolynomials we will only consider quasipolynomials of *delay type* (characterized by $\deg p_i < \deg p_m$, $i = 0, ..., m-1$) and of *neutral type* (characterized by $\deg p_i \leq \deg p_m$, $i = 0, ..., m-1$ and $\deg p_k = \deg p_m$ for at least one $k \leq m-1$).

For $n \geq 0$, $m > 0$ and any real vector $\tau = [\tau_0, \tau_1, \ldots, \tau_m]$ with ordered components $\tau_0 < \tau_1 < \cdots < \tau_m$, let $\mathcal{Q}_n^{m,\tau}(\mathbb{K})$ denote the vector space of quasipolynomials (3.1) with coefficients in $\mathbb{K}$, i.e.

$$\mathcal{Q}_n^{m,\tau}(\mathbb{K}) := \left\{ f\, ;\ f(z) = \sum_{j=0}^{m} \sum_{k=0}^{n} a_{kj} z^k e^{\tau_j z}\ ,\ a_{kj} \in \mathbb{K} \right\}. \tag{3.2}$$

A non-constant quasipolynomial $f(z) \in \mathcal{Q}_n^{m,\tau}(\mathbb{K})$ is called *Hurwitz stable* or simply *stable* if all its roots belong to the open complex left half plane. For fixed $n \geq 0$, $m > 0$, and fixed delays $\tau_0 < \tau_1 < \cdots < \tau_m$ let us denote by $\mathcal{H}_n^{m,\tau}(\mathbb{K})$ the *set of all Hurwitz stable quasipolynomials* $f(z) \in \mathcal{Q}_n^{m,\tau}(\mathbb{K})$ with $a_{nm} \neq 0$. Thus $\mathcal{H}_n^{m,\tau}(\mathbb{K}) \subset \mathcal{Q}_n^{m,\tau}(\mathbb{K}) \setminus \mathcal{Q}_{n-1}^{m,\tau}(\mathbb{K})$, and all the quasipolynomials in $\mathcal{H}_n^{m,\tau}(\mathbb{K})$ are of delay or of neutral type (as defined above, see [11]). For arbitrary quasipolynomials an analogue of the Hermite-Biehler Theorem can be proved and this result is known as Pontryagin's Theorem. However, Hurwitz type stability criteria are not available for quasipolynomials.

It is known that stable quasipolynomials enjoy the following *phase increasing property* [22], [18].

Theorem 6.
(i) If $f \in \mathcal{H}_n^{m,\tau}(\mathbb{C})$, *then*

$$\frac{\partial \arg(f(\imath\omega))}{\partial \omega} > \frac{\tau_0 + \tau_m}{2}, \quad \omega \in \mathbb{R}. \tag{3.3}$$

(ii) If $f \in \mathcal{H}_n^{m,\tau}(\mathbb{R})$, *then*

$$\frac{\partial \arg(f(\imath\omega))}{\partial \omega} > \frac{\tau_0 + \tau_m}{2} + \left| \frac{\sin(2\arg(f(\imath\omega)) - (\tau_0 + \tau_m)\omega)}{2\omega} \right|, \quad \omega > 0. \tag{3.4}$$

The above theorem shows that the phase increasing property of a quasipolynomial in $\mathcal{H}_n^{m,\tau}(\mathbb{C})$ depends upon the minimal delay τ_0 and the maximal delay τ_m. It was shown in [18] that the lower bounds in Theorem 6 are tight for both delay and neutral type quasipolynomials.

$\mathcal{H}_n^{m,\tau}(\mathbb{K})$ is a non-convex cone. However, in some directions it behaves like a convex set. Following [18] we define the concept of convex directions for quasipolynomials in the following way:

Definition 7. A quasipolynomial $g \in \mathcal{Q}_{n-1}^{m,\tau}(\mathbb{K})$ is called a *convex direction* for the set $\mathcal{H}_n^{m,\tau}(\mathbb{K})$ if, for *all* stable quasipolynomials $f \in \mathcal{H}_n^{m,\tau}(\mathbb{K})$, the stability of $f+g$ implies the stability of the whole segment of quasipolynomials $[f, f+g] = \{f+\mu g;\ \mu \in [0,1]\}$, i.e. if g satisfies, for all $f \in \mathcal{H}_n^{m,\tau}(\mathbb{K})$,

$$f, f+g \in \mathcal{H}_n^{m,\tau}(\mathbb{K}) \Rightarrow [f, f+g] \subset \mathcal{H}_n^{m,\tau}(\mathbb{K}).$$

We emphasize that convex directions for quasipolynomials are not defined relative to the whole class of quasipolynomials of given degree n but to the subclass of quasipolynomials of degree n having given time shifts $\tau_0, \tau_1, \ldots, \tau_m$.

As a simple consequence of Definition 7 we obtain that the convex directions g are characterized by the property that, for every Hurwitz quasipolynomial $f_0 \in \mathcal{H}_n^{m,\tau}(\mathbb{K})$ the *stability set* of the pair (f_0, g)

$$M(f_0, g) = \{\mu \geq 0;\ f_0 + \mu g\ \in \mathcal{H}_n^{m,\tau}(\mathbb{K})\} \tag{3.5}$$

is a real interval.

The following theorem gives a characterization of the convex directions for the set $\mathcal{H}_n^{m,\tau}(\mathbb{K})$.

Theorem 8 [18].

(i) A complex quasipolynomial $g \in \mathcal{Q}_{n-1}^{m,\tau}(\mathbb{C})$ is a convex direction for the set $\mathcal{H}_n^{m,\tau}(\mathbb{C})$ if and only if, for all $\omega \in \{w \in \mathbb{R} \mid g(\imath w) \neq 0\}$, the following condition is satisfied:

$$\frac{\partial \arg(g(\imath\omega))}{\partial \omega} \leq \frac{\tau_0 + \tau_m}{2} \tag{3.6}$$

(ii) A real quasipolynomial $g \in \mathcal{Q}_{n-1}^{m,\tau}(\mathbb{R})$ is a convex direction for the set $\mathcal{H}_n^{m,\tau}(\mathbb{R})$ if and only if for all $\omega \in \{w > 0 \mid g(\imath w) \neq 0\}$ the following condition is satisfied:

$$\frac{\partial \arg(g(\imath\omega))}{\partial \omega} \leq \frac{\tau_0 + \tau_m}{2} + \left|\frac{\sin(2\arg(g(\imath\omega)) - (\tau_0 + \tau_m)\omega)}{2\omega}\right| \tag{3.7}$$

From Theorem 8 it follows that if a quasipolynomial $g(z) \in \mathcal{Q}_{n-1}^{m,\tau}(\mathbb{K})$ is a convex direction then $cg(z)$ is also a convex direction for arbitrary constants $c \in \mathbb{K}$. Since the conditions (3.6) and (3.7) do not depend upon n we obtain – as in the case of Hurwitz polynomials – that a convex direction for $\mathcal{H}_n^{m,\tau}(\mathbb{K})$ is also a convex direction for $\mathcal{H}_{n'}^{m,\tau}(\mathbb{K})$ if $n' > \deg g$.

Remark 2. The above theorem characterizes those quasipolynomials $g \in \mathcal{Q}_{n-1}^{m,\tau}(\mathbb{K})$ for which the intersections $\{f + \mu g;\ \mu \geq 0\} \cap \mathcal{H}_n^{m,\tau}(\mathbb{K})$ are convex for *all* $f \in \mathcal{H}_n^{m,\tau}(\mathbb{K})$. As in the case of polynomials it is a different problem to characterize the convex directions for $\mathcal{H}_n^{m,\tau}(\mathbb{K})$ *relative to a given stable quasipolynomial* $f_0 \in \mathcal{H}_n^{m,\tau}(\mathbb{K})$, i.e. those $g \in \mathcal{Q}_{n-1}^{m,\tau}(\mathbb{K})$ for which the intersection $\{f_0 + \mu g;\ \mu \geq 0\} \cap \mathcal{H}_n^{m,\tau}(\mathbb{K})$ is convex (or, equivalently the stability set $M(f_0, g)$ (3.5) is an interval). This problem will be briefly discussed in Section 5.

4 Root Loci of Stable Polynomials

In this section we describe the root loci approach to the convex direction problem [16]. As starting point we use the following observation. The convex directions $q \in \mathcal{P}_{n-1}(\mathbb{K})$ for the set $\mathcal{S}_n(\mathbb{K}, \mathbb{C}_g)$ are characterized by this property: If, for any $p_0 \in \mathcal{S}_n(\mathbb{K}, \mathbb{C}_g)$ and some parameter value $\mu = \mu_0 > 0$, the polynomial $p_0 + \mu q$ possesses a root in the "bad" region $\mathbb{C}_b$ the same holds true for all $\mu \geq \mu_0$. In order to get a deeper understanding of this property we need to study the movement of the roots of $p_0 + \mu q$ as μ varies from 0 to ∞, i.e. the *root loci* of the polynomial pencil $p_0 + \mu q$.

Suppose that $p_0 \in \mathcal{S}_n(\mathbb{K}, \mathbb{C}_g)$ and $q \in \mathcal{P}_{n-1}(\mathbb{K})$ are of the form

$$p_0(z) = a_n z^n + a_{n-1} z^{n-1} + \cdots + a_1 z + a_0 \tag{4.1}$$

and

$$q(z) = b_m z^m + b_{m-1} z^{m-1} + \cdots b_1 z + b_0 \ , \tag{4.2}$$

respectively, where $0 \leq m < n$. In order to simplify the notation we make the following convention. Given $p_0(z)$, $q(z)$ of the form (4.1) and (4.2), we set $a_j = 0$ and $b_k = 0$ for all indices j and k for which these coefficients are not yet defined.

We will analyze the movement of the roots $z_j(\mu)$, $j = 1, \ldots, n$ of the polynomial

$$p_\mu(z) = p_0(z) + \mu q(z) \tag{4.3}$$

through the stability boundary $\partial\mathbb{C}_g \subset \mathbb{C}_b$ as μ varies from 0 to ∞.

For $\mu = 0$ all the roots of (4.3) lie in the open stability region $\mathbb{C}_g$ of the complex plane $\mathbb{C}$. When μ is increased the roots move continuously on the complex plane. More precisely, since $\deg p_0 = n$, there exist n continuous functions $z_j(\cdot) : \mathbb{R}_+ \mapsto \mathbb{C}$, $j = 1, ..., n$ such that, for every $\mu \geq 0$, $z_1(\mu), \ldots, z_n(\mu)$ are the roots of $p_\mu(z)$ (counting multiplicities), see [17, II.5.2]. Each of the functions $z_j(\cdot)$ is analytical in μ at every value μ_0 for which $z_j(\mu_0)$ is a simple root of $p_{\mu_0}(z)$. In this case we write $z_j'(\mu_0)$ for $(\partial z_j / \partial \mu)(\mu_0)$. With increasing μ the polynomial $p_\mu(z)$ loses stability at a parameter value μ_0 where one of the roots $z_j(\cdot)$ hits the boundary $\partial\mathbb{C}_g$. In order to characterize convex directions in terms of root loci we have to investigate under which conditions an unstable root of $p_\mu(z)$ may return to the stability region as μ is further increased.

We will consider the cases of Hurwitz and of Schur polynomials separately.

4.1 Root loci of Hurwitz stable polynomials

If q is a convex direction for the set $\mathcal{S}_n(\mathbb{C}, \mathbb{C}_-)$ of *complex* Hurwitz polynomials, the root loci of $p_\mu(z)$ can cross the stability boundary $\partial\mathbb{C}_- = \imath\mathbb{R}$ only in one direction: from left to right. However, the real case is different. If q is a convex direction for the set $\mathcal{S}_n(\mathbb{R}, \mathbb{C}_-)$ of *real* Hurwitz polynomials, roots of the polynomial $p_\mu(z)$ may return from instability to the stability region $\mathbb{C}_-$, but only through the origin. Therefore, in the real case, we must investigate the movement of the small roots of $p_\mu(z)$ for small variations of the parameter $\mu \geq 0$ around a parameter value $\mu_0 > 0$ such that $p_{\mu_0}(0) = 0$.

Suppose that the real polynomial $p_\mu(z)$ has a zero root of multiplicity $k \geq 1$ for some parameter value $\mu = \mu_0 > 0$. For μ varying in a small interval $[\mu_0 - \varepsilon, \mu_0 + \varepsilon]$ we want to determine the possible changes in the number of small roots of $p_0(z) + \mu q(z)$ with non-negative real parts. Note that $q(0) = b_0 \neq 0$ since otherwise $p_0(0) = 0$, contrary to the assumption that $p_0(z) \in \mathcal{S}_n(\mathbb{R}, \mathbb{C}_-)$. Moreover, there exists at most one value of $\mu_0 > 0$ for which $p_{\mu_0}(0) = 0$, namely $\mu_0 = -a_0/b_0$ (if $a_0/b_0 < 0$), see (4.1), (4.2).

We use the following notations. Given $k \in N$, let $u_0^{(k)}, u_1^{(k)}, \ldots, u_{k-1}^{(k)}$ denote the roots of $z^k - 1$

$$u_\nu^{(k)} = \exp\{\imath \frac{2\nu\pi}{k}\}, \qquad \nu = 0, 1, \ldots, k-1 \tag{4.4}$$

and $v_0^{(k)}, v_1^{(k)}, \ldots, v_{k-1}^{(k)}$ the roots of $z^k + 1$

$$v_\nu^{(k)} = \exp\{\imath \frac{(2\nu+1)\pi}{k}\}, \qquad \nu = 0, 1, \ldots, k-1 \tag{4.5}$$

A simple analysis shows that the absolute value of the difference between the number of roots of the form (4.4) with nonnegative real parts and the number of roots of the form (4.5) with nonnegative real parts does not exceed one [16]. Let us denote by $D(\varepsilon)$ the complex disk centered at $0 \in \mathbb{C}$ with radius $\varepsilon > 0$.

The following proposition [16] describes the behaviour of the roots of a general polynomial pencil p_μ in a small neighborhood of the origin for small variations of the parameter μ around $\mu_0 = -a_0/b_0$.

Proposition 9. *Let p_0, q be two real polynomials of the form (4.1) and (4.2), respectively, $a_0 \neq 0$, and suppose that $p_\mu(z) = p_0(z) + \mu q(z)$ has a root of multiplicity $k \geq 1$ at $z = 0$ for $\mu = \mu_0$. Then there exist $\varepsilon > 0$, $\delta > 0$ such that, for $s \in (-\varepsilon, \varepsilon) \setminus \{0\}$, p_{μ_0+s} has exactly k simple roots $z_\nu(s)$ in $D(\delta)$ and these roots have the following asymptotic behaviour as $|s| \to 0$:*

(i) If $b_0/d_k < 0$ then

$$z_\nu(s) = \begin{cases} s^{\frac{1}{k}} \left|\frac{b_0}{d_k}\right|^{\frac{1}{k}} u_j^{(k)} + o\left(s^{\frac{1}{k}}\right), & j = 0, 1, \ldots, k-1 \quad \text{for} \quad s > 0 \\ |s|^{\frac{1}{k}} \left|\frac{b_0}{d_k}\right|^{\frac{1}{k}} v_j^{(k)} + o\left(|s|^{\frac{1}{k}}\right), & j = 0, 1, \ldots, k-1 \quad \text{for} \quad s < 0 \end{cases}$$

(ii) If $b_0/d_k > 0$ then

$$z_\nu(z) = \begin{cases} s^{\frac{1}{k}} \left(\frac{b_0}{d_k}\right)^{\frac{1}{k}} v_j^{(k)} + o\left(s^{\frac{1}{k}}\right), & j = 0, 1, \ldots, k-1 \quad \text{for} \quad s > 0 \\ |s|^{\frac{1}{k}} \left(\frac{b_0}{d_k}\right)^{\frac{1}{k}} u_j^{(k)} + o\left(|s|^{\frac{1}{k}}\right), & j = 0, 1, \ldots, k-1 \quad \text{for} \quad s < 0 \end{cases}$$

where $d_k = a_k + \mu_0 b_k$.

Let $\mathcal{N}_+(p_\mu;\delta)$ denote the number of roots of $p_\mu(z)$ in $D(\delta)$ with nonnegative real parts (counting multiplicities). As a consequence of the proposition and the preceding considerations we obtain that under the conditions of Proposition 9 there exist $\varepsilon > 0$, $\delta > 0$ such that

$$|\mathcal{N}_+(p_{\mu^-};\delta) - \mathcal{N}_+(p_{\mu^+};\delta)| \leq 1, \qquad \mu^- \in (\mu_0 - \varepsilon, \mu_0), \quad \mu^+ \in (\mu_0, \mu_0 + \varepsilon)\,,$$

see [16]. Thus the net change in the number of small roots in the closed right half-plane as μ crosses the parameter value μ_0 is bounded by 1.

The above analysis provides the basis for the following characterization of convex directions in the case of *real* Hurwitz polynomials (statement (ii)).

Theorem 10 [16],[2].

(i) A polynomial $q \in \mathcal{P}_{n-1}(\mathbb{C})$ is a convex direction for the set $\mathcal{S}_n(\mathbb{C}, \mathbb{C}_-)$ of complex Hurwitz polynomials if and only if it satisfies the following condition for all polynomials $p_0 \in \mathcal{S}_n(\mathbb{C}, \mathbb{C}_-)$:

$(\mathrm{CD})_{\mathcal{S}_n(\mathbb{C},\mathbb{C}_-)}$ *If one of the roots $z_j(\mu)$, $j = 1, 2, \ldots, n$ of $p_\mu(z) = p_0(z) + \mu q(z)$, say $z_k(\mu)$, hits the imaginary axis $\imath\mathbb{R}$ for $\mu = \mu_0 > 0$ then $z_k(\mu_0)$ is a simple root of $p_0(z) + \mu_0 q(z)$ and $\mathrm{Re}\,\{z_k'(\mu_0)\} > 0$, i.e. as $\mu > 0$ increases the roots $z_j(\mu)$ can cross the imaginary axis only from left to right and with positive velocity.*

(ii) A polynomial $q \in \mathcal{P}_{n-1}(\mathbb{R})$ is a convex direction for the set $\mathcal{S}_n(\mathbb{R}, \mathbb{C}_-)$ of real Hurwitz polynomials if and only if it satisfies the following condition for all polynomials $p_0 \in \mathcal{S}_n(\mathbb{R}, \mathbb{C}_-)$:

$(\mathrm{CD})_{\mathcal{S}_n(\mathbb{R},\mathbb{C}_-)}$ *If one of the roots $z_j(\mu)$, $j = 1, 2, \ldots, n$ of $p_0(z) + \mu q(z)$, say $z_k(\mu)$, hits the punctured imaginary axis $\imath\mathbb{R} \setminus \{0\}$ for $\mu = \mu_0 > 0$ then $z_k(\mu_0)$ is a simple root of $p_0(z) + \mu_0 q(z)$ and $\mathrm{Re}\,\{z_k'(\mu_0)\} > 0$, i.e. as $\mu > 0$ increases the roots $z_j(\mu)$ can cross the punctured imaginary axis only from left to right and with positive velocity.*

Detailed proofs of (i) and (ii) can be found in [2] and [16], respectively. The proof of (i) is similar to the proof of the corresponding result for Schur polynomials which is given in the next subsection.

Theorem 10 shows that the difference between convex directions for complex and for real polynomials can be expressed in terms of root loci as follows. In the complex case a root $z_j(\mu)$ of $p_\mu(z)$ can never return to the stability domain $\mathbb{C}_-$ once it has left it. In the real case there is one hole in the imaginary axis through which the root loci of the real polynomial $p_\mu(z)$, $\mu \geq 0$ may return to the open left half-plane as μ increases, and this hole is at the origin. Moreover this hole can only be used once. In [16] it has been shown that, in the real case, the movement of the root loci from stability to instability (and back) obeys the following rules: The number $N_+(p_\mu)$ of unstable roots of $p_\mu(z) = p_0(z) + \mu q(z)$ is increasing as a function of $\mu \geq 0$ with one possible exception: $N_+(p_\mu)$ may decrease by 1 when μ passes through $\mu_0 = -a_0/b_0$ (if $b_0 \neq 0$ and $-a_0/b_0 > 0$). This will happen if and only if either $a_1 + \mu_0 b_1 < 0$ or $a_j + \mu_0 b_j = 0$ for $j = 0, 1, 2$ and $a_3 + \mu_0 b_3 > 0$.

The multiplicity of the zero root at the parameter value $\mu_0 = -a_0/b_0$ is at most 3. The jump sizes of the piecewise constant function $\mu \mapsto N_+(p_\mu)$ are multiples of 2 for every jump point $\mu \neq -a_0/b_0$.

Remark 3. It can be proved that condition $(\mathbf{CD})_{\mathcal{S}_n(\mathbb{C},\mathbb{C}_-)}$ is satisfied for a given pair of polynomials $q \in \mathcal{P}_{n-1}(\mathbb{C})$ and $p_0 \in \mathcal{S}_n(\mathbb{C},\mathbb{C}_-)$ if and only if for all $\mu > 0$, $\omega \in \mathbb{R}$

$$p_0(\imath\omega) + \mu q(\imath\omega) = 0 \;\Rightarrow\; \mathrm{Re}\left[\frac{p_0'(\imath\omega)}{p_0(\imath\omega)} - \frac{q'(\imath\omega)}{q(\imath\omega)}\right] > 0 \tag{4.6}$$

Similarly, condition $(\mathbf{CD})_{\mathcal{S}_n(\mathbb{R},\mathbb{C}_-)}$ is satisfied for $q \in \mathcal{P}_{n-1}(\mathbb{R})$ and $p_0 \in \mathcal{S}_n(\mathbb{R},\mathbb{C}_-)$ if and only if (4.6) holds for all $\mu > 0, \omega > 0$, see [16, Lemma 3.6].

4.2 Root loci of Schur stable polynomials

The next theorem is a counterpart of Theorem 10 for Schur polynomials. It characterizes the convex directions for the sets $\mathcal{S}_n(\mathbb{C},\mathbb{C}_1)$ and $\mathcal{S}_n(\mathbb{R},\mathbb{C}_1)$ in terms of root loci.

Theorem 11.
(i) A polynomial $q \in \mathcal{P}_{n-1}(\mathbb{C})$ is a convex direction for the set $\mathcal{S}_n(\mathbb{C},\mathbb{C}_1)$ of complex Schur polynomials if and only if it satisfies the following condition for all polynomials $p_0 \in \mathcal{S}_n(\mathbb{C},\mathbb{C}_1)$:

$(\mathbf{CD})_{\mathcal{S}_n(\mathbb{C},\mathbb{C}_1)}$ *If one of the roots $z_j(\mu)$, $j = 1,2,\ldots,n$ of $p_0(z) + \mu q(z)$, say $z_k(\mu)$, hits the unit circle $\{z \in \mathbb{C}\;;\; |z| = 1\}$ for $\mu = \mu_0 > 0$ then $z_k(\mu_0)$ is a simple root of $p_0(z) + \mu_0 q(z)$ and $|z_k(\mu_0)|' = (\partial|z_k(\mu)|/\partial\mu)(\mu_0) > 0$, i.e. as $\mu > 0$ increases the roots $z_j(\mu)$ of $p_0(z) + \mu q(z)$ can only cross the unit circle from the inside to the outside (with positive velocity).*

(ii) A polynomial $q \in \mathcal{P}_{n-1}(\mathbb{R})$ is a convex direction for the set $\mathcal{S}_n(\mathbb{R},\mathbb{C}_1)$ of real Schur polynomials if and only if it satisfies the following condition for all polynomials $p_0 \in \mathcal{S}_n(\mathbb{R},\mathbb{C}_1)$:

$(\mathbf{CD})_{\mathcal{S}_n(\mathbb{R},\mathbb{C}_1)}$ *If one of the roots $z_j(\mu)$, $j = 1,2,\ldots,n$ of $p_0(z) + \mu q(z)$, say $z_k(\mu)$, hits the punctured unit circle $\{z \in \mathbb{C}\;;\; |z| = 1,\; z \neq \pm 1\}$ for $\mu = \mu_0 > 0$ then $z_k(\mu_0)$ is a simple root of $p_0(z) + \mu_0 q(z)$ and $|z_k(\mu_0)|' = (\partial|z_k(\mu)|/\partial\mu)(\mu_0) > 0$, i.e. as $\mu > 0$ increases the roots $z_j(\mu)$ of $p_0(z)+\mu q(z)$ can only cross the punctured unit circle from the inside to the outside (with positive velocity).*

Proof. Suppose that $q(z)$ is a convex direction for the set $\mathcal{S}_n(\mathbb{C},\mathbb{C}_1)$ of complex Schur polynomials and $p_0 \in \mathcal{S}_n(\mathbb{C},\mathbb{C}_1)$. Let

$$p_0(e^{\imath\theta_0}) + \mu_0 q(e^{\imath\theta_0}) = 0 \tag{4.7}$$

for some $\mu_0 > 0$ and some $\theta_0 \in [0, 2\pi)$. Since $p_0 \in \mathcal{S}_n(\mathbb{C},\mathbb{C}_1)$ we have $p_0(e^{\imath\theta_0}) \neq 0$ and by (4.7) $q(e^{\imath\theta_0}) \neq 0$. Thus we can define two analytical argument functions

$\theta \mapsto \arg(p_0(e^{i\theta}))$, $\theta \mapsto \arg(q(e^{i\theta}))$ in a small neighborhood of θ_0. By the phase increasing property (2.10) of p_0 and the condition (2.8) at $\theta = \theta_0$ we obtain

$$\left(\arg(p_0(e^{i\theta_0})) - \arg(q(e^{i\theta_0}))\right)' = \mathrm{Re}\left[e^{i\theta_0}\left(\frac{p_0'(e^{i\theta_0})}{p_0(e^{i\theta_0})} - \frac{q'(e^{i\theta_0})}{q(e^{i\theta_0})}\right)\right] > 0\,. \quad (4.8)$$

If we assume, by contradiction, that $z_k(\mu_0) = e^{i\theta_0}$ is a multiple root of $p_{\mu_0}(z)$, i.e.

$$p_0(e^{i\theta_0}) + \mu_0 q(e^{i\theta_0}) = 0, \quad p_0'(e^{i\theta_0}) + \mu_0 q'(e^{i\theta_0}) = 0\,,$$

then

$$\mathrm{Re}\left[e^{i\theta_0}\left(\frac{p_0'(e^{i\theta_0})}{p_0(e^{i\theta_0})} - \frac{q'(e^{i\theta_0})}{q(e^{i\theta_0})}\right)\right] = 0$$

whence a contradiction to (4.8). Thus every root $z_k(\mu_0) = e^{i\theta_0}$ on the unit circle is a simple root of p_{μ_0} and in a small neighborhood of μ_0 in $\mathbb{R}$ there exists an analytic function $z_k(\mu)$ of μ satisfying

$$p_0(z_k(\mu)) + \mu q(z_k(\mu)) \equiv 0.$$

Differentiating this identity with respect to μ at μ_0 we obtain

$$\left[p_0'(e^{i\theta_0}) + \mu_0 q'(e^{i\theta_0})\right] z_k'(\mu_0) + q(e^{i\theta_0}) = 0\,.$$

Division by $-q(e^{i\theta_0}) = \mu_0^{-1} p_0(e^{i\theta_0})$ yields

$$\mu_0 e^{i\theta_0}\left(\frac{p_0'(e^{i\theta_0})}{p_0(e^{i\theta_0})} - \frac{q'(e^{i\theta_0})}{q(e^{i\theta_0})}\right)\frac{z_k'(\mu_0)}{z_k(\mu_0)} = 1\,. \quad (4.9)$$

Hence (4.8) implies

$$\mathrm{Re}\left[\frac{z_k'(\mu_0)}{z_k(\mu_0)}\right] > 0\,. \quad (4.10)$$

On the other hand $z_k(\mu) = r(\mu)e^{i\psi(\mu)}$ where $r(\mu) = |z_k(\mu)|$ and $\psi(\mu) = \arg(z_k(\mu))$ are analytical real-valued functions of μ in a neighborhood of μ_0. Thus

$$\frac{z_k'(\mu_0)}{z_k(\mu_0)} = \frac{r'(\mu_0)}{r(\mu_0)} + i\psi'(\mu_0)$$

and

$$\mathrm{Re}\left[\frac{z_k'(\mu_0)}{z_k(\mu_0)}\right] = \frac{r'(\mu_0)}{r(\mu_0)}.$$

Hence (4.10) implies $r'(\mu_0) = |z_k(\mu_0)|' > 0$. Therefore condition $\mathbf{(CD)}_{\mathcal{S}_n(\mathbb{C},\mathbb{C}_1)}$ is satisfied.

Conversely, suppose condition $\mathbf{(CD)}_{\mathcal{S}_n(\mathbb{C},\mathbb{C}_1)}$ holds. Then there are no holes in the unit circle through which the roots of $p_0(z) + \mu q(z)$ can enter the open unit disk C_1 from the outside when μ increases. Denote by μ_0 the smallest positive value of μ for which $p_\mu(z)$ loses the Schur stability property. Then at least one root remains outside the unit disk for all $\mu > \mu_0$, and $M(p_0, q)$ defined by (2.2) is an interval. Hence q is a convex direction for $\mathcal{S}_n(\mathbb{C}, \mathbb{C}_1)$.

(ii) The real case is complicated by the fact that there are two holes in the unit circle through which a root of the real polynomial $p_\mu(z) = p_0(z) + \mu q(z)$ can return into the unit disk after leaving it, namely $z = 1$ and $z = -1$. Thus the behaviour of the root loci at the exceptional points $z = \pm 1$ must be analyzed. For any $\delta > 0$, let $\mathbf{N}_{(+1)}(p_\mu;\delta)$ denote the number of roots of $p_\mu(z)$ of magnitude larger than 1 in the δ-neighborhood of $z = 1$ and $\mathbf{N}_{(-1)}(p_\mu;\delta)$ denote the number of such roots in the δ-neighborhood of $z = -1$. There is at most one value of μ for which $z = 1$ is a root of $p_\mu(z)$ and the same holds for $z = -1$. Analogously as in the Hurwitz case it can be shown (see [16]) that if $p_{\mu_1}(1) = 0$ and $p_{\mu_2}(-1) = 0$, there exist $\varepsilon > 0$ and $\delta > 0$ such that for all $\mu^- \in (\mu_1 - \varepsilon, \mu_1)$, $\mu^+ \in (\mu_1, \mu_1 + \varepsilon)$

$$|\mathbf{N}_{(+1)}(p_{\mu^-};\delta) - \mathbf{N}_{(+1)}(p_{\mu^+};\delta)| \le 1,$$

and for all $\mu_- \in (\mu_2 - \varepsilon, \mu_2)$, $\mu_+ \in (\mu_2, \mu_2 + \varepsilon)$

$$|\mathbf{N}_{(-1)}(p_{\mu_-};\delta) - \mathbf{N}_{(-1)}(p_{\mu_+};\delta)| \le 1,$$

respectively. Based on this fact the proof of (ii) can be carried out in a similar way as in the complex case replacing the unit circle by the punctured unit circle. For further details concerning the proof of (ii), see [16]. □

Remark 4. The previous proof shows that condition $(\mathbf{CD})_{\mathcal{S}_n(\mathbb{C},\mathbb{C}_1)}$ is satisfied for a given pair of polynomials $q \in \mathcal{P}_{n-1}(\mathbb{C})$ and $p_0 \in \mathcal{S}_n(\mathbb{C},\mathbb{C}_1)$ if and only if for all $\mu_0 > 0$, $\theta_0 \in [0, 2\pi)$

$$p_0(e^{\imath\theta_0}) + \mu_0 q(e^{\imath\theta_0}) = 0 \;\Rightarrow\; \mathrm{Re}\left[e^{\imath\theta_0}\left(\frac{p_0'(e^{\imath\theta_0})}{p_0(e^{\imath\theta_0})} - \frac{q'(e^{\imath\theta_0})}{q(e^{\imath\theta_0})}\right)\right] > 0 \qquad (4.11)$$

Similarly it can be shown that condition $(\mathbf{CD})_{\mathcal{S}_n(\mathbb{R},\mathbb{C}_1)}$ is satisfied for $q \in \mathcal{P}_{n-1}(\mathbb{R})$ and $p_0 \in \mathcal{S}_n(\mathbb{R},\mathbb{C}_1)$ if and only if (4.11) holds for all $\mu_0 > 0, \theta_0 \in (0,\pi)$, see [16, Lemma 5.6].

4.3 Relative convex directions for $\mathcal{S}_n(\mathbb{K},\mathbb{C}_-)$ and $\mathcal{S}_n(\mathbb{K},\mathbb{C}_1)$

In this section we discuss the problem of determining the set of relative convex directions for Hurwitz and Schur polynomials.

We begin with a general definition of the concept of relative convex direction.

Definition 12. Suppose X is a real or complex vector space and S a subset of X. Given $x_0 \in S$, a vector $y \in X$ is said to be a *convex direction for S relative to* x_0 if the intersection $R(x_0, y) \cap S$ of the ray $R(x_0, y) = \{x_0 + \alpha y;\ \alpha \ge 0\}$ with S is convex.
A vector $y \in X$ is said to be a *convex direction* for S if it is a convex direction for S relative to every $x \in S$.

The difference between the two concepts can be illustrated by a simple example. Consider the punctured Euclidean space $S = \mathbb{R}^n \setminus \{0\}$ (regarded as a subspace of the vector space $X = \mathbb{R}^n$). Then the set of convex directions for S is $\{0\}$, but for every $x_0 \in S$ the set of convex directions for S relative to x_0 is $\mathbb{R}^n \setminus (-\mathbb{R}_+ x_0)$.

Given any subset S in a vector space X two distinct convex direction problems arise:

Global Convex Direction Problem. Determine the set of all convex directions for S.

Relative Convex Direction Problem. For arbitrary $x_0 \in S$, determine the set of all convex directions for S relative to x_0.

Theorems 2 and 4 solve the Global Convex Direction Problem for the sets $\mathcal{S}_n(\mathbb{K}, \mathbb{C}_-) \subset \mathcal{P}_n(\mathbb{K})$ and $\mathcal{S}_n(\mathbb{K}, \mathbb{C}_1) \subset \mathcal{P}_n(\mathbb{K})$. On the other hand the Relative Convex Direction Problem for these sets is still unsolved. However, the proofs of the root loci characterizations in the previous subsections imply the following proposition (see [2] and [16]) which yields less conservative criteria than the global convex direction conditions stated in Theorems 2 and 4.

Proposition 13. *Let* $\mathbb{K} = \mathbb{R}$ *or* $\mathbb{K} = \mathbb{C}$.

(i) A polynomial $q \in \mathcal{P}_{n-1}(\mathbb{K})$ *is a convex direction for the set* $\mathcal{S}_n(\mathbb{K}, \mathbb{C}_-)$ *relative to a given polynomial* $p_0 \in \mathcal{S}_n(\mathbb{K}, \mathbb{C}_-)$ *if condition* $(\mathbf{CD})_{\mathcal{S}_n(\mathbb{K},\mathbb{C}_-)}$ *is satisfied (see Theorem 10).*

(ii) A polynomial $q \in \mathcal{P}_{n-1}(\mathbb{K})$ *is a convex direction for the set* $\mathcal{S}_n(\mathbb{K}, \mathbb{C}_1)$ *relative to a given polynomial* $p_0 \in \mathcal{S}_n(\mathbb{K}, \mathbb{C}_1)$ *if condition* $(\mathbf{CD})_{\mathcal{S}_n(\mathbb{K},\mathbb{C}_1)}$ *is satisfied (see Theorem 11).*

It has been shown by an example in [16] that there exist convex directions for $\mathcal{S}_n(\mathbb{K}, \mathbb{C}_-)$ relative to a given $p_0 \in \mathcal{S}_n(\mathbb{K}, \mathbb{C}_-)$ which do *not* satisfy $(\mathbf{CD})_{\mathcal{S}_n(\mathbb{K},\mathbb{C}_-)}$. Thus $(\mathbf{CD})_{\mathcal{S}_n(\mathbb{K},\mathbb{C}_-)}$ is a sufficient but not a necessary condition for $q \in \mathcal{P}_{n-1}(\mathbb{K})$ to be a convex direction for the set $\mathcal{S}_n(\mathbb{K}, \mathbb{C}_-)$ relative to the given $p_0 \in \mathcal{S}_n(\mathbb{K}, \mathbb{C}_-)$. In view of this it is surprising that – as we have seen in the preceding subsections – the condition $(\mathbf{CD})_{\mathcal{S}_n(\mathbb{K},\mathbb{C}_-)}$ is *necessarily* satisfied for all $p_0 \in \mathcal{S}_n(\mathbb{K}, \mathbb{C}_-)$ if $q \in \mathcal{P}_{n-1}(\mathbb{K})$ is a global convex direction for the set $\mathcal{S}_n(\mathbb{K}, \mathbb{C}_-)$. A similar comment applies to the Schur case.

The conditions $(\mathbf{CD})_{\mathcal{S}_n(\mathbb{K},\mathbb{C}_-)}$ and $(\mathbf{CD})_{\mathcal{S}_n(\mathbb{K},\mathbb{C}_1)}$ can be checked by graphical tests based on Nyquist plots.

Proposition 14.

(i) A polynomial $q \in \mathcal{P}_{n-1}(\mathbb{C})$ *satisfies* $(\mathbf{CD})_{\mathcal{S}_n(\mathbb{C},\mathbb{C}_-)}$ *for a given polynomial* $p_0 \in \mathcal{S}_n(\mathbb{C}, \mathbb{C}_-)$ *if and only if the Nyquist plot of* $h(z) = q(z)/p_0(z)$ *over* $\imath\mathbb{R}$ *crosses the negative real axis* $(-\infty, 0)$ *only in the clockwise direction, i.e. for every* $\omega \in \mathbb{R}$

$$h(\imath\omega) \in (-\infty, 0) \quad \Rightarrow \quad \frac{\partial \arg(h(\imath\omega))}{\partial\omega} < 0\,, \tag{4.12}$$

(ii) A polynomial $q \in \mathcal{P}_{n-1}(\mathbb{R})$ *satisfies* $(\mathbf{CD})_{\mathcal{S}_n(\mathbb{R},\mathbb{C}_-)}$ *for a given polynomial* $p_0 \in \mathcal{S}_n(\mathbb{R}, \mathbb{C}_-)$ *if and only if the Nyquist plot of* $h(z) = q(z)/p_0(z)$ *over the*

positive imaginary axis $\imath(0,\infty)$ crosses the negative real axis $(-\infty,0)$ only in the clockwise direction, i.e. for every $\omega > 0$

$$h(\imath\omega) \in (-\infty,0) \quad \Rightarrow \quad \frac{\partial \arg(h(\imath\omega))}{\partial\omega} < 0\,. \tag{4.13}$$

We omit the proof which can be found in [16] (for the real case) and prove instead the following counterpart for the Schur case.

Proposition 15.

(i) A polynomial $q \in \mathcal{P}_{n-1}(\mathbb{C})$ satisfies $(\mathbf{CD})_{\mathcal{S}_n(\mathbb{C},\mathbb{C}_1)}$ for a given polynomial $p_0 \in \mathcal{S}_n(\mathbb{C},\mathbb{C}_1)$ if and only if the Nyquist plot of $h(z) = q(z)/p_0(z)$ on the unit circle $\partial C_1 = \{e^{\imath\theta};\ \theta \in [0,2\pi)\}$ crosses the negative real axis $(-\infty,0)$ only in the clockwise direction, i.e. for all $\theta \in [0,2\pi)$

$$h(e^{\imath\theta}) \in (-\infty,0) \ \Rightarrow \ \frac{\partial \arg(h(e^{\imath\theta}))}{\partial\,\theta} < 0. \tag{4.14}$$

(ii) A polynomial $q \in \mathcal{P}_{n-1}(\mathbb{R})$ satisfies $(\mathbf{CD})_{\mathcal{S}_n(\mathbb{R},\mathbb{C}_1)}$ for a given polynomial $p_0 \in \mathcal{S}_n(\mathbb{R},\mathbb{C}_1)$ if and only if the the Nyquist plot of $h(z) = q(z)/p_0(z)$ on the upper half of the unit circle $\{e^{\imath\theta};\ \theta \in (0,\pi)\}$ crosses the negative real axis $(-\infty,0)$ only in the clockwise direction, i.e. for all $\theta \in (0,\pi)$

$$h(e^{\imath\theta}) \in (-\infty,0) \ \Rightarrow \ \frac{\partial \arg(h(e^{\imath\theta}))}{\partial\,\theta} < 0.$$

Proof. For every $\theta \in [0,2\pi)$, we have $h(e^{\imath\theta}) = q(e^{\imath\theta})/p_0(e^{\imath\theta}) \in (-\infty,0)$ if and only if the following equality holds with $\mu = -1/h(e^{\imath\theta}) > 0$:

$$p_0(e^{\imath\theta}) + \mu q(e^{\imath\theta}) = 0.$$

By (4.8), for all $\theta \in \{\phi \in [0,2\pi);\ q(e^{\imath\phi}) \neq 0\}$,

$$\frac{\partial \arg(h(e^{\imath\theta}))}{\partial\,\theta} = -\operatorname{Re}\left[e^{\imath\theta}\left(\frac{p_0'(e^{\imath\theta})}{p_0(e^{\imath\theta})} - \frac{q'(e^{\imath\theta})}{q(e^{\imath\theta})}\right)\right].$$

Hence (4.14) is satisfied if and only if, for all $\mu > 0,\ \theta \in [0,2\pi)$,

$$p_0(e^{\imath\theta}) + \mu q(e^{\imath\theta}) = 0 \ \Rightarrow \ \operatorname{Re}\left[e^{\imath\theta}\left(\frac{p_0'(e^{\imath\theta})}{p_0(e^{\imath\theta})} - \frac{q'(e^{\imath\theta})}{q(e^{\imath\theta})}\right)\right] > 0\,. \tag{4.15}$$

By the remark made after Theorem 11, condition $(\mathbf{CD})_{\mathcal{S}_n(\mathbb{C},\mathbb{C}_1)}$ is satisfied if and only if (4.15) holds. Thus (i) follows.
Statement (ii) can be proved similarly, making again use of the remark at the end of Subsection 4.2. □

5 Root Loci of Stable Quasipolynomials

In this section we apply the root loci approach in order to characterize convex directions for sets of stable quasipolynomials. To this end we investigate how the root loci of

$$f_\mu(z) = f_0(z) + \mu g(z), \quad \mu \geq 0$$

cross the imaginary axis if $f_0(z) \in \mathcal{H}_n^{m,\tau}(\mathbb{K})$ and $g(z)$ is a convex direction for the set $\mathcal{H}_n^{m,\tau}(\mathbb{K})$. Since $f_0(z) \in \mathcal{H}_n^{m,\tau}(\mathbb{K})$ all the roots of $f_0(z)$ lie in the open left half-plane. As μ is increased, the roots move continuously on the complex plane. Assume that, for some $\mu = \mu_0$, $f_{\mu_0}(z)$ has a zero $z = \imath\omega_0$ on the imaginary axis. Then there exists a continuous function $z_0(\mu)$ in a small neighborhood of μ_0 such that $f_\mu(z_0(\mu)) = 0$ and $z_0(\mu_0) = \imath\omega_0$. If $z_0(\mu_0) \in \imath\mathbb{R}$ is a simple root, then $z_0(\mu)$ is analytical. In this case we write $z_0'(\mu)$ for $(\partial z_0/\partial\mu)(\mu)$. We will see that, as μ increases, the roots of the quasipolynomials $f_\mu(z) = f_0(z) + \mu g(z)$ can move across the (punctured) imaginary axis $\imath\mathbb{R}$ only from left to right, if $g(z)$ is a convex direction for $\mathcal{H}_n^{m,\tau}(\mathbb{K})$ and $\mathbb{K} = \mathbb{C}$ (or $\mathbb{K} = \mathbb{R}$, respectively).

Consider a pair of quasipolynomials $f_0(z)$, $g(z)$ where $f_0(z) \in \mathcal{H}_n^{m,\tau}(\mathbb{K})$ is of the form

$$f_0(z) = \sum_{j=0}^{m}\sum_{k=0}^{n} a_{kj} z^k e^{\tau_j z}\,, \tag{5.1}$$

and $g(z) \in \mathcal{Q}_{n-1}^{m,\tau}(\mathbb{K})$ is of the form

$$g(z) = \sum_{j=0}^{m}\sum_{k=0}^{n-1} c_{kj} z^k e^{\tau_j z}\,. \tag{5.2}$$

To simplify the notations we make the following convention. Given $f_0(z)$, $g(z)$ of the form (5.1) and (5.2), respectively, we set $a_{ij} = 0$ and $c_{ij} = 0$ for all index pairs (i,j) for which these coefficients are not yet defined. In particular $c_{nj} = 0$, $j = 0, \ldots, m$. The main result in this section is the following characterization of convex directions for quasipolynomials of delay and of neutral type.

Theorem 16 [2],[1].

(i) A complex quasipolynomial $g(z) \in \mathcal{Q}_{n-1}^{m,\tau}(\mathbb{C})$ is a convex direction for the set $\mathcal{H}_n^{m,\tau}(\mathbb{C})$ if and only if it satisfies the following condition for all quasipolynomials $f_0(z) \in \mathcal{H}_n^{m,\tau}(\mathbb{C})$:

(CD)$_{\mathcal{H}_n^{m,\tau}(\mathbb{C})}$ *If one of the roots of $f_\mu(z) = f_0(z) + \mu g(z)$, say $z_0(\mu)$, hits the imaginary axis $\imath\mathbb{R}$ for $\mu = \mu_0 > 0$ then $z_0(\mu_0)$ is a simple root of $f_{\mu_0}(z)$ and* Re $\{z_0'(\mu_0)\} > 0$, *i.e. as μ increases the roots of $f_\mu(z)$ can only cross the imaginary axis from left to right and with positive velocity.*

(ii) A real quasipolynomial $g(z) \in \mathcal{Q}_{n-1}^{m,\tau}(\mathbb{R})$ is a convex direction for the set $\mathcal{H}_n^{m,\tau}(\mathbb{R})$ if and only if it satisfies the following condition for all quasipolynomials $f_0(z) \in \mathcal{H}_n^{m,\tau}(\mathbb{R})$:

(CD)$_{\mathcal{H}_n^{m,\tau}(\mathbb{R})}$ *If one of the roots of $f_\mu(z)$, say $z_0(\mu)$, hits the punctured* *imaginary axis* $\imath\mathbb{R}\setminus\{0\}$ *for* $\mu=\mu_0>0$ *then* $z_0(\mu_0)$ *is a simple root of* $f_{\mu_0}(z)$ *and* $\operatorname{Re}\{z_0'(\mu_0)\}>0$, *i.e. as* μ *increases the roots of* $f_\mu(z)$ *can only cross the punctured imaginary axis from left to right and with positive velocity.*

The proof of (i) follows the same lines as in the polynomial case, for details see [1]. The proof of (ii) requires a careful analysis of the behaviour of the root loci in a neigbourhood of the origin. The following counterpart to Proposition 9 has been proved in [2].

Proposition 17. *Let* $f_0(z)\in\mathcal{H}_n^{m,\tau}(\mathbb{R})$, $g(z)\in\mathcal{Q}_{n-1}^{m,\tau}(\mathbb{R})$ *be two real quasipolynomials of the form (5.1) and (5.2), respectively,* $g(0)=\sum_{j=0}^m c_{0j}\neq 0$, *and suppose that* $f_\mu(z)=f_0(z)+\mu g(z)$ *has a root of multiplicity* $k\geq 1$ *at* $z=0$ *for* $\mu=\mu_0$. *Then there exist* $\varepsilon>0$, $\delta>0$ *such that, for* $s\in(-\varepsilon,\varepsilon)\setminus\{0\}$, $f_{\mu_0+s}(z)$ *has exactly* k *simple roots in the complex disk* $D(\delta)$ *and these roots have the following asymptotic behaviour as* $|s|\to 0$*:*

(i) If $\left(\sum_{j=0}^m c_{0j}\right)/A_k(\mu_0)<0$ *then*

$$z_\nu(s)=\begin{cases} s^{\frac{1}{k}}\left|\dfrac{\sum_{j=0}^m c_{0j}}{A_k(\mu_0)}\right|^{\frac{1}{k}} u_\nu^{(k)}+o\left(s^{\frac{1}{k}}\right), & \nu=0,1,\dots,k-1 \ \ \textit{for } s>0\\ |s|^{\frac{1}{k}}\left|\dfrac{\sum_{j=0}^m c_{0j}}{A_k(\mu_0)}\right|^{\frac{1}{k}} v_\nu^{(k)}+o\left(|s|^{\frac{1}{k}}\right), & \nu=0,1,\dots,k-1 \ \ \textit{for } s<0\end{cases}$$

(ii) If $\left(\sum_{j=0}^m c_{0j}\right)/A_k(\mu_0)>0$ *then*

$$z_\nu(s)=\begin{cases} s^{\frac{1}{k}}\left(\dfrac{\sum_{j=0}^m c_{0j}}{A_k(\mu_0)}\right)^{\frac{1}{k}} v_\nu^{(k)}+o\left(s^{\frac{1}{k}}\right), & \nu=0,1,\dots,k-1 \ \ \textit{for } s>0\\ |s|^{\frac{1}{k}}\left(\dfrac{\sum_{j=0}^m c_{0j}}{A_k(\mu_0)}\right)^{\frac{1}{k}} u_\nu^{(k)}+o\left(|s|^{\frac{1}{k}}\right), & \nu=0,1,\dots,k-1 \ \ \textit{for } s<0\end{cases}$$

where

$$A_k(\mu_0)=\sum_{j=0}^m\left[(a_{kj}+\mu_0 c_{kj})+(a_{k-1j}+\mu_0 c_{k-1j})\frac{\tau_j}{1!}+\cdots+(a_{0j}+\mu_0 c_{0j})\frac{\tau_j^k}{k!}\right]$$

Let $\mathcal{N}_+(f_\mu;\delta)$ denote the number of roots of $f_\mu(z)$ in $D(\delta)$ with nonnegative real parts (counting multiplicities). As a consequence of the previous proposition we obtain that $\mathcal{N}_+(f_\mu;\delta)$ changes at most by one as μ crosses the value μ_0. More precisely, there exist $\varepsilon>0$, $\delta>0$ such that

$$\left|\mathcal{N}_+(f_{\mu^-};\delta)-\mathcal{N}_+(f_{\mu^+};\delta)\right|\leq 1\,,\quad \mu^-\in(\mu_0-\varepsilon,\mu_0)\,,\ \ \mu^+\in(\mu_0,\mu_0+\varepsilon)\qquad(5.3)$$

Therefore, as $\mu>0$ increases and some roots of $f_\mu(z)$ reach the closed right half-plane, not more than *one* of them may ever return to the open left half-plane through the origin $z=0$. Note that this can only happen at the parameter value $\mu_0=-f(0)/g(0))$ if $f(0)/g(0)<0$.

Based on these facts, statement (ii) in Theorem 16 can be proved similarly as in the polynomial case. For details see [1].

Remark 5. It follows from the proof of Theorem 16 (i) (see [2]) that the condition $(\mathbf{CD})_{\mathcal{H}_n^{m,\tau}(\mathbb{C})}$ is satisfied for a given pair of quasipolynomials $g \in \mathcal{Q}_{n-1}^{m,\tau}(\mathbb{C})$ and $f_0 \in \mathcal{H}_n^{m,\tau}(\mathbb{C})$ if and only if for all $\mu > 0$, $\omega \in \mathbb{R}$

$$f_0(\imath\omega) + \mu g(\imath\omega) = 0 \;\Rightarrow\; \mathrm{Re}\left[\frac{f_0'(\imath\omega)}{f_0(\imath\omega)} - \frac{g'(\imath\omega)}{g(\imath\omega)}\right] > 0. \tag{5.4}$$

Similarly, condition $(\mathbf{CD})_{\mathcal{H}_n^{m,\tau}(\mathbb{R})}$ is satisfied for $g \in \mathcal{Q}_{n-1}^{m,\tau}(\mathbb{R})$ and $f_0 \in \mathcal{H}_n^{m,\tau}(\mathbb{R})$ if and only if (5.4) holds for all $\mu > 0, \omega > 0$.

We conclude this chapter with some comments concerning the Relative Convex Direction Problem for quasipolynomials, see Subsection 4.3.

While the Global Convex Direction Problem for quasipolynomials has been solved in [18] (see Theorem 8) the Relative Convex Direction Problem is still open. However, as in the polynomial case the root loci approach yields a sufficient condition which is less conservative than the global convex direction condition in Theorem 8.

Proposition 18. *Let $\mathbb{K} = \mathbb{R}$ or $\mathbb{K} = \mathbb{C}$. A quasipolynomial $g(z) \in \mathcal{Q}_{n-1}^{m,\tau}(\mathbb{K})$ is a convex direction for the set $\mathcal{H}_n^{m,\tau}(\mathbb{K})$ relative to a quasipolynomial $f_0(z) \in \mathcal{H}_n^{m,\tau}(\mathbb{K})$ if the condition $(\mathbf{CD})_{\mathcal{H}_n^{m,\tau}(\mathbb{K})}$ stated in Theorem 16 is satisfied.*

A proof of this proposition is easily derived from the proof of the sufficiency statements in Theorem 16 (i),(ii), see [2], [1].

The conditions $(\mathbf{CD})_{\mathcal{H}_n^{m,\tau}(\mathbb{K})}$, $\mathbb{K} = \mathbb{R}, \mathbb{C}$ can be checked via a graphical test.

Proposition 19.

(i) A quasipolynomial $g(z) \in \mathcal{Q}_{n-1}^{m,\tau}(\mathbb{C})$ satisfies $(\mathbf{CD})_{\mathcal{H}_n^{m,\tau}(\mathbb{C})}$ for a given stable quasipolynomial $f_0(z) \in \mathcal{H}_n^{m,\tau}(\mathbb{C})$ if and only if the Nyquist plot of $h(z) = g(z)/f_0(z)$ on $\imath\mathbb{R}$ crosses the negative real axis $(-\infty, 0)$ only in the clockwise direction, i.e. for every $\omega \in \mathbb{R}$

$$h(\imath\omega) \in (-\infty, 0) \quad \Rightarrow \quad \frac{\partial \arg(h(\imath\omega))}{\partial\omega} < 0\,. \tag{5.5}$$

(ii) A quasipolynomial $g(z) \in \mathcal{Q}_{n-1}^{m,\tau}(\mathbb{R})$ satisfies $(\mathbf{CD})_{\mathcal{H}_n^{m,\tau}(\mathbb{R})}$ for a given stable quasipolynomial $f_0(z) \in \mathcal{H}_n^{m,\tau}(\mathbb{R})$ if and only if the Nyquist plot of $h(z) = g(z)/f_0(z)$ on $\imath(0,\infty)$ crosses the negative real axis $(-\infty, 0)$ only in the clockwise direction, i.e. (5.5) holds for all $\omega > 0$.

Using the previous remark the proof of this proposition can be carried through as in the polynomial case.

References

1. L. Atanassova, D. Hinrichsen and V.L. Kharitonov. On convex stability directions for real quasipolynomials. *International Series of Numerical Mathematics.*, Birkhäuser Verlag, Basel, 121: 43–52, 1996.
2. L. Atanassova, D. Hinrichsen and V.L. Kharitonov. Convex directions for complex Hurwitz stable polynomials and quasipolynomials. *Proc. of ECC*, Brussels, Belgium, 1997.
3. B.R. Barmish. *New Tools for Robustness of Linear Systems.* Macmillan Publishing Company, New York, 1994.
4. S. Basu, Boundary implications of stability and positivity properties of multidimensional systems. *Proc. IEEE*, 78: 614–636, 4, 1990.
5. S.P. Bhattacharyya, H. Chapellat, L.H. Keel, *Robust Control - The Parametric Approach.* Prentice Hall, Upper Saddle River, 1995.
6. N.K. Bose, *Applied Multidimensional Systems Theory.* Van Nostrand Reinhold, New York, 1982.
7. N.K. Bose, Robust multivariable scattering Hurwitz interval polynomials. *Linear Algebra and Its Applications*, 98: 123–136, 1988.
8. N.K. Bose. Argument conditions for Hurwitz and Schur polynomials from network theory. *IEEE Trans. Autom. Contr.* AC–39: 345–346, 1994.
9. N.K. Bose. Edge property from end points for scattering Hurwitz polynomials. *Automatica* 32: 655–657, 1996.
10. A.C. Bartlett, C.V. Hollot and H. Lin. Root location of an entire polytope of polynomials: it suffices to check the edges. *Math. Control, Signals and Systems*, 61–71, 1988.
11. R. Bellman and K.L. Cook. *Differential-Difference Equations.* Academic Press, New York, 1963.
12. M. Fu, A.W. Olbrot, and M.P. Polis. Edge theorem and graphical test for robust stability of neutral time-delay systems. *Automatica.* 17: 739-742, 1991.
13. Minyue Fu. A class of weak Kharitonov regions for robust stability of linear uncertain systems. *IEEE Trans. Autom. Contr.* AC–36:975–978, 1991.
14. Minyue Fu. Test of convex directions for robust stability. *Proc. 32-nd CDC*, San Antonio, Texas. 502–507, 1993.
15. J.K. Hale. *Theory of Functional Differential Equations.* Applied Mathematical Sciences, vol. 3, Springer Verlag, New York, 1977.
16. D. Hinrichsen and V.L. Kharitonov. On convex directions for stable polynomials. Technical Report 309, Inst. f. Dynamische Systeme, Universität Bremen, Bremen, 1994. To appear in: *Automatika i Telemekhanika.*
17. T. Kato. *Perturbation Theory for Linear Operators.* Springer Verlag, Berlin, 1980.
18. V. L. Kharitonov and A.P. Zhabko. Robust Stability of Time-Delay Systems. *IEEE Transactions on Automatic Control*, AC-39:2388–2397, 1994.
19. I.R. Petersen. A class of stability regions for which a Kharitonov like theorem holds. *IEEE Trans. Autom. Contr.* AC–34:1111–1115, 1989.
20. A. Rantzer. Stability conditions for polytopes of polynomials. *IEEE Transactions on Automatic Control*, AC-37:79–89, 1992.
21. E. Schwengeler. *Geometrisches über die Verteilung der Nullstellen spezieller ganzer Functionen (Exponentialsummen).* Ph.D. thesis, Zürich, 1925.
22. N.G. Tchebotareff and N.N. Meiman. *The Routh-Hurwitz Problem for Polynomials and Entire Functions.* Proc. of Mathematical Institute, Acad. Sci. USSR, 1949 (in Russian).

Delay-Independent Stability of Linear Neutral Systems: A Riccati Equation Approach

Erik I. Verriest[1] and Silviu-Iulian Niculescu[2]

[1] School of Electrical and Computer Engineering, Georgia Institute of Technology, Atlanta, GA 30332-0250, USA
[2] Laboratoire de Mathématiques Appliquées, Ecole Nationale Supérieure de Techniques Avancées, 32, Blvd. Victor, 75739 Paris, France

Abstract. This chapter focuses on the problem of asymptotic stability of a class of linear neutral systems described by differential equations with delayed state. The delay is assumed unknown, but constant. *Sufficient* conditions for *delay-independent* asymptotic stability are given in terms of the existence of symmetric and positive definite solutions of a *continuous* Riccati algebraic matrix equation coupled with a *discrete* Lyapunov equation.

1 Introduction

The stability of time-delay systems is a problem of practical and theoretical interests since the existence of a delay in a physical system may induce instability or poor performance. In certain control problems, one encounters linear hyperbolic differential equations with mixed initial and derivative boundary conditions, see, e.g. processes including steam or water pipes, loss-less transmission lines. Using a technique proposed in Hale and Lunel [3], these systems can be easily described by functional differential equations of *neutral* type.

A different example is proposed by Niculescu and Brogliato [7], where the effect of force measurements delays on the stability of manipulators in contact with a rigid environment is considered. The closed-loop system is represented by a linear time-invariant neutral equation. In this case, the time-delay may be a *cause* of *possible bouncing* of the robot's tip on the environment. The effect of small delays on the stability properties of some closed-loop neutral systems have been considered in [6] and the references therein.

In this chapter, one considers a particular class of time-delay systems described by linear neutral differential equations. We are interested in giving conditions for *delay-independent* stability conditions (which do not carry any information on the delay size). A guided tour of the general corresponding methods for linear systems with delayed states could be found in [9]. For some backgrounds on the stability of functional differential equations of neutral type, see e.g. Hale and Lunel [3], Kolmanovskii and Myshkis [5].

The approach adopted here is based on the Lyapunov's second method and makes use of an appropriate Lyapunov-Krasovskii functional. *Sufficient* delay-independent stability conditions are given in terms of some appropriate Riccati matrix equation coupled with a Lyapunov one.

The chapter is organized as follows: in Section 2, the main results are given. Singular value tests in terms of an $\mathcal{H}_\infty$ norm of some transfer functions are proposed in Section 3. A formulation in terms of LMI is given in Section 4. Some concluding remarks end the chapter.

2 Main Results

Consider the following class of linear neutral systems:

$$\dot{x}(t) - C\dot{x}(t-\tau) = Ax(t) + Bx(t-\tau) \tag{2.1}$$

with the initial condition

$$x(t_0+\theta) \;=\; \phi(\theta),\ \forall\, \theta \in [-\tau, 0]; \quad (t_0, \phi) \in \mathbb{R}^+ \times \mathcal{C}^v_{n,\tau}, \tag{2.2}$$

where $x(t) \in \mathbb{R}^n$ is the state, $\tau > 0$ is the delay and C, A and B are constant matrices of appropriate dimension.

We have the following result:

Theorem 1. *The neutral system (2.1)-(2.2) is delay-independent asymptotically stable if*

(i) A is a Hurwitz stable matrix;
(ii) C is a Schur-Cohn stable matrix;
(iii) there exist two symmetric and positive definite matrices $R > 0$ and $Q > 0$ such that the following Riccati equation has a symmetric and positive definite solution $P > 0$:

$$\begin{aligned} A^TP + PA + S + Q + \\ +\left[P(AC+B) + SC\right]R^{-1}\left[C^TS + (B^T + C^TA^T)P\right] &= 0, \end{aligned} \tag{2.3}$$

where $S > 0$ is the symmetric and positive definite solution of the Lyapunov discrete equation:

$$C^TSC - S + R \;=\; 0. \tag{2.4}$$

The proof of the Theorem is given in Appendix B and makes use of the following Lyapunov-Krasovskii functional candidate:

$$\begin{aligned} V(x_t) &= (x(t) - Cx(t-\tau))^T P (x(t) - Cx(t-\tau)) \\ &\quad + \int_{-\tau}^{0} x(t+\theta) S x(t+\theta) d\theta. \end{aligned} \tag{2.5}$$

Notice that since C is a Schur Cohn stable matrix, then the Lyapunov equation (2.4) has always a symmetric and positive definite solution $S > 0$ for every positive definite matrix $R > 0$.

In conclusion, the delay-independent stability problem of the neutral system (2.1)-(2.2) is transformed into the *existence* of a symmetric and positive-definite solution of the "parametrized" Riccati equation (2.3), where the papRameters are given by a *couple* of positive-definite matrices satisfying the discrete Lyapunov equation (2.4).

Remark 1. The Schur Cohn stability of the matrix C ensures the *stability* of the operator $\mathcal{D} : \mathcal{C}_{n,\tau} \mapsto \mathbb{R}^n$:

$$\mathcal{D}(\phi) = \phi(0) - C\phi(-\tau),$$

which is a *necessary* condition to have the stability of the neutral differential equation (2.1)-(2.2).

Notice also that the Hurwitz stability of the matrix A is a *necessary* condition for the existence of a symmetric positive definite solution to the Riccati equation (2.4), but is not a *sufficient* one.

Remark 2. A similar result has been proposed by Slemrod and Infante [10] using a particular Lyapunov-Krasovskii candidate (2.5) with:

$$P = I_n \qquad S = \frac{1}{2}[A + A^T - B^T C - C^T B],$$

such that the matrix S is symmetric ans positive-definite. Notice that their result uses some particular "constraints" on the system's matrices A, B and C.

The Riccati equation (2.3) is similar to the Riccati equations encountered in the LQG theory, but with a negative sign in the quadratic term. Indeed, if we consider the system:

$$\dot{\xi}(t) = A\xi(t) + (AC + B)u(t),$$

with the quadratic index:

$$\mathcal{J} = \int_0^\infty \begin{bmatrix} x(t) \\ u(t) \end{bmatrix}^T \cdot \begin{bmatrix} Q+S & SC \\ C^T S & R \end{bmatrix} \cdot \begin{bmatrix} x(t) \\ u(t) \end{bmatrix} dt,$$

the corresponding LQG Riccati equation is:

$$A^T X + XA - [X(AC+B) + SC]R^{-1}[(B^T + C^T A^T)X + C^T S] + Q + S = 0.$$

A different Lyapunov-Krasovskii functional has been proposed by Verriest in [12]:

$$\begin{aligned} V(x(t), x_t, \dot{x}_t) &= x(t)^T P_1 x(t) + \int_{-\tau}^0 x(t+\theta)^T P_2 x(t+\theta) d\theta \\ &\quad + \int_{-\tau}^0 \dot{x}(t+\theta)^T P_3 \dot{x}(t+\theta) d\theta, \end{aligned} \tag{2.6}$$

where P_i ($i = \overline{1,3}$) are symmetric and positive definite matrices satisfying some appropriate Riccati inequalities (see [12]).

The form of the Lyapunov functional (2.6) includes "information" on the derivatives $\dot{x}_t$. A proper norm for this asymptotic stability case is given by:

$$\|x_t\|_{c1} = \sup_{-\tau\leq\theta\leq 0} \{\|x(t+\theta)\|, \|\dot{x}(t+\theta)\|\}.$$

Some connections between the stability results obtained using the *norms* $\|\cdot\|_c$ and $\|\cdot\|_{c1}$ could be found in Els'golts' and Norkin [2]. For the sake of simplicity, we do not consider this approach here.

In the case of a scalar neutral system:

$$\dot{x}(t) - c\dot{x}(t-\tau) = ax(t) + bx(t-\tau) \tag{2.7}$$

with a, b and $c \in \mathbb{R}$, *Theorem 1* becomes:

Corollary 2. *The scalar neutral system (2.7) is delay-independent asymptotically stable if*

(i) $a < 0$,
(ii) $| c |< 1$,
(iii) $| b |<| a |$.

Notice that this result "approaches" the *necessary and sufficient* condition obtained in [4] using a frequential domain approach (the exact condition is given by *(i)-(iii)*, but with *(iii)* changed in $| b |\leq| a |$).

Consider now a more general form for the system (2.1) with A and B continuous time-varying matrices, i.e.

$$\dot{x}(t) - Cx(t-\tau) = A(t)x(t) + B(t)x(t-\tau) \tag{2.8}$$

In this case, *Theorem 1* may be rewritten as follows:

Theorem 3. *The neutral system (2.8)-(2.2) is delay-independent uniformly asymptotically stable if*

(i) C is a Schur-Cohn stable matrix;
(ii) there exist two symmetric and positive definite matrices $R > 0$ and $Q > 0$ such that the following Riccati equation has a symmetric solution $P(t)$ satisfying $p_m I_n \leq P(t) \leq p_M I_n$ for some positive real numbers p_m and p_M:

$$\dot{P} + A^T P + PA - [P(AC+B) + SC] \cdot$$
$$\cdot (R+2S)^{-1} \left[C^T S + (B^T + C^T A^T)P\right] + Q + S = 0,$$

where $S > 0$ is the symmetric and positive definite solution of the Lyapunov discrete equation:

$$C^T SC - S + R = 0.$$

The proof of the *Theorem 2* follows the same ideas as the proof of *Theorem 1*, but with a time-varying matrix $P(t)$ in the Lyapunov-Krasovskii functional (2.5). Notice that the existence of p_m and p_M allows that the corresponding Lyapunov functional candidate is *positive-definite* and has an *infinitesimal upper bound.*

It is easy to see that for *constant* matrices $A(t)$ and $B(t)$, one completely *recovers* the results given in *Theorem 1*. Furthermore, if $C = 0$ one recovers the results proposed by Verriest in [13].

3 Singular Value Test for Delay-Independent Asymptotic Stability

Theorem 1 gives a sufficient condition to guarantee the asymptotic stability independently of the delay size of the neutral system (2.1)-(2.2) in terms of the *existence* of *symmetric* and *positive definite* solutions to a continuous Riccati equation (2.3) and to a discrete Lyapunov equation.

Notice that to the Riccati equation (2.3), one can associate the *Hamiltonian* matrix:

$$H = \begin{bmatrix} A-(B+AC)R^{-1}C^TS & (AC+B)R^{-1}(AC+B)^T \\ -(Q+S)+SCR^{-1}C^TS & -\left[A-(B+AC)R^{-1}C^TS\right]^T \end{bmatrix} \quad (3.1)$$

In order to have a symmetric and positive definite solution to the Riccati equation (2.3), one needs that the associated Hamiltonian matrix *has no eigenvalues on the imaginary axis*, or equivalently (via the bounded rel lemma) that the *associated transfer matrix*

$$\begin{aligned} G(s) &= (Q+S-SCR^{-1}C^TS)^{\frac{1}{2}} \times \\ &\quad \times \left(sI_n - A-(B+AC)R^{-1}C^TS\right)^{-1}(AC+B)R^{-\frac{1}{2}} \end{aligned} \quad (3.2)$$

satisfies the norm condition:

$$\sup_{\omega\geq 0} \|G(j\omega)\| < 1. \quad (3.3)$$

Notice that in the case $C = 0$, the Lyapunov equation (2.4) becomes the equality $S = R$ and the condition (3.3) for the transfer (3.2) *recovers* the singular value test proposed in [11] for delay-independent asymptotic stability of linear systems with delayed state:

$$\sup_{\omega\geq 0} \|(Q+S)^{\frac{1}{2}}(j\omega I_n - A)^{-1}BS^{-\frac{1}{2}}\| < 1.$$

4 LMI Formulation

Theorem 1 can be easily converted into an *LMI feasibility problem* (see [1] and the references therein). We have the following result:

Corollary 4. *The neutral system (2.1)-(2.2) is delay-independent asymptotically stable if there exist two symmetric and positive definite matrices $P > 0$ and $S > 0$ such that the following LMIs hold:*

$$\begin{bmatrix} A^TP + PA + S & P(AC+B) + SC \\ C^TS + (B^T + C^TA^T)P & C^TSC - S \end{bmatrix} < 0, \tag{4.1}$$

$$C^TSC - S < 0. \tag{4.2}$$

Furthermore, if $C = 0$, i.e. the system (2.1) becomes a retarded one:

$$\dot{x}(t) = Ax(t) + A_d x(t-\tau), \tag{4.3}$$

Theorem 1 still holds, i.e.:

Corollary 5 [8, 1]. *The linear system with delayed state (4.3)-(2.2) is delay-independent asymptotically stable if there exist two symmetric and positive definite matrix $P > 0$ and $S > 0$ such that the following LMI holds:*

$$\begin{bmatrix} A^TP + PA + S & PB \\ B^TP & -S \end{bmatrix} < 0, \tag{4.4}$$

5 Concluding Remarks

The problem of stability of a class of linear systems described by differential equations of neutral type has been considered. Sufficient delay-independent conditions are given in terms of some algebraic Riccati matrix equations combined with appropriate Lyapunov equations. The approach adopted here is based on the Lyapunov's second method. Two numerical tools for the analysis have been also considered: a singular value test and linear matrix inequality (LMI) techniques. The proposed results can be easily extended to multiple delays case. Furthermore, in the particular case of linear systems with delayed state, one covers previous results form the literature [11].

A Stability Theory

Consider the following functional differential equation of neutral type:

$$\frac{d}{dt}\left[\mathcal{D}(x_t)\right] = f(x_t), \tag{A.1}$$

with an appropriate initial condition:

$$x(t_0 + \theta) = \phi(\theta),\ \forall\, \theta \in [-\tau, 0]; \quad (t_0, \phi) \in \mathbb{R}^+ \times \mathcal{C}^v_{n,\tau}, \tag{A.2}$$

where $\mathcal{D} : \mathcal{C}_{n,\tau} \to \mathbb{R}^n$, $\mathcal{D}(\phi) = \phi(0) - C\phi(-\tau)$ and $x(t) \in \mathbb{R}^n$. We say that the operator $\mathcal{D}$ is *stable* if the zero solution of the corresponding homogeneous difference equation is uniformly asymptotically stable. For our choice, this condition is replaced by the *Schur-Cohn* stability of the matrix C. For a general framework, see e.g. Hale and Lunel [3].

If $V : \mathbb{R} \times \mathcal{C}_{n,\tau} \to \mathbb{R}^n$ is continuous and $x(t_0, \phi)$ is the solution of the neutral differential equation (A.1) through the (t_0, ϕ) defined by (A.2), we define:

$$\dot{V}(t_0, \phi) = \limsup_{h \to 0+} \frac{1}{h} \left[V(t+h, x_{t+h})(t_0, \phi) - V(t_0, \phi) \right].$$

We have the following result:

Theorem 6 [3]. *Suppose $\mathcal{D}$ is stable, $f : \mathbb{R} \times \mathcal{C}_{n,\tau} \to \mathbb{R}^n$ takes bounded sets of $\mathcal{C}_{n,\tau}$ in bounded sets of $\mathbb{R}^n$ and suppose $u(s)$, $v(s)$ and $w(s)$ are continuous, nonnegative and nondecreasing functions with $u(s)$, $v(s) > 0$ for $s \neq 0$ and $u(0) = v(0) = 0$.*

If there is a continuous function $V : \mathbb{R} \times \mathcal{C}_{n,\tau} \to \mathbb{R}^n$ such that

(i) $u(\|\mathcal{D}(\phi)\|) \leq V(t, \phi) \leq v(\|\phi\|_c)$,
(ii) $\dot{V}(t, \phi) \leq -w(\|\phi(0)\|)$

then the solution $x = 0$ of the neutral equation (A.1)-(A.2) is uniformly stable.

If $u(s) \to \infty$ as $s \to \infty$ the solutions are uniformly bounded.

If $w(s) > 0$ for $s > 0$, then the solution $x = 0$ is uniformly asymptotically stable.

The same conclusions hold if the upper bound on $\dot{V}(t, \phi)$ is given by $-w(\|\phi(0)\|)$.

B Proof of Theorem 1

Let us consider the following Lyapunov-Krasovskii functional candidate:

$$\begin{aligned} V(x_t) &= (x(t) - Cx(t-\tau))^T P (x(t) - Cx(t-\tau)) \\ &\quad + \int_{-\tau}^{0} x(t+\theta) S x(t+\theta) d\theta, \end{aligned} \tag{B.1}$$

where P and S are the solutions of the Riccati equation (2.3) and respectively of the Lyapunov equation (2.4) and let introduce the operator $\mathcal{D} : \mathcal{C}_{n,\tau} \to \mathbb{R}^n$:

$$\mathcal{D}(\phi) = \phi(0) - C\phi(-\tau), \quad \phi \in \mathcal{C}_{n,\tau}. \tag{B.2}$$

It is easy to see that the functional V satisfies the condition:

$$u(| D\phi |) \leq V(\phi) \leq v(\|\phi\|_c), \tag{B.3}$$

where $u(s) = \lambda_{min}(P)s^2$ and $v(s) = [\lambda_{max}(P) + \tau\lambda_{max}(S)]s^2$. The derivative of $V(\cdot)$ along the trajectory of the neutral system (2.1) is given by:

$$\begin{aligned}\dot{V}(x_t) &= (Ax(t) + Bx(t-\tau))^T P(x(t) - Cx(t-\tau)) \\ &\quad + (x(t) - Cx(t-\tau))^T P(Ax(t) + Bx(t-\tau)) \\ &\quad + x(t)^T Sx(t) - x(t-\tau)^T Sx(t-\tau) \end{aligned} \tag{B.4}$$

Simple computation allows to rewrite the equation (B.4) as follows

$$\begin{aligned}\dot{V}(x_t) &= (x(t) - Cx(t-\tau))^T (A^T P + PA + S)(x(t) - Cx(t-\tau)) \\ &\quad + (x(t) - Cx(t-\tau))^T PACx(t-\tau) \\ &\quad + x(t-\tau)^T C^T A^T P(x(t) - Cx(t-\tau)) \\ &\quad + (x(t) - Cx(t-\tau))^T SCx(t-\tau) + x(t-\tau)^T C^T S(x(t) - Cx(t-\tau)) \\ &\quad + x(t-\tau)^T C^T SCx(t-\tau) - x(t-\tau)^T Sx(t-\tau) \\ &\quad + (x(t) - Cx(t-\tau))^T PBx(t-\tau) \\ &\quad + x(t-\tau)^T B^T P(x(t) - Cx(t-\tau)). \end{aligned} \tag{B.5}$$

Since S is the positive definite solution of the Lyapunov equation (2.4) and using the operator form (B.2), the relation (B.5), follows:

$$\begin{aligned}\dot{V}(x_t) &= \mathcal{D}(x_t)^T (A^T P + PA + S)\mathcal{D}(x_t) + \mathcal{D}(x_t)^T (PAC + PB + SC)x(t-\tau) \\ &\quad + x(t-\tau)^T (C^T S + C^T A^T P + B^T P)\mathcal{D}(x_t) \\ &\quad - x(t-\tau)^T Rx(t-\tau) \end{aligned} \tag{B.6}$$

Since P is the symmetric and positive definite solution of the Riccati equation (2.3) and using the Schur complement property, we have:

$$\begin{aligned}\dot{V}(x_t) &\leq -\mathcal{D}(x_t)^T Q\mathcal{D}(x_t) \\ &\quad - \left[(C^T A^T P + B^T + C^T S)\mathcal{D}(x_t) - Rx(t-\tau)\right]^T R^{-1} \times \\ &\quad \times \left[(C^T A^T P + B^T + C^T S)\mathcal{D}(x_t) - Rx(t-\tau)\right] \\ &\leq -\mathcal{D}(x_t)^T Q\mathcal{D}(x_t) \end{aligned} \tag{B.7}$$

The inequalities (B.3) and (B.7) allow to conclude the uniform asymptotic stability of the trivial solution of the neutral differential equation (2.1) (see Appendix A or [3], *Theorem 8.1*, pp. 292-293).

Furthermore, the negativity of the Lyapunov functional candidate does not use any information about the delay size and in conclusion, the asymptotic stability property holds for any positive delay.

References

1. Boyd, S., El Ghaoui, L., Feron, E. and Balakrishnan, V.: *Linear Matrix Inequalities in System and Control Theory*, SIAM Studies in Applied Mathematics, **15**, 1994.

2. Els'golts', L. E. and Norkin, S. B., *Introduction to the Theory and Application of Differential Equations with Deviating Arguments,* Math. in Science Eng., **105**, Academic Press, New York, 1973.
3. Hale, J. K. and Verduyn Lunel, S. M.: *Introduction to Functional Differential Equations* , Applied Math. Sciences, **99**, Springer-Verlag, New York, 1991.
4. Hale, J. K., Infante, E. F. and Tseng, F. S. P.: Stability in Linear Delay Equations. J. Math. Anal. Appl., **105** (1985) 533-555.
5. Kolmanovskii, V. B. and Myshkis, A.: *Applied Theory of Functional Differential Equations*, Math. and Its Appl., Kluwer Academic Publ., Dordrecht, 1992.
6. Logemann, H. and Townley, S.: The effect of small delays in the feedback loop on the stability of neutral systems. *Syst. & Contr. Lett.* **27** (1996) 267-274.
7. Niculescu, S. I. and Brogliato, B.: On force measurements time-delays in control of constrained manipulators. Proc. IFAC Syst. Struct. & Contr., Nantes, France (1995) 266-271.
8. Niculescu, S. I.: *On the stability and stabilization of linear systems with delayed state*, (in French) Ph.D. Dissertation, L.A.G., Institut National Polytechnique de Grenoble (France), February 1996.
9. Niculescu, S. I., Verriest, E. I., Dugard L. and Dion, J. M.: Stability and robust stability of time-delay systems: A guided tour. this monography (Chapter 1), 1996.
10. Slemrod, M. and Infante, E. F.: Asymptotic stability criteria for linear systems of differential difference equations of neutral type and their discrete analogues. J. Math. Anal. Appl. **38** (1972) 399-415.
11. Verriest, E. I., Fan, M. K. H. and Kullstam, J.: Frequency domain robust stability criteria for linear delay systems. Proc. 32nd IEEE CDC, San Antonio, Texas, U.S.A. (1993) 3473-3478.
12. Verriest, E. I.: Riccati type conditions for robust stability of delay systems. Proc. MTNS'96, St. Louis, U.S.A. (1996).
13. Verriest, E. I.: Robust stability of time-varying systems with unknown bounded delays. Proc. 33rd IEEE Conf. Dec. Contr., Lake Buena Vista, Florida, U.S.A. (1994) 417-422.

Robust Stability and Stabilization of Time-Delay Systems via Integral Quadratic Constraint Approach

M. Fu[1], H. Li[2] and S.-I. Niculescu[3]

[1] Dept. Electrical & Computer Engineering, The University of Newcastle, NSW 2308 Australia
[2] Laboratoire d'Automatique de Grenoble, ENSIEG, BP 46, 38402 Saint Martin d'Hères, France
[3] Laboratoire de Mathématiques Appliquées, Ecole Nationale Supérieure de Techniques Avancées, 32, Blvd. Victor, 75739 Paris, France

Abstract. In this chapter, we consider two problems associated with time-delay systems: robust stability analysis and robust stabilization. We first obtain two results for robust stability using the integral quadratic constraint approach and the linear matrix inequality technique. Both results give an estimate of the maximum time-delay which preserves robust stability. The first stability result is simpler to apply while the second one gives a less conservative robust stability condition. We then apply these stability results to solve the associated robust stabilization problem using static state feedback. Our results provide new design procedures involving linear matrix inequalities.

1 Introduction

Consider a time-delay system described by

$$\dot{x}(t) = A_0 x(t) + A_d x(t-\tau) + B_u u(t) \tag{1.1}$$

where $x(t) \in \mathbb{R}^n$ is the state, $u(t) \in \mathbb{R}^m$ is the control input, τ is an unknown constant time delay, A_0, A_d and B_u are constant matrices.

The system above has been analyzed by many researchers. Two types of robust stability conditions have been reported in the literature: the so-called *delay independent* conditions and *delay-dependent* conditions. In comparison, the delay independent conditions are simpler to apply, but the delay-dependent conditions are less conservative in general. With the recent advances in convex optimization (see, e.g., [2]), the focus of the current research is towards finding less conservative delay-dependent conditions by allowing more complex convex optimization. See [13] for a review of robust stability results.

One of the goals in this chapter is to provide new conditions under which the robust stability of the autonomous system of (1.1) is guaranteed. Our work is based on two ingredients: 1) a sufficient condition for robust stability expressed in the frequency domain; and 2) the integral quadratic constraint (IQC) approach to robustness analysis. Two stability results are presented. Both results are expressed in terms of linear matrix inequalities (LMIs), and they give an estimate of the maximum time-delay which preserves robust stability. The first stability result is simpler to apply while the second one is less conservative. We point out that the stability results in this chapter generalize those in [13].

After derived the two stability results, we then apply these results to solve the associated robust stabilization problem for the system (1.1) using state feedback control. We also provide explicit formula for controllers. Finally, we show several examples which demonstrate the applications as well as the advantages of the results obtained in this chapter.

2 Preliminaries

Several preliminary results are required for robust stability analysis of the autonomous system of (1.1). Throughout this chapter, we denote $A = A_0 + A_d$.

Lemma 1. *The autonomous system of (1.1) is asymptotically stable if A is asymptotically stable and that*

$$\mathcal{A}(j\omega,\tau) := j\omega I - A - \tau\rho_1(j\omega\tau)A_dA_0 - \tau\rho_2(j\omega\tau)A_dA_d \tag{2.1}$$

is nonsingular for all $\omega \in \mathbb{R}$, where

$$\rho_1(jv) = -e^{-jv/2}\frac{\sin(v/2)}{(v/2)}, \quad \rho_2(jv) = \rho_1(jv)e^{-jv}. \tag{2.2}$$

Proof. It is well-known that the autonomous system of (1.1) is asymptotically stable if and only if

$$\hat{\mathcal{A}}(j\omega,\tau) = j\omega I - A_0 - A_de^{-j\omega\tau}$$

is nonsingular for all $\omega \in \mathbb{R}$.

Suppose $\mathcal{A}(j\omega,\tau)$ is nonsingular, we need to show that $\hat{\mathcal{A}}(j\omega,\tau)$ is nonsingular. Let x be such that $\hat{\mathcal{A}}(j\omega,\tau)x = 0$. We need to show that $x = 0$. To see this, we note

$$\begin{aligned} 0 &= (j\omega I - A_0 - A_de^{-j\omega\tau})x \\ &= (j\omega I - A - A_d(e^{-j\omega\tau} - 1))x \\ &= (j\omega I - A - \tau\rho_1(j\omega\tau)A_dj\omega)x \\ &= (j\omega I - A - \tau\rho_1(j\omega\tau)A_d(A_0 + A_de^{-j\omega\tau}))x \\ &= \mathcal{A}(j\omega,\tau)x \end{aligned} \tag{2.3}$$

So x must be zero due to the nonsingularity of $\mathcal{A}(j\omega,\tau)$. □

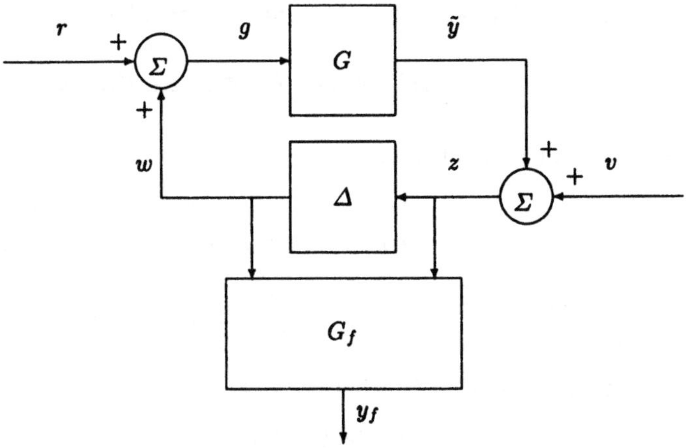

Fig. 1. Interconnected Feedback System

Consider the interconnected system in Figure 1 which is also described by the following equations:

$$\begin{aligned} \dot{x}(t) &= Ax(t) + Bg(t) \\ \tilde{y}(t) &= Cx(t) + Dg(t) \\ z(t) &= \tilde{y}(t) + v(t) \\ g(t) &= r(t) + w(t) \\ w(t) &= \Delta(z(t)) \end{aligned} \tag{2.4}$$

where $\Delta(\cdot) \in \underline{\Delta}$ which is a set of linear or nonlinear dynamic operators to be specified later. Denote

$$G(s) = C(sI - A)^{-1}B + D \tag{2.5}$$

and assume A to be asymptotically stable in the preliminaries and stability analysis sections.

The feedback block $\Delta(\cdot)$ is assumed to satisfy an IQC which is constructed via a *filter* given as follows:

$$\begin{aligned} \dot{x}_f &= A_f x_f + B_f u_f, \quad x_f(0) = 0 \\ y_f &= C_f x_f + D_f u_f \\ u_f &= \begin{bmatrix} z \\ w \end{bmatrix} \end{aligned} \tag{2.6}$$

where A_f is asymptotically stable. Denote the transfer function of the filter by

$$G_f(s) = C_f(sI - A_f)^{-1}B_f + D_f \tag{2.7}$$

The IQC used in this chapter is then described by the following inequality:

$$\int_0^T y_f' \tilde{\Phi} y_f \geq 0, \quad \text{as } T \to \infty, \quad \forall\, \Delta \in \underline{\Delta}, z \in \mathcal{L}_2[0,\infty), \tag{2.8}$$

where $\tilde{\Phi}$ is a constant symmetric matrix.

Remark 1. The definition above does not require $w \in \mathcal{L}_2[0,\infty)$. But if this is the case, then the IQC (2.6)-(2.8) becomes, following the Parseval Theorem,

$$\int_{-\infty}^{+\infty} [z^*(j\omega) \;\; w^*(j\omega)]\Phi(j\omega) \begin{bmatrix} z(j\omega) \\ w(j\omega) \end{bmatrix} d\omega \geq 0, \;\; \forall\, \Delta \in \underline{\Delta} \tag{2.9}$$

where $z(j\omega), w(j\omega)$ are Fourier transforms of $z(t), w(t)$, respectively, and

$$\Phi(s) = G_f^*(s)\tilde{\Phi}G_f(s) \tag{2.10}$$

The following results serve the foundation of the IQC approach.

Theorem 2. (The IQC Theorem) *[19, 16, 15] Given a connected set of operators $\underline{\Delta}$, containing the zero operator, for the feedback block of the system (2.4), the system is absolutely stable if there exists some $\Phi(s)$ of the form (2.10) and a constant $\epsilon > 0$ such that both (2.8) and the following condition are satisfied:*

$$[G^*(j\omega) \;\; I]\Phi(j\omega) \begin{bmatrix} G(j\omega) \\ I \end{bmatrix} + \epsilon I \leq 0, \;\; \forall\, \omega \in (-\infty,\infty) \tag{2.11}$$

Further, for causal and asymptotically stable linear time-invariant (LTI) $\Delta(\cdot)$, (2.8) is equivalent to the following:

$$[I \;\; \Delta^*(j\omega)]\Phi(j\omega) \begin{bmatrix} I \\ \Delta(j\omega) \end{bmatrix} \geq 0, \;\; \forall\, \omega \in (-\infty,\infty),\; \Delta \in \underline{\Delta} \tag{2.12}$$

That is, the system (2.4) is absolutely stable if there exists $\Phi(s)$ of the form (2.10) such that (2.11) and (2.12) hold.

Lemma 3. (KYP Lemma) *[1, 17] Given $A \in \mathbb{R}^{n\times n}$, $B \in \mathbb{R}^{n\times k}$ and symmetric $\Omega \in \mathbb{R}^{(n+k)\times(n+k)}$, there exists a symmetric matrix $P \in \mathbb{R}^{n\times n}$ such that*

$$\begin{bmatrix} A^TP + PA & PB \\ B^TP & 0 \end{bmatrix} + \Omega < 0 \tag{2.13}$$

if and only if there exists some constant $\epsilon > 0$ such that

$$[B^T((j\omega I - A)^{-1})^* \;\; I]\Omega \begin{bmatrix} (j\omega I - A)^{-1}B \\ I \end{bmatrix} + \epsilon I \leq 0, \;\; \forall \;\; \omega \in (-\infty,\infty) \tag{2.14}$$

Further, if A is Hurwitz and the top-left $n \times n$ submatrix of Ω is positive semidefinite, then (2.13) implies $P > 0$.

We also recall the following two linear matrix inequality results:

Theorem 4. (Positive Real Parrot Theorem) *[2, 11, 10, 7] Given a symmetric matrix $\Psi \in \mathbb{R}^{m\times m}$ and two matrices U, V of column dimension m. There exists a matrix Θ of compatible dimensions such that*

$$\Psi + U^T\Theta^T V + V^T\Theta U < 0 \tag{2.15}$$

if and only if

$$U_\perp^T \Psi U_\perp < 0 \tag{2.16}$$
$$V_\perp^T \Psi V_\perp < 0 \tag{2.17}$$

where $U_\perp$ (resp. $V_\perp$) is any matrix whose columns form a basis of the null space of U (resp. V).

Lemma 5. *[6] Given matrices $A, B_1, B_2, C_1, C_2, X_d, Q = Q^T, W_1 = W_1^T, W_2 = W_2^T$ of appropriate dimensions, suppose $W_1 > 0$, $W_2 > 0$. Then there exists a matrix K of appropriate dimension such that*

$$\left[\begin{array}{cc|c} QA^T + AQ & QC_1^T & B_1 \\ C_1Q & -W_1 & 0 \\ \hline B_1^T & 0 & -W_2 \end{array}\right] + \left[\begin{array}{c} B_2 \\ X_d \\ \hline 0 \end{array}\right] K \left[\begin{array}{cc|c} I & 0 & 0 \end{array}\right] + \left[\begin{array}{c} I \\ 0 \\ \hline 0 \end{array}\right] K^T \left[\begin{array}{cc|c} B_2^T & X_d^T & 0 \end{array}\right] < 0 \tag{2.18}$$

holds if and only if the following LMI holds:

$$\left[\begin{array}{c|c} \mathcal{N} & 0 \\ \hline 0 & I \end{array}\right]^T \left[\begin{array}{cc|c} QA^T + AQ & QC_1^T & B_1 \\ C_1Q & -W_1 & 0 \\ \hline B_1^T & 0 & -W_2 \end{array}\right] \left[\begin{array}{c|c} \mathcal{N} & 0 \\ \hline 0 & I \end{array}\right] < 0 \tag{2.19}$$

where $\mathcal{N}$ is any matrix whose columns form a basis of the null space of $[B_2^T \ X_d^T]$.

Denote K_1 the solution of the following formula:

$$\left[\begin{array}{c} K_1 \\ \hline * \end{array}\right] = -\left[\begin{array}{c|c} 0 & X_d^T \\ \hline X_d & -W_1 \end{array}\right]^+ \left[\begin{array}{c} B_2^T \\ \hline C_1Q \end{array}\right], \tag{2.20}$$

where $^+$ denote the pseudoinverse. Further let K_2 be any solution of the LMI

$$\Psi(K_1) + B_2(I - X_d^+ X_d)K_2 + K_2^T(I - X_d^+ X_d)B_2^T < 0, \tag{2.21}$$

where

$$\begin{aligned} \Psi(K_1) &= QA^T + AQ + K_1^T B_2^T + B_2 K_1 \\ &\quad + \left[\begin{array}{c} C_1Q + X_dK_1 \\ B_1^T \end{array}\right]^T \left[\begin{array}{cc} W_1^{-1} & 0 \\ 0 & W_2^{-1} \end{array}\right] \left[\begin{array}{c} C_1Q + X_dK_1 \\ B_1^T \end{array}\right]. \end{aligned} \tag{2.22}$$

Suppose (2.19) holds. Then a desired K for (2.18) is given by $K = K_1$ if $(I - X_d^+ X_d)B_2^T = 0$ or $K = K_1 + (I - X_d^+ X_d)K_2$ otherwise.

3 Stability Analysis

Consider the time-delay system (1.1). For stability analysis, we assume that $u(t) \equiv 0$. Express

$$A_d = HE, \quad H \in \mathbb{R}^{n\times q}, \quad E \in \mathbb{R}^{q\times n} \tag{3.1}$$

where $q \leq n$, and H and E are of full rank. Define

$$B^T = \begin{bmatrix} H^T \\ H^T \end{bmatrix}, \quad C = \begin{bmatrix} C_1 \\ C_2 \end{bmatrix} = \begin{bmatrix} EA_0 \\ EA_d \end{bmatrix}, \quad C_\tau = \tau C, \quad D = 0, \tag{3.2}$$

and $\underline{\Delta}$ being the set of LTI operators with Fourier transform given by

$$\Delta(j\omega) = \lambda \operatorname{diag}\{\rho_1(j\omega\tau)I_q, \rho_2(j\omega\tau)I_q\} \tag{3.3}$$

for some $\lambda \in [0, 1]$, where $\rho_1(\cdot)$ and $\rho_2(\cdot)$ are defined in (2.2).

Using Lemma 1, we know that the system (1.1) is robustly stable if the following system is robustly stable:

$$\begin{aligned} \dot{x}(t) &= Ax(t) + Bg(t) \\ \tilde{y}(t) &= C_\tau x(t) + Dg(t) \\ z(t) &= \tilde{y}(t) + v(t) \\ g(t) &= r(t) + w(t) \\ w(t) &= \Delta(z(t)) \end{aligned} \tag{3.4}$$

Following the IQC Theorem, we assert that the system (1.1) is robustly stable if there exists some IQC, or equivalently, $\Phi(s)$ as in (2.10) such that (2.11) and (2.12) hold. Note that the notion of absolute stability coincides with the notion of robust stability for a linear uncertain block Δ. In the rest of this section, we study two IQCs which give two robust stability conditions.

The first IQC is a simple constant D-scaling used in the analysis of structured singular value. More precisely, we take

$$G_f(s) = \operatorname{diag}\{I_{2q}, I_{2q}\} \tag{3.5}$$

and

$$\tilde{\Phi} = \tau^{-1} \operatorname{diag}\{\Lambda_1, \Lambda_2, -\Lambda_1, -\Lambda_2\} \tag{3.6}$$

for some $q\times q$ symmetric and positive-definite $\Lambda_i, i = 1, 2$, which are to be chosen. Denote

$$\Lambda = \operatorname{diag}\{\Lambda_1, \Lambda_2\} \tag{3.7}$$

The resulting IQC has

$$\Phi(s) = \tau^{-1} \operatorname{diag}\{\Lambda, -\Lambda\} \tag{3.8}$$

It is straightforward to verify that (2.12) holds because $\rho_i(\cdot)$ are contractive. The sufficient condition (2.11) for robust stability becomes

$$\begin{bmatrix} B^T((j\omega I - A)^{-1})^* & I_{2q} \end{bmatrix} \begin{bmatrix} \tau C^T \Lambda C & 0 \\ 0 & -\tau^{-1}\Lambda \end{bmatrix} \begin{bmatrix} (j\omega I - A)^{-1}B \\ I_{2q} \end{bmatrix} + \epsilon I \leq 0 \tag{3.9}$$

for all ω.

Using the KYP Lemma, the above is equivalent to the existence of $P = P^T > 0$ such that the following LMI holds:

$$\begin{bmatrix} A^TP + PA + \tau C_1^T\Lambda_1 C_1 + \tau C_2^T \Lambda_2 C_2 & PH & PH \\ H^TP & -\tau^{-1}\Lambda_1 & 0 \\ H^TP & 0 & -\tau^{-1}\Lambda_2 \end{bmatrix} < 0$$

Multiplying τ to the second and third row and column blocks, which does not alter the validity of the LMI, the above becomes

$$\Pi(\tau) = \begin{bmatrix} A^TP + PA + \tau C_1^T\Lambda_1 C_1 + \tau C_2^T \Lambda_2 C_2 & \tau PH & \tau PH \\ \tau H^TP & -\tau\Lambda_1 & 0 \\ \tau H^TP & 0 & -\tau\Lambda_2 \end{bmatrix} < 0 \quad (3.10)$$

Note that $\Pi(\tau)$ is affine in P, Λ_1 and Λ_2.

We summarize the analysis above as follows:

Theorem 6. *The autonomous time-delay system of (1.1) is robustly stable for all $0 < \tau \le \bar{\tau}$ if there exist $n \times n$ symmetric and positive definite matrices Λ_1, Λ_2 and P such that the LMI*

$$\Pi(\bar{\tau}) < 0 \quad (3.11)$$

holds, where $\Pi(\tau)$ is defined in (3.10).

Proof. Suppose (3.11) holds. It follows from the analysis above that the system (1.1) is robustly stable for $\bar{\tau}$. The conclusion that the above also implies the robust stability for all $0 \le \tau \le \bar{\tau}$ follows from the fact that the $\Pi(\tau)$ is convex in τ. More precisely, $\Pi(\tau) < 0$ when τ is sufficiently small, due to (3.11) and $\Lambda_i > 0$. The rest follows from the convexity of $\Pi(\tau)$. □

The second IQC we use to study the robust stability of the system (1.1) will be more involved but give a less conservative condition. Let

$$f(s) = c_f(sI - a_f)^{-1}b_f + d_f \quad (3.12)$$

be any asymptotically stable SISO filter with the following property:

$$|f(jv)| \ge \left|\frac{\sin(v)}{v}\right|, \quad \forall v \in \mathbb{R} \quad (3.13)$$

Denote the diagonal transfer matrix

$$F(s) = f(s)I_{2q} = C_f(sI - A_f)^{-1}B_f + D_f \quad (3.14)$$

We will discuss how to choose $f(s)$ later.

Now define

$$G_f(s) = \text{diag}\{F(s\tau), I_{2q}\} \quad (3.15)$$

and $\tilde{\Phi}$ as in (3.6). This yields

$$y_f(s) = \begin{bmatrix} F(s\tau)z(s) \\ w(s) \end{bmatrix} \tag{3.16}$$

$$\Phi(s) = G_f^*(s)\tilde{\Phi}G_f(s) = \tau^{-1}\ \mathrm{diag}\{F^*(s\tau)\Lambda F(s\tau), -\Lambda\} \tag{3.17}$$

Subsequently, condition (2.12) is automatically satisfied because

$$\begin{aligned} &\tau[I_{2q}\quad \Delta^*(j\omega)]\Phi(j\omega)\begin{bmatrix} I_{2q} \\ \Delta(j\omega) \end{bmatrix} \\ = \;& F^*(j\omega\tau)\Lambda F(j\omega\tau) - \lambda^2 \mathrm{diag}\{\rho_1^*(j\omega\tau)\rho_1(j\omega\tau)\Lambda_1, \rho_2^*(j\omega\tau)\rho_2(j\omega\tau)\Lambda_2\} \\ = \;& \Lambda^{1/2}(F^*(j\omega\tau)F(j\omega\tau) - \lambda^2\mathrm{diag}\{\rho_1^*(j\omega\tau)\rho_1(j\omega\tau), \rho_2^*(j\omega\tau)\rho_2(j\omega\tau)\})\Lambda^{1/2} \\ \geq \;& 0 \end{aligned}$$

Therefore, a sufficient condition for robust stability of system (1.1) is the condition (2.11) which, in our case, becomes

$$G_\tau^*(j\omega)F^*(j\omega\tau)\tau^{-1}\Lambda F(j\omega\tau)G_\tau(j\omega) - \tau^{-1}\Lambda + \epsilon I_{2q} \leq 0,\quad \forall\, \omega \in (-\infty, \infty) \tag{3.18}$$

for some (sufficiently small) $\epsilon > 0$, where $G_\tau(s) = C_\tau(sI - A)^{-1}B$.

Our next step is to convert the frequency domain condition (3.18) into the state space. To this end, we denote by $\bar{C}_\tau(sI - \bar{A}_\tau)^{-1}\bar{B}_\tau$ a state-space realization of $F(s\tau)G_\tau(s)$. Then, it is straightforward to verify that

$$\bar{A}_\tau = \begin{bmatrix} \tau^{-1}A_f & B_f C \\ 0 & A \end{bmatrix},\quad \bar{B} = \begin{bmatrix} 0 \\ B \end{bmatrix},\quad \bar{C}_\tau = [\, C_f \quad D_f C_\tau \,] \tag{3.19}$$

Condition (3.18) can be rewritten as

$$[\bar{B}^T((j\omega I - \bar{A}_\tau)^{-1})^*\ I]\ \mathrm{diag}\{\tau^{-1}\bar{C}_\tau^T\Lambda\bar{C}_\tau, -\tau^{-1}\Lambda\}\begin{bmatrix} (j\omega I - \bar{A}_\tau)^{-1}\bar{B} \\ I \end{bmatrix} + \epsilon I \leq 0$$

for all $\omega \in (-\infty, \infty)$.

Applying the KYP Lemma, the above is equivalent to the existence of some $\bar{P} = \bar{P}^T > 0$ such that the following linear matrix inequality holds:

$$\bar{\Pi}(\tau) = \begin{bmatrix} \bar{A}_\tau^T\bar{P} + \bar{P}\bar{A}_\tau + \tau^{-1}\bar{C}_\tau^T\Lambda\bar{C}_\tau & \bar{P}\bar{B} \\ \bar{B}^T\bar{P} & -\tau^{-1}\Lambda \end{bmatrix} < 0 \tag{3.20}$$

The above analysis is summarized as follows:

Theorem 7. *The autonomous time-delay system of (1.1) is robustly stable for all $\tau \leq \bar{\tau}$ if there exist an asymptotically stable filter $f(s)$, and symmetric and positive-definite matrices Λ_1, Λ_2 and $\bar{P}$ such that the following LMI holds:*

$$\bar{\Pi}(\bar{\tau}) < 0 \tag{3.21}$$

where $\bar{\Pi}(\cdot)$ is defined in (3.20).

Proof. The proof is very similar to that of Theorem 6, so the details are omitted. The only step worth of discussion is the fact that $\bar{\Pi}(\bar{\tau}) < 0$ implies $\bar{\Pi}(\tau) < 0$ for all $0 < \tau \leq \bar{\tau}$. This step is a bit tedious but not too difficult to verify too. □

4 Stabilization

Consider the time-delay system (1.1). Our objective is to design a static state feedback controller such that the closed-loop system of (1.1) is uniformly asymptotically stable. The results derived in this section are based on Theorems 6 and 7.

Let a desired controller be given in the following form:

$$u(t) = Kx(t) \tag{4.1}$$

where K is the gain matrix to be designed.

With the controller (4.1), the closed-loop system of (1.1) is as follows:

$$\dot{x}(t) = (A_0 + B_u K)x(t) + A_d x(t-\tau) \tag{4.2}$$

Applying Theorem 6 and Lemma 5 to the system (4.2), we obtain the following result:

Theorem 8. *There exists a state feedback controller (4.1) such that the closed-loop time-delay system of (1.1) with this controller is robustly stable for all* $0 < \tau \leq \bar{\tau}$ *if there exist* $n \times n$ *symmetric and positive definite matrices* Γ_1, Γ_2 *and* Q *such that the LMI*

$$\left[\begin{array}{c|c} \mathcal{N}_B & 0 \\ \hline 0 & I \end{array}\right]^T \Pi_c(\bar{\tau}) \left[\begin{array}{c|c} \mathcal{N}_B & 0 \\ \hline 0 & I \end{array}\right] < 0 \tag{4.3}$$

holds, where

$$\Pi_c(\tau) = \left[\begin{array}{cc|ccc} QA^T + AQ & QC_1^T & QC_2^T & H\Gamma_1 & H\Gamma_2 \\ C_1 Q & -\tau^{-1}\Gamma_1 & 0 & 0 & 0 \\ \hline C_2 Q & 0 & -\tau^{-1}\Gamma_2 & 0 & 0 \\ \Gamma_1 H^T & 0 & 0 & -\tau^{-1}\Gamma_1 & 0 \\ \Gamma_2 H^T & 0 & 0 & 0 & -\tau^{-1}\Gamma_2 \end{array}\right],$$

and $\mathcal{N}_B$ *is any matrix whose columns form a basis of the null space of* $[B_u^T \;\; B_u^T E^T]$.

Further, suppose (4.3) holds. Let K_1 *be the solution of the following formula:*

$$\left[\begin{array}{c} K_1 \\ \hline * \end{array}\right] = -\left[\begin{array}{c|c} 0 & B_u^T E^T \\ \hline EB_u & -\bar{\tau}^{-1}\Gamma_1 \end{array}\right]^+ \left[\begin{array}{c} B_u^T \\ \hline C_1 Q \end{array}\right], \tag{4.4}$$

and K_2 *be any solution of the LMI*

$$\Psi(K_1) + B_u(I - (EB_u)^+ EB_u)K_2 + K_2^T(I - (EB_u)^+ EB_u)B_u^T < 0, \tag{4.5}$$

where

$$\Psi(K_1) = QA^T + AQ + K_1^T B_u^T + B_u K_1 + \bar{\tau}\begin{bmatrix} C_1Q + EB_uK_1 \\ C_2Q \\ H^T \\ H^T \end{bmatrix}^T \begin{bmatrix} \Gamma_1^{-1} & 0 & 0 & 0 \\ 0 & \Gamma_2^{-1} & 0 & 0 \\ 0 & 0 & \Gamma_1 & 0 \\ 0 & 0 & 0 & \Gamma_2 \end{bmatrix} \begin{bmatrix} C_1Q + EB_uK_1 \\ C_2Q \\ H^T \\ H^T \end{bmatrix} \tag{4.6}$$

Then, a desired controller gain matrix K is given by $K = K_1Q^{-1}$ if

$$(I - (EB_u)^+ EB_u)B_u^T = 0$$

or $K = (K_1 + (I - (EB_u)^+EB_u)K_2)Q^{-1}$ otherwise.

Proof. Applying Theorem 6 to the closed-loop system (4.2), we find that this system is robustly stable for all $0 < \tau \leq \bar{\tau}$ if there exist $n \times n$ symmetric and positive definite matrices Λ_1, Λ_2 and P such that the matrix inequality

$$\left[\begin{array}{c|cc} (A + B_uK)^T P + P(A + B_uK) + \bar{\tau}C_2^T \Lambda_2 C_2 & \bar{\tau}PH & \bar{\tau}PH \\ +\bar{\tau}(C_1 + EB_uK)^T \Lambda_1 (C_1 + EB_uK) & & \\ \hline \bar{\tau}H^TP & -\bar{\tau}\Lambda_1 & 0 \\ \bar{\tau}H^TP & 0 & -\bar{\tau}\Lambda_2 \end{array}\right] < 0 \tag{4.7}$$

holds. Using Schur complements, (4.7) can be rewritten as

$$\left[\begin{array}{cc|ccc} A^TP + PA & \bar{\tau}C_1^T & \bar{\tau}C_2^T & \bar{\tau}PH & \bar{\tau}PH \\ \bar{\tau}C_1 & -\bar{\tau}\Lambda_1^{-1} & 0 & 0 & 0 \\ \hline \bar{\tau}C_2 & 0 & -\bar{\tau}\Lambda_2^{-1} & 0 & 0 \\ \bar{\tau}H^TP & 0 & 0 & -\bar{\tau}\Lambda_1 & 0 \\ \bar{\tau}H^TP & 0 & 0 & 0 & -\bar{\tau}\Lambda_2 \end{array}\right] + \left[\begin{array}{c} PB_u \\ \bar{\tau}EB_u \\ \hline 0 \\ 0 \\ 0 \end{array}\right] K[I \; 0 \mid 0 \; 0 \; 0] + \left[\begin{array}{c} I \\ 0 \\ \hline 0 \\ 0 \\ 0 \end{array}\right] K^T[B_u^TP \; \bar{\tau}B_u^TE^T \mid 0 \; 0 \; 0] < 0 \tag{4.8}$$

Define $Q = P^{-1}$, $\Gamma_1 = \Lambda_1^{-1}$ and $\Gamma_2 = \Lambda_2^{-1}$, respectively. Multiplying

$$\mathrm{diag}\{Q, \bar{\tau}^{-1}I, \bar{\tau}^{-1}I, \bar{\tau}^{-1}\Gamma_1, \bar{\tau}^{-1}\Gamma_2\}$$

to both sides, the inequality above is equivalent to

$$\Pi_c(\bar{\tau}) + \left[\begin{array}{c} B_u \\ EB_u \\ \hline 0 \\ 0 \\ 0 \end{array}\right] (KQ)[I \; 0 \mid 0 \; 0 \; 0] + \left[\begin{array}{c} I \\ 0 \\ \hline 0 \\ 0 \\ 0 \end{array}\right] (KQ)^T[B_u^T \; B_u^TE^T \mid 0 \; 0 \; 0] < 0 \tag{4.9}$$

Then, the results in the theorem are obtained by applying Lemma 5. □

Corresponding to Theorem 7 for stability analysis, we can obtain the following less conservative result for robust stabilization:

Theorem 9. *There exists a state feedback controller (4.1) such that the closed-loop time-delay system of (1.1) with this controller is robustly stable for all $\tau \le \bar{\tau}$ if there exist symmetric and positive definite matrices $\bar{\Gamma}$, Q_f and Q such that the LMIs*

$$\left[\begin{array}{c|c} \mathcal{N}_f & 0 \\ \hline 0 & I \end{array}\right]^T \bar{\Pi}_c(\bar{\tau}) \left[\begin{array}{c|c} \mathcal{N}_f & 0 \\ \hline 0 & I \end{array}\right] < 0 \tag{4.10}$$

$$\begin{bmatrix} A_f Q_f + Q_f A_f^T & Q_f C_f^T \\ C_f Q_f & -\bar{\Gamma} \end{bmatrix} < 0 \tag{4.11}$$

hold, where

$$\bar{\Pi}_c(\tau) = \left[\begin{array}{ccc|c} QA^T + AQ & QC^T B_f^T & QC^T D_f^T & B\bar{\Gamma} \\ B_f CQ & \tau^{-1}(A_f Q_f + Q_f A_f^T) & \tau^{-1} Q_f C_f^T & 0 \\ D_f CQ & \tau^{-1} C_f Q_f & -\tau^{-1}\bar{\Gamma} & 0 \\ \hline \bar{\Gamma} B^T & 0 & 0 & -\tau^{-1}\bar{\Gamma} \end{array}\right]. \tag{4.12}$$

and $\mathcal{N}_f$ is any matrix whose columns form a basis of the null space of the matrix $[B_u^T \ \ X_d^T]$ with

$$X_d = \begin{bmatrix} B_f \\ D_f \end{bmatrix} \begin{bmatrix} EB_u \\ 0 \end{bmatrix}. \tag{4.13}$$

Further, suppose (4.10)-(4.11) hold. Denote

$$X_c = \begin{bmatrix} B_f \\ D_f \end{bmatrix} C \tag{4.14}$$

and

$$W_f = \begin{bmatrix} Q_f A_f^T + Q_f A_f & Q_f C_f^T \\ C_f Q_f & -\bar{\Gamma} \end{bmatrix}, \tag{4.15}$$

Let K_1 be the solution of the following formula:

$$\left[\begin{array}{c} K_1 \\ \hline * \end{array}\right] = -\left[\begin{array}{c|c} 0 & X_d^T \\ \hline X_d & \bar{\tau}^{-1} W_f \end{array}\right]^+ \left[\begin{array}{c} B_u^T \\ \hline X_c Q \end{array}\right], \tag{4.16}$$

and let K_2 be any solution of the LMI

$$\Psi(K_1) + B_u(I - X_d^+ X_d)K_2 + K_2^T(I - X_d^+ X_d)B_u^T < 0, \tag{4.17}$$

where

$$\begin{aligned} \Psi(K_1) &= QA^T + AQ + K_1^T B_u^T + B_u K_1 \\ &\quad + \bar{\tau} \begin{bmatrix} X_c Q + X_d K_1 \\ B^T \end{bmatrix}^T \begin{bmatrix} W_f^{-1} & 0 \\ 0 & \bar{\Gamma} \end{bmatrix} \begin{bmatrix} X_c Q + X_d K_1 \\ B^T \end{bmatrix}. \end{aligned} \tag{4.18}$$

Then, a desired controller gain matrix K *is given by* $K = K_1 Q^{-1}$ *if*

$$(I - X_d^+ X_d) B_u^T = 0$$

or $K = (K_1 + (I - X_d^+ X_d) K_2) Q^{-1}$ *otherwise.*

Proof. The proof of this theorem follows the same line as that for Theorem 8. Namely, we apply Theorem 7 and Lemma 5 to the closed-loop system (4.2). We first use Schur complements to rewrite the robust stability condition in Theorem 8, assuming $K = 0$. That is, $\bar{\Pi}(\bar{\tau}) < 0$ if and only if

$$\left[\begin{array}{cc|c} \bar{A}_\tau^T \bar{P} + \bar{P}\bar{A}_\tau & \bar{C}_\tau^T & \bar{P}\bar{B} \\ \bar{C}_\tau & -\bar{\tau}\Lambda^{-1} & 0 \\ \hline \bar{B}^T \bar{P} & 0 & -\bar{\tau}^{-1}\Lambda \end{array}\right] < 0$$

Furthermore, we take $\bar{P} = \mathrm{diag}\{P_f, P\}$ since there is no interaction between the closed-loop system and the filter. Let $\bar{Q} = \bar{P}^{-1} = \mathrm{diag}\{Q_f, Q\}$ and $\bar{\Gamma} = \Lambda^{-1}$. Multiplying $\mathrm{diag}\{\bar{Q}, \bar{\tau}^{-1} I, \bar{\Gamma}\}$ to the both sides, the above inequality is converted into

$$\left[\begin{array}{ccc|c} \tau^{-1}(A_f Q_f + Q_f A_f^T) & B_f C Q & \tau^{-1} Q_f C_f^T & 0 \\ Q C^T B_f^T & Q A^T + A Q & Q C^T D_f^T & B\bar{\Gamma} \\ \tau^{-1} C_f Q_f & D_f C Q & -\tau^{-1}\bar{\Gamma} & 0 \\ \hline 0 & \bar{\Gamma} B^T & 0 & -\tau^{-1}\bar{\Gamma} \end{array}\right] < 0.$$

Swapping the first two rows and columns, which does not affect the inequality, we further convert the above into

$$\bar{\Pi}_c(\bar{\tau}) < 0 \tag{4.19}$$

where $\bar{\Pi}_c(\bar{\tau})$ is defined in (4.12).

Now let static state feedback be used, i.e., A_0 becomes $A_0 + B_u K$. Subsequently, A becomes $A + B_u K$ and C becomes

$$C + \begin{bmatrix} E B_u K \\ 0 \end{bmatrix}.$$

Hence, the robust stability condition (4.19) becomes the following robust stabilization condition:

$$\bar{\Pi}_c(\bar{\tau}) + \left[\begin{array}{c} B_u \\ X_d \\ \hline 0 \end{array}\right] (KQ) \left[\begin{array}{cc|c} I & 0 & 0 \end{array}\right] + \left[\begin{array}{c} I \\ 0 \\ \hline 0 \end{array}\right] (KQ)^T \left[\begin{array}{cc|c} B_u^T & X_d^T & 0 \end{array}\right] < 0$$

Then, it is tedious but straightforward to prove the characterization of K by applying Theorem 7 and Lemma 5. □

Remark 2. If we take the filter $f(s) = 1$ and further constrain Λ to be $\Lambda = \mathrm{diag}\{\Lambda_1, \Lambda_2\}$, then it is obvious that Theorem 9 reduces to Theorem 8. To see this, we may select $B_f = 0$, $C_f = 0$ and $A_f = -I$, then clearly (4.10) reduces to (4.3) while (4.11) is trivially satisfied.

5 Examples

Before providing illustrative examples, we address the problem of finding a suitable filter $f(s)$. First, we note that $f(s)$ is a SISO transfer function, and that the constraint on $f(s)$ (3.13) is independent of the system (1.1). This means that once a "good" $f(s)$ is found, it can be applied to various time-delay systems of the form (1.1). The complexity of $f(s)$ is mainly determined by the degree of $f(s)$. A second order example is given below:

$$f(s) = \frac{2(s+0.9)}{(s+0.8)(s+2.216)} \tag{5.1}$$

with its Bode plot given in Figure 2. Also plotted in Figure 2 is $|\sin(\omega)/\omega|$ to justify (3.13).

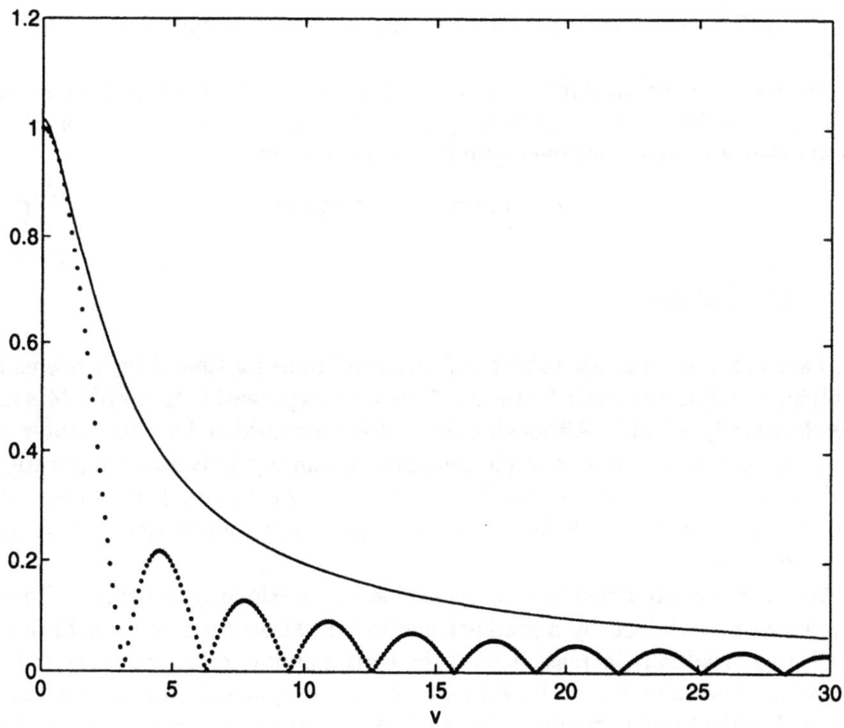

Fig. 2. Example of $f(s)$. Solid line: $|f(jv)|$; Dotted line: $|\sin(v)/v|$

Example 1: Consider the autonomous system of (1.1) with

$$A_0 = \begin{bmatrix} -2 & 0 \\ 0 & 0.25 \end{bmatrix}, \quad A_d = \begin{bmatrix} -1 & 0 \\ -0.1 & -0.85 \end{bmatrix} \tag{5.2}$$

Using Theorem 7 and the $f(s)$, the maximum τ is obtained to be $\tau_{\max} = 0.9848$.

Obviously, the conservatism of $\tau_{\max}$ depends on the filter $f(s)$. It is found in simulation that second order filters usually outperform first order ones. Also, higher order filters can be used to obtain slightly larger $\tau_{\max}$.

Using Theorem 6, the maximum τ is obtained to be $\tau_{\max} = 0.6417$.

As comparisons, we notice that the maximum τ using the results in [13, 12] is $\tau_{\max} = 0.58$ while the optimal τ for the system with the given parameters is $\tau_o = 1.54$[12].

Example 2: Consider the system (1.1) with

$$A_0 = \begin{bmatrix} 2 & 0 \\ 1.75 & 0.35 \end{bmatrix}, \quad A_d = \begin{bmatrix} -1 & 0 \\ -0.1 & -0.25 \end{bmatrix}, \quad B_u = \begin{bmatrix} 1 \\ 1 \end{bmatrix}. \tag{5.3}$$

Since $A_0 + A_d$ is unstable and (A_0, B_u) is not controllable, the system (1.1) with the above given parameters can't be stabilized independent of the time-delay using state feedback controller.

Using Theorem 9 and the $f(s)$, the maximum τ is obtained to be $\tau_{\max} = 0.984$.

To avoid numerical difficulties, we synthesize the state feedback controller using $\tau_{\max} = 0.92$ instead. Following the explicit K formula in Theorem 9, we obtain that a desired controller gain matrix is given by

$$K = [-1.7063 \quad -1.2815]. \tag{5.4}$$

6 Conclusion

We have obtained two new robust stability conditions for time-delay systems by applying the IQC approach. These conditions are expressed in terms of LMIs and therefore easily solvable. Although a single delay is considered in this chapter, we stress that an extension to multiple delays can be simply derived. As applications of these new robust stability results, robust stabilization problems using static state feedback control have been tackled. Explicit controller formulas have also been provided.

We have not explained how to determine the maximal time delay $\bar{\tau}$. Generally, $\bar{\tau}$ can be obtained by a gradient method. First we set $\bar{\tau}$ to be sufficiently small, then gradually increase it until the corresponding robust stability or stabilization conditions are no longer feasible. A fine gradient can be adopted in the final critical region to obtain larger $\bar{\tau}$. Alternatively, we can use a bisection method. That is, we start with any lower bound and and a upper bound for $\bar{\tau}$. Then, choose the initial $\bar{\tau}$ to be the average of the bounds and test for the solvability of robust stability or stabilization conditions. The bounds will be improved according to the outcome of the test. This procedure is repeated until the gap between the bounds is sufficiently small.

References

1. B. D. O. Anderson, "A system theory criterion for positive real matrices," *SIAM J. Contr. Optimization*, vol. 6, no. 2, pp. 171-192, 1967.
2. S. Boyd, L. El Ghaoui, E. Feron and V. Balakrishnan, *Linear Matrix Inequalities in System and Control Theory*, Philadelphia: SIAM, 1994.
3. S. Dasgupta, G. Chockalingam, B. D. O. Anderson and M. Fu, "Lyapunov functions for uncertain systems with applications to the stability to time varying systems," *IEEE Trans. Circ. Syst.*, vol. 41, pp. 93-106, 1994.
4. M. Fu and S. Dasgupta: "The integral quadratic constraint approach to robustness analysis: An overview," *IFAC Workshop on Robust Control*, Napa Valley, San Fransisco, 1996.
5. M. Fu and S. Dasgupta, "Parametric Lyapunov functions for uncertain systems: the multiplier approach," presented at *MNTS*, St. Louis, June, 1996.
6. P. Gahinet, "Explicit controller formulas for LMI-based H_∞ synthesis," *Automatica*, vol 32, pp. 1007-1014, 1996.
7. P. Gahinet and P. Apkarian, "A linear matrix inequality approach to $\mathcal{H}_\infty$ control," *Int. J. Robust and Nonlinear Contr.*, vol. 4, pp. 421-428, 1994.
8. P. Gahinet, P. Apkarian and M. Chilali, "Affine parameter-dependent Lyapunov functions for real parameter uncertainty," *Proc. 33rd IEEE Conf. Decision and Control*, pp. 2026-2031, Lake Buena Vista, FL, Dec., 1994.
9. T. Iwasaki, S. Hara and T. Asai, "Well-posedness theorem: a classification of LMI/BMI-reducible robust control problems," *Int. Symp. Intelligent Robotics Systems*, Bangalore, India, Nov. 1995.
10. T. Iwasaki and R. E. Skelton, "A complete solution to the general $\mathcal{H}_\infty$ problem: LMI existence conditions and state space formulas," *Proc. American Control Conf.*, pp. 605-609, San Fransisco, CA, June 2-4, 1993.
11. J. H. Ly, M. G. Safonov and F. Ahmad, "Positive real Parrot theorem with applications to LMI controller synthesis," *Proc. American Control Conf.*, pp. 50-52, Baltimore, MD, June 1994.
12. S.-I. Niculescu, *On the stability and stabilization of linear systems with delayed state* (in French), Ph.D. thesis, Laboratoire d'Automatique de Grenoble, INPG, February 1996.
13. S.-I. Niculescu, J.-M. Dion, and L. Dugard: "Delays-dependent stability for linear systems with two delays: a convex optimization approach," Internal Report, Laboratoire d'Automatique de Grenoble, 1995.
14. A. Packard, K. Zhou, P. Pandey and G. Becker, "A collection of robust control problems leading to LMI's," *Proc. 30th IEEE Conf. Decision and Control*, pp. 1245-1250, 1991.
15. A. Rantzer and A. Megretski, "System analysis via integral quadratic constraints," *Proc. 33rd IEEE Conf. Decision and Control*, pp. 3062-3067, Lake Buena Vista, FL, Dec., 1994.
16. M. G. Safonov, *Stability and Robustness of Multivariable Feedback Systems*, MIT Press, 1980.
17. J. Willems, *The Analysis of Feedback Systems*, MIT Press, 1971.

18. V. A. Yakubovich, "Frequency conditions of absolute stability of control systems with many nonlinearities," *Automatica i Telemekhanica*, vol. 28, pp. 5-30, 1967.
19. V. A. Yakubovich, "S-procedure in nonlinear control theory," *Vestnik Leningrad Universiteta, Ser. Matematika*, pp. 62-77, 1971.

Graphical Test for Robust Stability with Distributed Delayed Feedback

Erik I. Verriest

School of Electrical and Computer Engineering, Georgia Institute of Technology, Atlanta, GA 30332-0250

Abstract. The performance of a nominally designed state feedback control for a linear systems is analyzed in the case that the information, available at time t for feedback, consists of a functional of the state over the interval $[t-T, t]$. Sufficient conditions are given for the stability and asymptotic stability, *independent* of the matrix valued weight functions on the delay-perturbed state. These sufficient conditions, obtained via the Lyapunov-Krasovskii theory, revolve around the existence of some positive definite matrix functions satisfying certain Riccati-type differential equations. Connections are made with the theory of robust control and its frequency domain criteria. New graphical criteria akin to the Nyquist criterion are derived to obtain the delay perturbation margin.

This chapter applies the theory of stability of linear differential delay systems with distributed delays. In earlier work [3, 4, 8, 9, 7, 14, 17, 19, 20, 21, 22], the theory of Lyapunov functionals was exploited to obtain sufficient conditions for the stability of time-invariant and time-varying differential delay systems, independent of the delay time. The conditions that were obtained were of an algebraic nature since they involved the existence of a triple of positive definite matrices satisfying a certain Riccati equation. This Riccati equation has a positive sign in its quadratic term, and is thus of the type encountered in robust control theory [12]. In fact, connections between the stability of the delay differential equation, and the theory of robust control were obtained in [9, 18]. Similar techniques were used to analyze the stochastic stability of delay systems with crisp (i.e., non-distributed) delays in [3, 4]. Since the existence proofs are constructive, they can be exploited to *prescribe* a whole class of stabilizing gains [10].

In the present chapter the performance degradation and stability margins for delay perturbations of a nominal state feedback control are analysed. Preliminary ideas were explored in [17]. Consider thus a multivariable time invariant linear system

$$\dot{x} = Ax + Bu \tag{0.1}$$

where $A \in \mathbb{R}^{n \times n}$, $B \in \mathbb{R}^{n \times m}$. The input u is m-dimensional, and $x \in \mathbb{R}^n$. It is assumed that for this system, a *nominal* state feedback gain K was designed for which the closed loop system

$$\dot{x} = (A - BK)x + Bu \tag{0.2}$$

is stable. We shall analyze the robustness of this feedback gain in case the available information at time t contains only a "fraction", λ, of the instantaneous state $x(t)$, and the remaining part, $(1-\lambda)$, is given by a functional of the state history over the interval $[t-T,t]$. This contrasts with common robust design techniques which consider perturbations with bounds on the transfer function ([15]). As in many situations such frequency domain information may not be available, the Lyapunov-Krasovskii theory presents a viable alternative, based on a time domain bound. Of course, in the presence of additional frequency domain bounds, one should expect more precise statements. Some Nyquist type graphical robust stability tests are also provided for such cases.

More precisely, we assume that the signal fed into the control gain matix K is

$$y(t) = \lambda x(t) + \int_0^T \mu(\tau)x(t-\tau)\,d\tau, \qquad \int_0^T |\mu(\tau)|\,d\tau = 1-\lambda. \tag{0.3}$$

The normalized weight $\mu_0(t) \stackrel{\text{def}}{=} \frac{\mu(t)}{1-\lambda}$ will be referred to as the ***shape*** of the delay perturbation, and the smallest λ for which stability can be guaranteed, the ***delay perturbation margin.*** With this definition, $\lambda = 1$ actually means that the nominal solution may not be robust, as stability can only be guaranteed (recall that we only have ***sufficient*** conditions) for $\mu = 0$. It could therefore happen that a slight delay perturbation of the feedback signal destabilizes the closed loop.

The actual system dynamics in closed loop are modeled by the functional differential equation

$$\dot{x}(t) = (A - \lambda BK)x(t) - BK\int_0^T \mu(\tau)x(t-\tau)\,d\tau. \tag{0.4}$$

In function of the parameter λ, we will set up ***sufficient*** conditions for the asymptotic stability of this system, independent of T and the precise shape of $\mu(\tau)$, except for the single constraint that $\int_0^T |\mu(t)|\,dt = 1-\lambda$. This problem is significant, as many continuum systems (infinite dimensional) are approximated by finite dimensional ones, and therefore what is fed back is in fact the infinite dimensional exact state. The propagation delays in vibrating plates and beams [1] provide specific examples. In such problems the local vectorfield is a smooth superposition of the states at all previous instants in some interval. For instance, in viscoelastic structures the stress-strain states of the materials are modeled by such equations [6]: e.g., the Piola-Kirchoff stress function in the equation of motion of a one-dimensional viscoelastic body which moves longitudinally has a distributed delay integral of the history of the displacement. Many other applications in man-machine systems, process control, remote control and robotics, involve delays due to transportation lags and conduction or communication times which may be distributed due to reflection at boundaries whose distance smoothly varies with the angle (line of sight). If in a feedback controller, the actual computation times are taken into account, one is also forced to consider delay differential (or difference) models [2]. Likewise, sampled data systems provide another instance where delays - here periodically time varying - are introduced. Distributed delay is further prevalent in population dynamics. In all these applications, robustness with respect to the shape information is important (this includes the robustness with respect to the maximal delay T), especially since the precise shape information is seldom known in practice [6, Section 1.4].

To facilitate the analysis, it is first assumed that the shape function $\mu(t)$ as well as the effective interference time T are known. As will be apparent, the sufficient conditions derived only require partial information. This is then exploited to obtain ***robust*** stability conditions.

In Section 2, some background material on ***retarded functional differential equations*** and the Lyapunov-Krasovskii theory is introduced. Sufficient conditions for stability in terms of Riccati-like equations are derived in Section 3. In Section 4 the results are reinterpreted as robust stability conditions and frequency domain criteria are derived via the positive real lemma. In Section 5, the stability margins for the delay perturbed controller are derived. A simple graphical test for robust stability is given in Section 6, and its use as a tool to determine a bound on the stability margin is illustrated. Some examples are collected in Section 7 to illustrate the concepts.

Throughout, the following notation will be used: If M is an arbitrary matrix, then M' denotes its transpose, and if M is an invertible (square) matrix we shall use M^{-T} for $(M^{-1})^T$.

1 Retarded Functional Differential Equations

The delay differential equation is usually represented as a functional differential equation [5]. We take $C([-T,0],\mathbb{R}^n)$, the Banach space of continuous functions $[-T,0] \to \mathbb{R}^n$, with the norm $\|\phi\| = \sup_t |\phi(t)|$, as the natural state space for such systems [6]. Here, $|\phi(t)|$ denotes the Euclidean norm of $\phi(t) \in \mathbb{R}^n$. Let the initial data $\phi \in C([-T,0],\mathbb{R}^n)$ for the problem be given.

For $t \geq t_0 - T$, let $x(t;t_0,\phi)$ denote its solution at time t with the initial data ϕ, specified at time t_0, i.e., $x(t_0+\theta,t_0,\phi) = \phi(\theta)$ for $\theta \in [-T,0]$. Because of the time-invariance of the dynamical system, $x(t;t_0,\phi) = x(t-t_0,0,\phi)$ for all $t > t_0$. As customary, $x(t+\theta)$ for $\theta \in [-T,0]$ is denoted by $x_t(\theta)$, which is the state (infinite dimensional) of the delay system. Equation (0.4) is then of the general form

$$\dot{x}(t) = F(x_t) \quad , \quad x_{t_0} = \phi \quad , \quad F(0) = 0, \tag{1.1}$$

with $F : \mathcal{D}_T = C([-T,0],\mathbb{R}^n) \to \mathbb{R}^n$, linear in the argument. It is well known that this retarded functional Cauchy problem has a unique solution, defined for all $t \in [t_0 - T, \infty)$, and which depends continuously on F and ϕ [5, 6]. Its equilibrium solution is $x_t \equiv 0$. Let

$$Q_\delta = \{\phi \in C[-T, 0] \mid \|\phi\| < \delta\}.$$

The following definitions are standard in the theory of time-invariant delay differential equations. [6, p. 98]:

Definition 1. The equilibrium solution, $x_t \equiv 0$, of the delay-differential equation (0.4) is said to be

1. *(uniformly) stable* if for any $\epsilon > 0$, there is a $\delta = \delta(\epsilon) > 0$, such that $|x(t; 0, \phi)| \leq \epsilon$ for any initial function $\phi \in Q_\delta$, and $\forall t > 0$. In the opposite case, it is called *unstable.*
2. *(uniformly) asymptotically stable* if it is (uniformly) stable, and there is a $K > 0$ such that for any $\gamma > 0$ there is a $T(\gamma, K) > 0$ such that $|x(t; 0, \phi)| \leq \gamma$, $\forall t \geq T(\gamma, K)$ and $\phi \in Q_K$.
3. *(uniformly) exponentially stable* if there are constants $K > 0, C_1 > 0, C_2 > 0$, such that for any $\phi \in Q_K$, the solution $x(t; 0, \phi)$ of the system (0.4) satisfies the inequality

$$|x(t; 0, \phi)| \leq C_1 \|\phi\| e^{-C_2 t}, \qquad 0 \leq t \leq \infty$$

These definitions are the straight forward extensions of the various stability concepts in the finite dimensional case. Note that because of the time-invariance of the dynamical equation, the above definitions for stability, asymptotic stability and exponential stability are in fact automatically *uniform.*

The main results of this chapter hinge on the following theorems, taken from the book of Kolmogorov and Myshkis [6] and repeated here to make the chapter self contained: Let Ω be the class of scalar nondecreasing functions $\alpha \in C([0, \infty), \mathbb{R})$ such that $\alpha(r) > 0$ for $r > 0$, and $\alpha(0) = 0$.

Definition 2. Let $V : Q_K \to \mathbb{R}$ be a continuous functional such that $V(0) \equiv 0$. The functional $V : \psi \to V(\psi)$ is called *positive definite* (*negative definite*) if there is a function $\alpha \in \Omega$ such that $V(\psi) \geq \alpha(|\psi(0)|)$ (respectively, $V(\psi) \leq -\alpha(|\psi(0)|)$) for all $\psi \in Q_K$. It is said to have *infinitesimal upper bound* if $|V(\psi)| \leq \alpha(\|\psi\|)$, for all $\psi \in Q_K$.

Theorem 3 (Theorem 1.2, p. 103 [6]). *If there is a continuous positive definite functional $V : Q_K \to \mathbb{R}$ with derivative $\dot{V} \leq 0$, then the trivial solution of (1.1) is (uniformly) stable.*

Theorem 4 (Theorem 1.1, p. 103 [6]). *Assume that for some $T > 0$, there exists a positive definite continuous functional $(\psi \to V(\psi)) : Q_K \to \mathbb{R}$ which has infinitesimal upper bound and whose derivative $\dot{V}$ is a negative definite functional on Q_K. Then the trivial solution of (1.1) is (uniformly) asymptotically stable.*

Theorem 5 (Theorem 1.3, p. 103 [6]). *A necessary and sufficient condition for the (uniform) exponential stability of the trivial solution of (1.1) is that there exists a continuous functional* $V : Q_K \to \mathbb{R}$ *such that for some positive constants* C_i; $i = 1 \cdots 4$, *and* $\psi, \eta \in Q_K$:

$$C_1\|\psi\| \leq V(\psi) \leq C_2\|\psi\|, \tag{1.2}$$

$$\dot{V}(\psi) \leq -C_3\|\psi\|, \tag{1.3}$$

$$|V(\psi) - V(\eta)| \leq C_4\|\psi - \eta\|. \tag{1.4}$$

Notes and Comments If one wants to prove stability in its various forms for equations with *unbounded* delays, then instead of the subset Q_K of $C([-T,0],\mathbb{R}^n)$, one needs to work with the *metric space* of continuous functions $\phi : (-\infty, 0] \to \mathbb{R}^n$. We shall not consider this case here.
One obtains *global* stability in its various guises, if the the set Q_K may be replaced by *all* of $C([-T,0],\mathbb{R}^n)$, i.e., the bound K may be arbitrary large.

2 Riccati-type Equations as Sufficient Conditions

We derive here a sufficient condition for stability, which is facilitated by an auxiliary key-lemma, proven in [14]. To make this chapter self-contained, it is here repeated. Consider the general distributed delay system (the nondistributed case was treated in [3, 8])

$$\dot{x}(t) = Ax(t) + \int_0^T B(\tau)x(t-\tau)\,d\tau. \tag{2.1}$$

Lemma 6. *Given a symmetric matrix* $P \in \mathbb{R}^{n\times n}$ *and a differentiable matrix valued function* $Q(t) = Q'(t)$, *satisfying* $Q(T) = 0$ *and* $\frac{dQ(t)}{dt} = -\mathcal{C}(t)'\mathcal{C}(t)$, *define the Lyapunov-Krasovskii functional* $V : C([-T,0],\mathbb{R}^n) \to \mathbb{R}$ by:

$$V(\psi) = \psi(0)'P\psi(0) + \int_{-T}^0 \psi(\theta)'Q(-\theta)\psi(\theta)\,d\theta. \tag{2.2}$$

If $\mathcal{L}$ *is the directional derivative along solutions* $x(t)$ *of (2.1), then the inequality*

$$\mathcal{L}V(x_t) \leq -x(t)'Rx(t), \tag{2.3}$$

holds, where

$$-R = A'P + PA + C'PC + Q(0) + P\int_0^T B(\tau)\mathcal{C}(\tau)^{-1}\mathcal{C}(\tau)^{-T}B(\tau)'\,d\tau\, P. \tag{2.4}$$

Proof. Consider the *Lyapunov-Krasovskii functional* (2.2). Its directional derivative along trajectories of (2.1) is

$$\frac{dV(x_t)}{dt} = \dot{x}(t)'Px(t) + x(t)'P\dot{x}(t) + x(t)'Q(0)x(t) - x(t-T)'Q(T)x(t-T) +$$

$$+\int_{t-T}^{t} x(\tau)' \frac{\partial Q(t-\tau)}{\partial t} x(\tau)\, d\tau \tag{2.5}$$

Substituting (2.1) with the expression for $\dot{Q}$ and 'completing the squares', the above expression yields for the directional derivative $\mathcal{L}$,

$$\mathcal{L}V = x(t)'Mx(t) - x(t-T)'Q(T)x(t-T) - \int_0^T \xi(t,\tau)\xi'(t,\tau)\, d\tau.$$

where

$$M = A'P + PA + Q(0) + P\left[\int_0^T B(\tau)C(\tau)^{-1}C(\tau)^{-T}B(\tau)'\, d\tau\right] P,$$

and

$$\xi(t,\tau) = [x(t-\tau)'C(\tau)' - x(t)'PB(\tau)C(\tau)^{-1}].$$

Since by assumption $Q(T) = 0$, and the second term is negative semi-definite,

$$\mathcal{L}V(t) \leq -x(t)'R(t)x(t) \tag{2.6}$$

follows. □

One then easily establishes the following

Theorem 7. *Consider the system (2.1). If there exists a pair of positive definite (symmetric) matrices P, and R, and a matrix function $C(\cdot) : [0,T] \to \mathbb{R}^{n\times n}$ such that $C(t)$ is nonsingular for all $t \in [0,T]$, and*

$$A'P + PA + R + \int_0^T [C(\tau)C(\tau)' + PB(\tau)C(\tau)^{-1}C(\tau)^{-T}B(\tau)'P]\, d\tau = 0 \tag{2.7}$$

then the system is uniformly stable.

Proof. Let $Q(t) = \int_t^T C(\tau)'C(\tau)\, d\tau$. From condition (2.7) and Lemma 6, the existence of a positive definite Lyapunov-Krasovskii functional V is guaranteed, with

$$\mathcal{L}V(x_t) \leq 0. \tag{2.8}$$

The stability of the trivial solution follows then by Theorem 3. □

Theorem 8. *Consider the system (2.1). If there exists a pair of positive definite (symmetric) matrices P and R, and a nonsingular matrix function $C(\cdot) : [0,T] \to \mathbb{R}^{n\times n}$, satisfying*

$$\left\| \int_0^T C'(t)C(t)\, dt \right\| < \infty, \tag{2.9}$$

such that

$$A'P+PA+R+\int_0^T [C(\tau)C(\tau)' + PB(\tau)C(\tau)^{-1}C(\tau)^{-T}B(\tau)'P]\, d\tau = 0, \tag{2.10}$$

then the system (2.1) is uniformly asymptotically stable.

Proof. We already established in Theorem 7 the positive definiteness of the functional (2.2) with $Q(t) = \int_t^T C(\tau)'C(\tau)\,d\tau$. Bounding the other way:

$$\begin{aligned} V(\psi) &\leq \psi(0)'P\psi(0) + \int_{-T}^{0} \psi(\theta)'Q(-\theta)\psi(\theta)\,d\theta \\ &\leq |\psi(0)|^2\lambda_{\max}(P) + \max_{\theta}|\psi(\theta)|^2 \int_{-T}^{0} \lambda_{\max}(Q(-\theta))\,d\theta \\ &\leq \|\psi\|^2 \left[\lambda_{\max}(P) + \int_0^T \lambda_{\max}(Q(\theta))\,d\theta\right]. \end{aligned} \tag{2.11}$$

By the choice of Q:

$$\lambda_{\max}(Q(\theta)) \leq \lambda_{\max}\left[\int_0^T C'C\,d\tau\right]. \tag{2.12}$$

Hence we get in (2.11)

$$V(\psi) \leq \|\psi\|^2 \left[\lambda_{\max}(P) + T\lambda_{\max}\left(\int_0^T C'C\,d\tau\right)\right], \tag{2.13}$$

showing that under the given conditions V also has infinitesimal upper bound. Finally, by Lemma 6, we have now:

$$\begin{aligned} \mathcal{L}(x_t) &\leq -x_t(0)'Rx_t(0) \\ &\leq -|x_t(0)|^2[\lambda_{\min}(R)] \end{aligned} \tag{2.14}$$

Thus, under the given conditions the trivial solution is asymptotically stable by Theorem 4. □

3 Robust Stability and Frequency Domain Criteria

The basic conditions obtained in the previous section involve an algebraic Riccati equation of the form $A'P + PA + R + Q_1 + PQ_2P = 0$. The difference with the Riccati equations in the LQG theory is that the quadratic term appears with the plus sign. This type of Riccati equation is known to appear in robust control theory. In fact, the specific *time dependence* of the weight matrix $B(\cdot)$ and the exact support $[0, T]$ of the distributed delay are immaterial in the sufficient conditions that were derived. This makes room for *robust* stability conditions. Indeed, the theorems stated in section 3 may be rephrased as robust (asymptotic) stability theorems. In particular we have the following definition, which allows to generalize the above to robust stability criteria:

Definition 9. Let $\mathcal{B} \in \mathbb{R}^{n\times n}$ be symmetric and positive definite. The distributed-delay differential system (2.1) is *$\mathcal{B}$-robustly stable* (respectively *robustly asymptotically stable*) if the equilibrium solution is stable (respectively asymptotically stable) for *all* bounded delay intervals $[0,T]$, and weight matrix functions $B(\cdot) : [0,T] \to \mathbb{R}^{n\times n}$ satisfying

$$\min_{\beta(\cdot)>0} \int_0^T \frac{B(t)B(t)'}{\beta(t)}\, dt \int_0^T \beta(t)\, dt \leq \mathcal{B}. \tag{3.1}$$

Definition 9 selects one particular functional of the exact form of the delay distribution matrix $B(\cdot)$. Robust stability is then expressed for a class of weights in terms of the evaluation of this functional, and the precise shape of $B(\cdot)$ can be 'forgotten'. Also it is assumed that the delay is bounded, but the precise value of the bound T is immaterial.

Theorem 10. *The distributed delay system (2.1) is $\mathcal{B}$-robustly asymptotically stable if there exists a positive definite matrix P such that*

$$A'P + PA + I + P\mathcal{B}P < 0. \tag{3.2}$$

Proof. It follows from (3.2) that the conditions of Theorem 8 are satisfied with $\mathcal{C}(t) = \sqrt{\beta(t)}I > 0$, such that $\int_0^T \mathcal{C}(t)'\mathcal{C}(t) = \int_0^T \beta(\tau)\, d\tau = 1$, since then also in Definition 9

$$P\mathcal{B}P \geq P \int_0^T \frac{B(\tau)B(\tau)'}{\beta(\tau)}\, d\tau\, P.$$

□

Corollary 11. *The distributed delay system (2.1) is $\mathcal{B}$-robustly asymptotically stable if A is stable, and*

$$\sup_{\omega} \|[j\omega I - A]^{-1}\mathcal{B}^{1/2}\| < 1. \tag{3.3}$$

Proof. This is a direct consequence of the strict bounded real lemma [12, 13] applied to (3.2), and provides an extension to the frequency domain robust stability criterion for delay systems, developed in [9]. □

4 Stabilization with Delayed Feedback

We return now to the model for the delay perturbed control, (0.4), and apply the robust stability criterion obtained in the previous section. The A and $B(\tau)$ in the general theory are respectively replaced by $A - \lambda BK$ and $BK\mu(\tau)$.

4.1 Sufficient Condition

The bound of Definition 9 gives with the above $B(\tau)$,

$$\begin{aligned}\mathcal{B} &= \min_{\beta(\cdot)>0} \int_0^T BKK'B\frac{\mu(\tau)^2}{\beta(\tau)}\,d\tau \int_0^T \beta(\tau)\,d\tau \\ &= BKK'B'\left(\int_0^T |\mu(\tau)|)\,d\tau\right)^2 \\ &= BKK'B'(1-\lambda)^2, \qquad (4.1)\end{aligned}$$

since $\beta(\tau) = |\mu(\tau)|$ is the minimizer of $\int_0^T \frac{\mu(\tau)^2}{\beta(\tau)}\,d\tau \int_0^T \beta(\tau)\,d\tau$. Consequently, the frequency domain criterion (3.3) from the Corollary 11 is then, identifying $\mathcal{B}^{1/2}$ with the form $(1-\lambda)BK$,

$$\sup_\omega \|[j\omega I - A + \lambda BK)^{-1}BK\| < \frac{1}{1-\lambda} \qquad (4.2)$$

Some standard matrix simplifications lead to

$$\begin{aligned}\|(sI - A + \lambda BK)^{-1}BK\| &= \|[(sI-A)(I+\lambda(sI-A)^{-1}BK)]^{-1}BK\| \\ &= \|[I+\lambda[(sI-A)^{-1}BK]^{-1}(sI-A)^{-1}BK\| \\ &= \|[I+\lambda K_0(s)]^{-1}K_0(s)\| \qquad (4.3)\end{aligned}$$

where $K_0(s) \stackrel{\text{def}}{=} (sI-A)^{-1}BK$ is the nominal return ratio. Restricting $s = j\omega$, we have thus proven the following :

Theorem 12. *Consider the open loop system $\dot{x} = Ax + Bu$, with a stabilizing nominal state feedback control gain K. If the feedback control is perturbed by delay as in (3), then the system is robustly asymptotically stable for all values of λ in the interval $[0,1]$ for which*

$$\sup_\omega (1-\lambda)\|[I+\lambda K_0(j\omega)]^{-1}K_0(j\omega)\| \le 1 \quad , \quad K_0(s) = (sI-A)^{-1}BK$$

is satisfied.

Multiplying both sides by λ, the criterion can also be restated as

$$\sup_\omega \|[I + K(j\omega)]^{-1}K(j\omega)\| \le \frac{\lambda}{1-\lambda}, \qquad (4.4)$$

where $K(j\omega) = \lambda K_0(j\omega)$. The left hand side is the transfer matrix from reference to control input due only to the ***instantaneous*** state information, thus with the reduced gain λK, and the delay part removed from the loop.

The criterion of Theorem 12 can be expressed in the alternative form by applying Woodbury's (matrix inversion) lemma to the left hand side:

$$[I + \lambda K_0(s)]^{-1} K_0(s) = (sI - A)^{-1} B [I + \lambda \widetilde{K}(s)]^{-1} K. \tag{4.5}$$

Note that $\widetilde{K}(s) \overset{def}{=} K(sI - A)^{-1}B$ is an $m \times m$ matrix. The criterion is:

$$\sup_{\omega}(1 - \lambda) \| (j\omega I - A)^{-1} B [I + \lambda \widetilde{K}(j\omega)]^{-1} K \| \leq 1. \tag{4.6}$$

In the single input case, using lower case k and b respectively for K and B, the left hand side of Equation (4.5) reduces to:

$$[I + \lambda K_0(s)]^{-1} K_0(s) = \frac{(sI - A)^{-1} bk}{1 + \lambda \tilde{k}(s)}, \tag{4.7}$$

with the robust stability criterion:

$$\sup_{\omega} (1 - \lambda) \frac{\| (j\omega I - A)^{-1} b \| \, \|k\|}{|1 + \lambda \tilde{k}(j\omega)|} \leq 1. \tag{4.8}$$

4.2 Alternative Criterion and a Necessary Condition

Based on results by Maciejowski in [15], we may proceed differently:

Let $G_0(s) = (sI - A)^{-1}B$ be the open loop transfer function from input to state. Consider the delayed feedback as a perturbation of the nominal static feedback gain. The exact transfer function of the feedback part, i.e. the "gain" system from state $x(t)$ to the feedback control $Ky(t)$, is, with (3):

$$\begin{aligned} F(s) &= [\lambda + (1 - \lambda) m_0(s)] K \\ &= [1 + (1 - \lambda)(m_0(s) - 1)] K \\ &= [1 + \Delta(s)] K \\ &= [1 + (1 - \lambda) \Delta_0(s)] K \end{aligned} \tag{4.9}$$

where $m_0(s)$ is the Laplace transform of the compactly supported normalized ($\int_0^T |\mu_0(t)| \, dt = 1$) scalar shape function $\mu_0(t)$. Defining $\mu_0(t)$ for $t > T$ to be zero, the extension has Laplace transform

$$m_0(s) = \int_0^{\infty} \mu_0(t) e^{-st} \, dt = \int_0^{T} \mu_0(t) e^{-st} \, dt. \tag{4.10}$$

The delay perturbation term is the convolution of the state with $\mu(t) = (1 - \lambda)\mu_0(t)$. Since the support of μ is finite, its Laplace transform is not rational in general. (E.g. if $\mu_0(t) = \frac{1}{T}$ in $[0, T]$, then $m_0(s) = \frac{1 - e^{-sT}}{sT}$.) We shall now restrict attention to the class of shapes $\mu_0(t)$ for which $G_0(s) = (sI - A)^{-1}B$

and $G(s) = G_0(s)(1 + \Delta(s))$ have the same number of unstable poles. So since the loop is stable for $\Delta = 0$, it will remain stable provided the number of encirclements of -1 by the characteristic loci of G_0F remains unchanged. This means that for any permissible F and any ω.

$$\det[I + G(j\omega)F(j\omega)] \neq 0. \tag{4.11}$$

Adapting the derivation in [15, p. 115] slightly, it is easily seen that this is implied by

$$\begin{aligned} \| G_0(j\omega)K[I + G_0(j\omega)K]^{-1} \| &< \frac{1}{\| \Delta(j\omega) \|} \\ &= \frac{1}{(1-\lambda)} \frac{1}{\|\Delta_0(j\omega)\|}. \end{aligned} \tag{4.12}$$

It is important to note that the sufficient condition for stability is also necessary if all permissible perturbations may actually occur. The problem with this approach is that the precise information on $\mu(t)$, necessary to compute $\|\Delta_0(j\omega)\|$, may not be available. The L_1-norm $\int_0^\infty |\mu(t)|\, dt$ may be an easier statistic to bound. One case however is quite interesting: this is the case where the system has a single input, analyzed in the next section.

5 Single Input Case: Frequency Response Tests

The stabilizability by feedback of the state is analyzed, in case the state is perturbed by delays (e.g., due to multipath). The criteria derived in the previous section lead to interesting graphical methods, permitting to derive the stability margins (margin on λ) in a straightforward way. A set of conditions based on Theorem 8 are first derived for the case where only a time domain bound is known. Next, invoking Rouché's theorem, criteria are derived in case additional bounds in the Laplace domain are known. It is shown that the delay perturbation stability margin can be derived from a single Nyquist plot (of $k(j\omega)$), or more precisely, its location with respect to a fixed circle bundle in the complex plane. This is reminiscent of the classical closed loop frequency response analysis with the M-circles.

5.1 Frequency Sweep

To fix the ideas, assume that a deterministic finite dimensional, single input *unstable* system

$$\dot{x} = Ax + bu \tag{5.1}$$

is controlled by feeding back the available state information $y(t)$ over a gain k. As typical in control involving a communication link, let this available state information be a *delay perturbed* version of the instantaneous state, i.e.,

$$y(t) = \lambda x(t) + \int_0^T \mu(\tau)x(t-\tau)\, d\tau, \tag{5.2}$$

with "total weight"

$$\int_0^T |\mu(\tau)|\, d\tau = 1 - \lambda, \text{ for } \lambda \in [0,1]. \tag{5.3}$$

Applying the Corollary 11 in Section 4 to these dynamics, one finds that the delay perturbed system is asymptotically stable if the matrix $A - \lambda bk$ is Hurwitz (i.e. stable), and

$$\sup_\omega \| \, [j\omega I - A + \lambda bk]^{-1} bk \, \| < \frac{1}{1-\lambda}; \tag{5.4}$$

This leads to the following robust stability margin.

Theorem 13. *The scalar system (5.1-5.2) with feedback $u = ky$ is robustly stable with delay margin λ_0 if $\forall \lambda \in [\lambda_0, 1]$, the matrix $A - \lambda bk$ is Hurwitz stable and*

$$\lambda_0 = \sup_\omega [\, M(j\omega) - \sqrt{M(j\omega)^2 - N(j\omega)}\,], \tag{5.5}$$

where

$$M(s) = \frac{l(s)^2 + \operatorname{Re} \tilde{k}(s)}{l(s)^2 - |\tilde{k}(s)|^2} \tag{5.6}$$

$$N(s) = \frac{l(s)^2 - 1}{l(s)^2 - |\tilde{k}(s)|^2}, \tag{5.7}$$

and

$$l(s) = \|k\| \, \|(sI - A)^{-1} b\| \tag{5.8}$$

Proof. From (5.4), or equivalently (4.8), robust stability holds for λ such that

$$\forall s = j\omega: \quad \|k\| \, \|(sI - A)^{-1} b\| \, (1 - \lambda) \le |1 + \lambda \tilde{k}(s)|. \tag{5.9}$$

With $l(s)$ as defined in (5.8) this gives

$$\lambda^2 [l(s)^2 - |\tilde{k}(s)|^2] - 2\lambda [l(s)^2 + \operatorname{Re} \tilde{k}(s)] + [l(s)^2 - 1] \le 0. \tag{5.10}$$

The coefficient of λ^2 in (5.10) is nonnegative since

$$\begin{aligned} |\tilde{k}(s)|^2 &= |k(sI - A)^{-1} b|^2 \\ &\le \|k\|^2 \|(sI - A)^{-1} b\|^2 = l(s)^2. \end{aligned} \tag{5.11}$$

With $M(s)$ and $N(s)$ as defined in the theorem statement, the inequality (5.10) is equivalent to

$$\lambda^2 - 2\lambda M(s) + N(s) < 0. \tag{5.12}$$

This quadratic polynomial in (5.12) has real roots iff the discriminant condition,

$$M(s)^2 - N(s) > 0 \tag{5.13}$$

holds. By assumption the closed loop is asymptotically stable in the (delay free) nominal case. Hence for $\lambda = 1$ the inequality (5.4) holds for all ω and one has

$$1 - 2M(s) + N(s) < 0. \tag{5.14}$$

By continuity, there must exist a neighborhood for $\lambda = 1$ for which the discriminant condition (5.10) holds. In view of (5.14), the discriminant of (5.12) satisfies,

$$M(s)^2 - N(s) > M(s)^2 - 2M(s) + 1 = [M(s) - 1)]^2] > 0, \tag{5.15}$$

This implies the existence of real roots λ_+ and λ_- in a neighborhood of 1 (Condition (5.4) is necessary in the delay free, i.e. $\lambda = 1$, case.) Consequently, $\lambda_- \leq 1 \leq \lambda_+$, with

$$\lambda_+(s) = M(s) + \sqrt{M(s)^2 - N(s)} > M(s) + |M(s) - 1| > 1 \tag{5.16}$$
$$\lambda_-(s) = M(s) - \sqrt{M(s)^2 - N(s)} > M(s) - |M(s) - 1| < 1. \tag{5.17}$$

The set of values for λ in $[0, 1]$ for which robust stability holds is consequently bounded by $\sup_\omega \lambda_-(j\omega)$, the margin λ_0 for the delay perturbation. □

Comments One can plot a frequency sweep of $\lambda_-(j\omega)$ and determine $\sup \lambda_-(j\omega)$ graphically, thus obtaining the margin λ_0 for the delay perturbation. Note that the quadratic form (5.10) degenerates to a linear form if $l(s)^2 = |\tilde{k}(s)|^2$. This will be illustrated in the examples below.

5.2 Criteria Based on Rouché's Theorem

Consider now the more direct alternative approach which requires some Laplace domain information about $\mu(t)$ in addition to (5.3):

The closed loop characteristic equation for the single input case is first rewritten as

$$\det\left[sI - A + \lambda bk + (1 - \lambda)m_0(s)bk\right] = 0, \tag{5.18}$$

which can be brought to two equivalent forms, using again $a(s) = \det(sI - A)$ and $\tilde{k}(s) = k(sI - A)^{-1}b$:

$$a(s)\left[1 + \tilde{k}(s)(1 + (1 - \lambda)(m_0(s) - 1))\right] = 0, \tag{5.19}$$

and

$$a(s)\left[1 + \lambda\tilde{k}(s) + (1 - \lambda)m_0(s)\tilde{k}(s)\right] = 0. \tag{5.20}$$

These two forms will be treated separately with different simplifying assumptions, as we shall not assume full knowledge of $m_0(s)$. Invoking Rouché's theorem, either form leads to a sufficient condition for robust stability with delay margin and results in a graphical test.

Recall from classical control theory (see a brief review, e.g. in [15]) that the loci where the closed loop gain $\left|\frac{k(s)}{1+k(s)}\right|$ is constant are the so-called M-circles in the $k(s)$-complex plane: they form a circle bundle with centers at 0 (closed loop gain 0) and at -1 (closed loop gain infinite). Let $\mathcal{M}_m$ be the domain $\left\{s \mid \left|\frac{s}{1+s}\right| \leq m\right\}$ bounded by the M-circle M_m. This is the domain containing the origin.

For the first result of this type, introduce also the transfer functions $\Delta(s)$ and $\Delta_0(s)$ defined by $\Delta(s) = (1-\lambda)\Delta_0(s) = (1-\lambda)(m_0(s)-1)$.

Theorem 14. *The system (1-3) is robustly stable with delay margin λ_0 if there exist $\delta_0 > 0$ and $\kappa > 0$ such that*

$$\lambda_0 = 1 - \frac{1}{\delta_0 \kappa} > 0 \tag{5.21}$$

and

i) $\tilde{k}(j\omega) = k(j\omega I - A)^{-1}b$ encircles the point $-1+j0$ n_p times counterclockwise (CCW), where n_p is the number of unstable open loop poles (provided there are no hidden modes).
ii) $m_0(j\omega)$ lies in the disk C_{1,δ_0}, centered at $s = 1 + j0$ and with radius δ_0.
iii) $\tilde{k}(j\omega)$ lies inside the domain $\mathcal{M}_\kappa$ (which is the outside *of the bounding M-circle).*

Proof. Condition i) is simply the Nyquist criterion and expresses stability for the nominal (delay free) closed loop system. By Rouché's theorem, the number of zeros in the right half plane of $a(s)[1+\tilde{k}(s)]$ and (5.19) are equal if

$$|\tilde{k}(s)\Delta_0(s)|(1-\lambda) < |1+\tilde{k}|. \tag{5.22}$$

Hence the delay perturbed system is stable if

$$\left|\frac{\tilde{k}(s)}{1+\tilde{k}(s)}\right| < \frac{1}{|\Delta_0(s)|\,(1-\lambda)}. \tag{5.23}$$

With the confinement condition ii), $|\Delta_0(s)| < \delta_0$ is satisfied, so that stability holds for all $\lambda \in [0,1]$ for which the Nyquist plot of $\tilde{k}(s)$ is confined to the domain $\mathcal{M}_\kappa$. With all conditions satisfied one may conclude that the nominal feedback is also stable with delay perturbation λ_0, assuming of course the asymptotic stability of the nomimal feedback. Since $\frac{1}{1-\lambda}$ is monotonically increasing, it follows then that stability holds for all $\lambda_0 \leq \lambda \leq 1$. □

With a uniformly distributed delay $\mu_0(t) = \frac{1}{T}$ in $[0,T]$, one obtains $m_0(s) = \frac{1-e^{-sT}}{sT}$ and its Nyquist plot is contained in the disk C_{1,δ_0} with $\delta_0 = 1.259591$. The extremal point on the Nyquist plot occurs for $\omega T = 4.08557$.

The statement and derivation of the second criterion are very similar: This time, use the form (5.20) for the characteristic equation. Thus if

$$|(1-\lambda)\tilde{k}(s)m_0(s)| < |1+\lambda\tilde{k}(s)| \tag{5.24}$$

then it follows again from Rouché's theorem that the perturbed closed loop will be stable if $a(s)[1+\lambda\tilde{k}(s)]$ has no zeros with positive real part; i.e., $a(s)[1+\lambda\tilde{k}(s)+(1-\lambda)\tilde{k}(s)m_0(s)]$ and $a(s)[1+\lambda\tilde{k}(s)]$ have the same number of zeros in the right half plane. These are the poles of the closed loop system. Since for $s = j\omega$

$$|m_0(s)| \leq \int_0^T |\mu_0(t)e^{-st}|\,dt = \int_0^T |\mu_0(t)|\,dt = 1, \tag{5.25}$$

condition (5.24) is satisfied if

$$(1-\lambda)|\tilde{k}(s)| < |1+\lambda\tilde{k}(s)| \tag{5.26}$$

or, equivalently

$$\left|\frac{\lambda\tilde{k}(s)}{1+\lambda\tilde{k}(s)}\right| < \frac{\lambda}{1-\lambda} \tag{5.27}$$

Finally, note that the set of $\lambda\tilde{k}(s)$ for which the above inequality holds is the domain $\mathcal{M}_{\frac{\lambda}{1-\lambda}}$ bounded by an M-circle $M_{\frac{\lambda}{1-\lambda}}$. The corresponding domain for $\tilde{k}(s)$ is obtained by the scaling $\frac{1}{\lambda}$, i.e.,

$$\tilde{k}(s) \in \frac{1}{\lambda}\mathcal{M}_{\frac{\lambda}{1-\lambda}} \stackrel{\text{def}}{=} \mathcal{L}_\lambda. \tag{5.28}$$

The right hand side, being a conformal transformation of the M-bundle, constitutes another circle bundle of 'L-circles' (L for lambda). The boundary of $\mathcal{L}_\lambda$ is a circle with center $(\frac{\lambda}{1-2\lambda}, 0)$ and radius $R_\lambda = \frac{1-\lambda}{|1-2\lambda|}$. This bundle is tangent to the vertical through -1. The $\mathcal{L}$-bundle is shown in Figure 1 for $\lambda = 0$ to $\lambda = 1$ in increments of 0.5. We conclude:

Theorem 15. *The system (1-3) is robustly stable with margin λ_0 if:*

i) $\tilde{k}(j\omega) = k(j\omega I - A)^{-1}b$ encircles the critical point $-\frac{1}{\lambda} + j0$, n_p times CCW, where n_p is the number of unstable open loop poles (provided there are no huidden modes).

ii) $\tilde{k}(j\omega)$ lies inside the domain $\mathcal{L}_{\lambda_0}$.

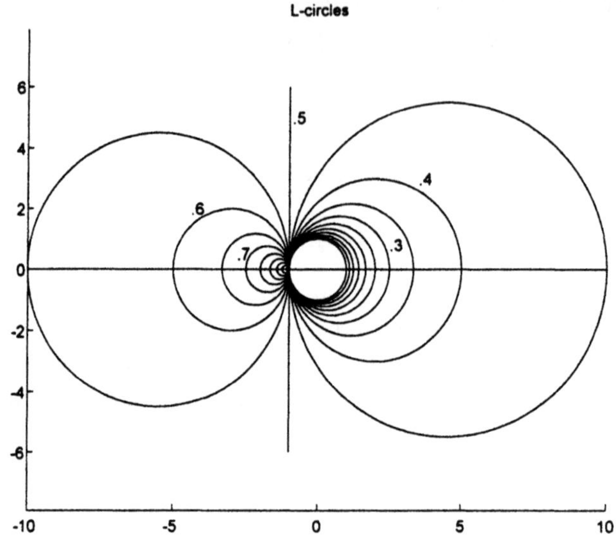

Fig. 1. The $\mathcal{L}$-bundle

Remarks: i) The first condition in the theorem again guarantees the asymptotic stability of the nominally designed closed loop system (with state feedback gain k).
ii) Unlike the previous theorem, this theorem in fact does *not* assume further (Laplace domain) conditions on $\mu(t)$, since the bound (5.25) is automatic. However supplemental information about μ_0 may enable to set a more precise bound in (5.25). With this information one could relax condition iii) somewhat as was done in Theorem 14.
iii) Recall that a stability margin λ_0 for the delay perturbation means that the system is stabilized for *all* delay perturbations with $\lambda \geq \lambda_0$.
iv) Note finally that in order to *guarantee* stabilizability of an otherwise unstable system with this method, it is essential that the input $y(t)$ contains a *delay-free* version of the state $x(t)$, i.e., it is essential that $\lambda > 0$. This is an artifact of the sufficiency of our condition, it may not be necessary.

6 Examples

We illustrate in this section the ease with which the various criteria can be used.

Example 1. Consider the multivariable second order system with

$$A = \begin{bmatrix} 1 & 1 \\ 0 & -1 \end{bmatrix}, \; B = \begin{bmatrix} 1 & 0 \\ 0 & 1 \end{bmatrix}, \text{ with } K = \begin{bmatrix} 3 & 0 \\ 1 & -1 \end{bmatrix}. \tag{6.1}$$

The nominal closed loop dynamics are given by

$$A_{FB} = \begin{bmatrix} -2 & 1 \\ -1 & 0 \end{bmatrix}. \tag{6.2}$$

which has a double pole at -1. Notice that (A_{FB}, B) is in the observer canonical form. The norm of $(1-\lambda)[I+\lambda K_0(s)]^{-1}K_0(s)$ is displayed as a two dimensional surface in function of delay perturbation λ and frequency ω. Figure 2 shows the 3D plot. For the same system, Figure 3 gives the contours for the norm $(1-\lambda)\,\|\,(j\omega I - A)^{-1}B[I+\lambda\widetilde{K}(s)]^{-1}K\,\|$. The critical level is of course level 1. It is readily seen that for $\lambda > \lambda_0 = 0.681$ the norm remains below 1, hence indicating robust stability. The delay margin can be found interactively using a mathematics package. One gets $\lambda_0 = 0.681$. We have also computed the delay

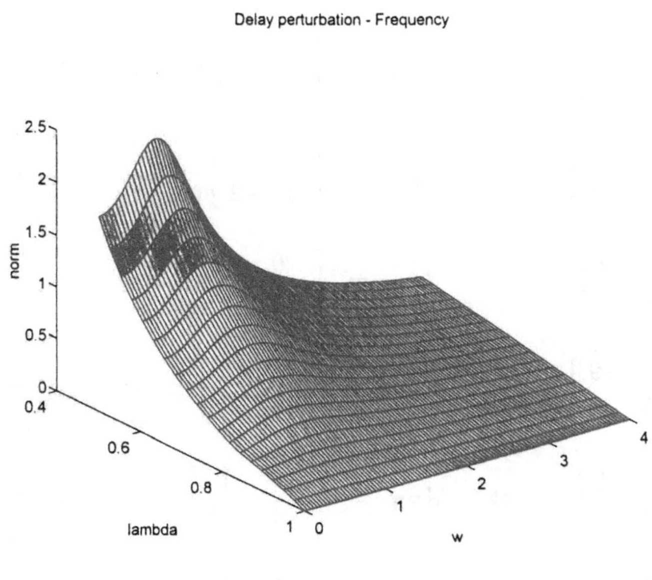

Fig. 2. 3-D plot of the norm as function of λ and ω for Example 1

margin for some different feedback gains, but all such that the closed loop has a

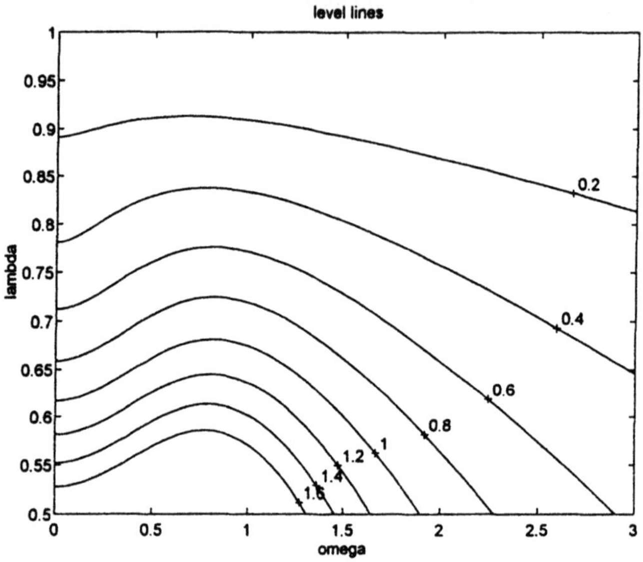

Fig. 3. Level lines for the norm as function of λ and ω for Example 1

double pole at -1:

$$
\begin{array}{lllll}
K = \begin{bmatrix} 3 & 2 \\ -1 & -1 \end{bmatrix} & \Rightarrow & A_{FB} = \begin{bmatrix} -2 & -1 \\ 1 & 0 \end{bmatrix} & \Rightarrow & \lambda_0 = 0.708 \\
K = \begin{bmatrix} 1 & 2 \\ -1 & 1 \end{bmatrix} & \Rightarrow & A_{FB} = \begin{bmatrix} 0 & -1 \\ 1 & -2 \end{bmatrix} & \Rightarrow & \lambda_0 = 0.891 \\
K = \begin{bmatrix} 1 & 0 \\ 1 & 1 \end{bmatrix} & \Rightarrow & A_{FB} = \begin{bmatrix} 0 & 1 \\ -1 & -2 \end{bmatrix} & \Rightarrow & \lambda_0 = 0.834 \\
K = \begin{bmatrix} 2 & -9 \\ 0 & 0 \end{bmatrix} & \Rightarrow & A_{FB} = \begin{bmatrix} -1 & 10 \\ 0 & -1 \end{bmatrix} & \Rightarrow & \lambda_0 = 0.911 \\
K = \begin{bmatrix} 2 & 11 \\ 0 & 0 \end{bmatrix} & \Rightarrow & A_{FB} = \begin{bmatrix} -1 & -10 \\ 0 & -1 \end{bmatrix} & \Rightarrow & \lambda_0 = 0.924 \\
K = \begin{bmatrix} 2 & 0 \\ 0 & 0 \end{bmatrix} & \Rightarrow & A_{FB} = \begin{bmatrix} -1 & 1 \\ 0 & -1 \end{bmatrix} & \Rightarrow & \lambda_0 = 0.750 \\
K = \begin{bmatrix} 2 & 1 \\ 0 & 0 \end{bmatrix} & \Rightarrow & A_{FB} = \begin{bmatrix} -1 & 0 \\ 0 & -1 \end{bmatrix} & \Rightarrow & \lambda_0 = 0.764
\end{array}
\tag{6.3}
$$

The last four cases correspond with a single feedback ($m = 1$). It is obvious that the the additional degrees of freedom can be exploited to obtain the best possible delay margin (the smaller λ_0 value).

Example 2. Consider a single input system. For comparison, take the system from example 1, but with input matrix

$$b = \begin{bmatrix} 1 \\ 0 \end{bmatrix}. \tag{6.4}$$

We compute the delay perturbation margin $\lambda_0 = \sup_\omega \lambda_-(j\omega)$. In this case, with feedback gain (k_1, k_2) but constrained by the requirement that the closed loop poles are at -1, we get:

$$\det \begin{bmatrix} 1-k_1 & 1-k_2 \\ 0 & -1 \end{bmatrix} = (s-1)^2 \Rightarrow \begin{cases} k_1 = 2 \\ k_2 \text{ arbitrary.} \end{cases} \tag{6.5}$$

Then the matrix $A - \lambda bk$ has eigenvalues $1 - 2\lambda$ and -1, and $A - \lambda bk$ is Hurwitz for $\lambda > 0.5$. Using Theorem 13, we also derive

$$\begin{aligned} l(j\omega) &= \sqrt{\frac{k_2^2 + 4}{1 + \omega^2}} \\ k(j\omega) &= \frac{-2(1 + j\omega)}{1 + \omega^2} \end{aligned}$$

From these, it is straightforward to compute

$$\lambda_-(j\omega) = \frac{2 + k_2^2 - \sqrt{4 + k_2^2(1 + \omega^2)}}{k_2^2}. \tag{6.6}$$

Obviously the supremum is attainable for $\omega = 0$, and gives

$$\lambda_-(0) = \frac{2 + k_2^2 - \sqrt{4 + k_2^2}}{k_2^2}, \tag{6.7}$$

a delay margin $\lambda_0(k_2)$ which is symmetrical in k_2. This delay margin λ_0 is given as function of k_2 in Figure 4. Note that for $k_2 = 0$ the quadratic form (5.10) degenerates to a linear one: $\frac{3-4\lambda-\omega^2}{1+\omega^2} < 0$, yielding $\lambda > 0.75$. It is clear from Figure 4 that this gives the best robustness (smallest λ_0) margin.

Example 3. Let $\mu_0(t) = \beta e^{-\beta t}$, and $T \to \infty$. Since T is no longer bounded, the delay margin based on the Lyapunov-Krasovskii theory is strictly speaking not applicable. In this case the shape function is $m_0(j\omega) = \frac{\beta}{\beta + j\omega}$, and $|\Delta_0(j\omega)| = |m_0(j\omega) - 1| = \frac{\omega}{\sqrt{\beta^2 + \omega^2}} < 1$. The condition ii) in theorem 14 (or the more general statement (37)) is satisfied. By Theorem 14 stability is deduced if

$$\overline{\sigma}\{G_0(j\omega)K[I + G_0(j\omega)K]^{-1}\} < \frac{\sqrt{\beta^2 + \omega^2}}{\omega} \frac{1}{1 - \lambda}. \tag{6.8}$$

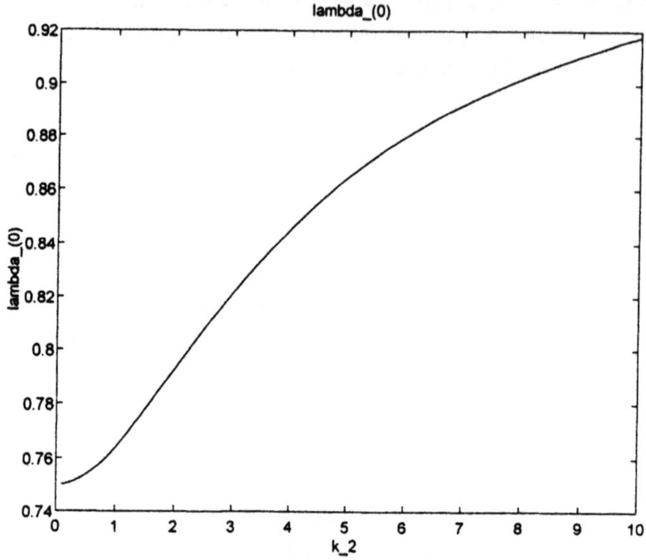

Fig. 4. Delay margin λ_0 as function of the gain k_2 in Example 2

As a simple illustration, consider the simple scalar case with $A = 1$, $B = 1$ and nominal gain $K = 2$. The left hand side of condition (6.8) equals $\frac{2}{\sqrt{1+\omega^2}}$, and the condition can be rearranged to

$$4(1-\lambda)^2 < \frac{(\beta^2+\omega^2)(1+\omega^2)}{\omega^2}. \tag{6.9}$$

The right hand side of the above is maximal for $\omega^2 = \beta$, thus

$$4(1-\lambda)^2 \leq (1+\beta)^2. \tag{6.10}$$

This gives the delay perturbation margin $\lambda_0 = \frac{1-\beta}{2}$. Note that if $\beta \to 0$, stability remains guaranteed for $\lambda > \frac{1}{2}$. We emphasize that we have here assumed that the weight $\mu(t)$ was exactly known, which is quite unrealistic. Using the more conservative Theorem 14 for the same problem we can only guarantee robust stability for $\lambda > \frac{3}{4}$. If the graphical method of Theorem 15 is used with the $\mathcal{L}_\lambda$-domains, note that the Nyquist plot of $k(j\omega) = \frac{2}{j\omega-1}$ coincides with the L-circle for $\lambda = \frac{3}{4}$. Again the margin $\lambda_0 = .75$ guarantees robust stability for *all* perturbations confined to $\mathcal{L}_{3/4}$.

Now for this simple case the explicit solution may be obtained through Laplace transforming the perturbed equation. Because of the infinite delay the Laplace transform of $x(t)$ is rational. Indeed, with $x(\theta) = 0$ for $\theta < 0$, but $x(0) \neq 0$, the system equation

$$\dot{x} = x - 2\lambda x - 2(1-\lambda)\int_0^\infty \beta e^{-\beta\tau} x(t-\tau)\, d\tau, \tag{6.11}$$

transforms to

$$sX(s) - x_0 = (1 - 2\lambda - 2(1 - \lambda)\frac{\beta}{s + \beta}X(s). \tag{6.12}$$

from which in turn

$$X(s) = \frac{s + \beta}{s^2 + (2\lambda - 1 + \beta)s + \beta}x_0. \tag{6.13}$$

Clearly, the solution is stable for

$$\lambda > \frac{1 - \beta}{2} \tag{6.14}$$

The limit for $\beta \to 0$ gives the exact stability margin $(\lambda_0)_{\text{exact}} = \frac{1}{2}$. This illustrates that the margins obtained by the graphical methods may still be quite conservative. If one formally applies the criterion from theorem 12, the bound $\lambda_0 = \frac{3}{4}$ results as well.

Example 4 [17]. Consider the double integrator $\dot{y} = v$, $\dot{v} = u$, with nominal feedback $k_1 v + y$. The Nyquist plots for $k_1 = 1$, 1.5 and 2 are superimposed on the $\mathcal{L}$-bundle in Figure 5. The larger k_1, the lower the guaranteed margin λ_0, and the more robust with respect to delay perturbations. In this example $\lambda_0 = \frac{1}{2}$ is a limit to the guaranteed performance. As the conditions are only sufficient, the actual margin may still be greater.

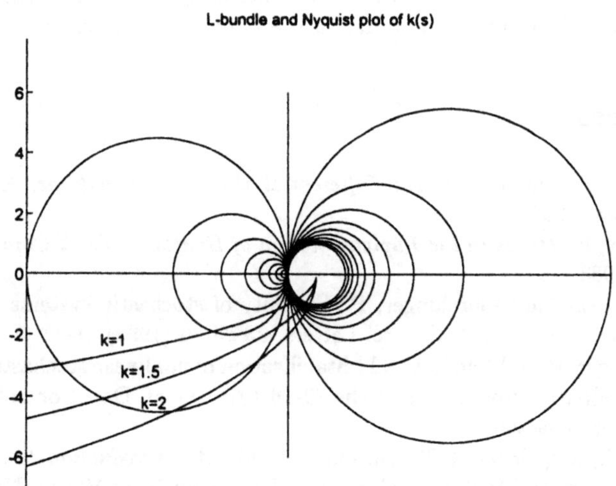

Fig. 5. $\mathcal{L}$-bundle and Nyquist plots for Example 4

7 Conclusions

We have studied the stability properties of a linear system with distributed delay parameters. Our main results are sufficient conditions for stability and asymptotic stability in terms of solvability of certain Riccati equations. Via the positive real lemma, frequency domain criteria were derived as well. The conditions were such that the precise form of the weight matrix function and the effective extend of the delay are immaterial. These results led then to conditions for *robust* stability of distributed delay systems. In turn, these were applied to the problem of stabilizability of systems with delay perturbation in the feedback in case time domain information (bounds) is known for the weight function. With additional information in the frequency domain another criterion was derived in terms of a containment conditions on the Nyquist plot of the total loop transfer function. If all perturbations can occur theses conditions are also necessary. The graphical tests are easily performed, as only one plot needs to be generated and compared against a certain circle bundle (reparametrized M-circles, or the newly introduced L-circles). In all instances, it was shown that as long as a pure delay free state is present in the corrupted signal, stabilizability is ensured whenever the system is stabilizable with instantaneous state feedback, This proves another robustness result with respect to delayed feedback. It can be shown that if the weight matrix function $B(\cdot)$ is *impulsive*, the conditions of robust stability for systems with multiple delays are retrieved [8, 9, 10]. Future research will involve incorporation of the multivariable Nyquist criteria [15, 16] in case the shape factor is a matrix, instead of simply a scalar. Also the effect of stochastic perturbations should be investigated as stochasticity is another way of expressing uncertainty and nonreproducibility in the behavior of a system.

References

1. Bellman, R. and Cooke, K.: *Differential Difference Equations*, Academic Press, New York, 1963
2. Moroney, P.: *Issues in the Implementation of Digital Feedback Compensators*, MIT Press, 1983
3. Verriest, E.I. and Florchinger, P.: Stability of stochastic systems with uncertain time delays. Systems & Control Letters. **23** No. 6 (1994) 41-47
4. Florchinger, P. and Verriest, E.I.: Stabilization of nonlinear stochastic systems with delay feedback. Proceedings of the 32-nd IEEE Conf. Dec. Control. San Antonio, TX. (1993) 859-860
5. Hale,J.K., Magelhães, L.T. and Oliva, W.M.: *An Introduction to Infinite Dimensional Dynamical Systems - Geometric Theory*. Springer-Verlag, New York, 1984
6. Kolmanovskii, V. and Myshkis, A.: *Applied Theory of Functional Differential Equations*. Mathematics and Its Applications, Kluwer Academic Publishers, Dordrecht, 1992
7. Verriest, E.I.: Robust stability of time-varying systems with unknown bounded delays. Proceedings of the 34th IEEE Conference on Decision and Control. Lake Buena Vista, FL (1994) 417-422

8. Verriest, E.I. and Ivanov, A.F.: Robust stability of systems with delayed feedback. Circuits, Systems and Signal Processing **13**(2)/**13**(3) (1994) 213-222
9. Verriest, E.I., Fan, M.K.H. and Kullstam, J.: Frequency domain robust stability criteria for linear delay systems. Proceedings of the 32nd Conference on Decision and Control, San Antonio, TX (1993) 3473-3478
10. Verriest, E.I.: Stabilization of deterministic and stochastic systems with uncertain time delays. Proceedings of the 33nd Conference on Decision and Control, Lake Buena Vista FL. (1994) 3829-3834
11. Wonham, W.M.: On a matrix Riccati equation of stochastic control. SIAM J. Control **4** No. 4 (1968) 681-690
12. Hewer, G.: Existence theorems for positive semidefinite and sign indefinite stabilizing solutions of H_∞ Riccati equations. SIAM J. Control and Optimization **31** No. 1 (1993) 16-29
13. Willems, J.C.: Least squares stationary optimal control and the algebraic Riccati equation. IEEE Trans. Automatic Control **16** (1971) 621-634
14. Verriest, E.I.: Stability and stabilization of stochastic systems with distributed delays. Proceedings of the 34-th IEEE Conference on Decision and Control, New Orleans, LA (1995) 2205-2210
15. Maciejowski, J.M.: *Multivariable Feedback Design*, Addison-Wesley, 1989
16. Postlethwaite, I. and MacFarlane, A.G.J.: *A Complex Variable Approach to the Analysis of Linear Multivariable Feedback Systems*, Springer-Verlag, Lecture Notes in Control and Information Sciences **12** 1979
17. Verriest, E.I.: Stability of systems with distributed delays. Proceedings of the IFAC Conference on System Structure and Control, Nantes, France (1995) 294-299.
18. Niculescu, S.-I.: Sur la stabilité et la stabilisation des systèmes linéaires à états retardés. Ph.D. Dissertation Institut National Polytechnique de Grenoble, Grenoble, France 1996.
19. Verriest, E.I.: Robust stability and stabilization of deterministic and stochastic time-delay systems. Actes du Colloque, Analyse et Commande des Systèmes avec Retards, Ecole des Mines de Nantes, Nantes, France (1996) 197-206.
20. Verriest, E.I. and Aggoune, W.: Stability of nonlinear differential delay systems. Proceedings of the CESA'96 IMACS Multiconference, Symposium on Modelling, Analysis and Simulation, Lille, France (1996) 790-795.
21. Verriest, E.I.: Riccati type conditions for robust stability of delay systems. Presented at the MTNS-96, St. Louis (1996).
22. Verriest, E.I. and Fan, M.K.H.: Robust stability of nonlinearly perturbed delay systems. Proceedings of the 35-th IEEE Conference on Decision and Control, Kobe, Japan (1996) 2090-2091.

Numerics of the Stability Exponent and Eigenvalue Abscissas of a Matrix Delay System

James Louisell*

Department of Mathematics
University of Southern Colorado
Pueblo, CO 81001, USA
Fax : (719) 549 - 2732

Abstract. The author presents a method for determining the stability exponent and eigenvalue abscissas of a linear delay system. The method is based on examining the endpoint values of the solution to a functional equation occurring in the Lyapunov theory of delay equations. The question of existence of the solution to this functional equation is examined in more detail than in the previous delay systems literature. Numerical examples are given, including one in which we show that the decay rate of a feedback system can be improved by delay feedback.

1 Introduction

In this chapter the author develops a method for very accurate determination of the stability exponent for any delay - differential system of the form (*) $x'(t) = A_0x(t)+A_1x(t-h)$. Here $h > 0$ is any positive real number, and $A_0, A_1 \in \mathbb{R}^{n\times n}$. In addition, we will show how to find the other eigenvalue abscissas in the order in which they naturally appear. For simplicity, the theorems are given for the case in which the system has a single delay h as above. Nonetheless, except for notation and computational size, there is nothing to keep the development from being carried over to the case of multiple commensurate delays.

Theorems giving criteria for the stability of linear autonomous delay - differential systems date at least to the work of Pontryagin [18]. For extensive bibliographies as well as many basic theorems on stability, one can consult the books by Hale and Lunel [6], Bellman and Cooke [1], and Diekman *et al* [4]. One need hardly mention that the systems and control literature is replete with theorems on the stability of functional differential equations [11, 12, 14, 17, 19]. It is something of a wonder, then, that the study of the precise stability *exponent* in delay - differential systems has received such scant attention [16]. Up until now, there has been very little in the way of analytic work leading directly to a procedure for determining the stability or growth exponent, or any other eigenvalue abscissas, such as given in this chapter.

Our approach will be based on the analysis of a quadratic functional which is analogous in the theory of delay equations to the quadratic functional $R(x) = x^T Px$ in ordinary differential equations, where P is the matrix

* This work was supported by the NSF (USA) under Grant Number DMS - 9500565.

$P = \int_0^\infty e^{(tA^T)} e^{(tA)} dt$ for any asymptotically stable matrix A, and otherwise P is the solution of the matrix Lyapunov equation $A^*P + PA + I = 0$. Authors such as Infante and Castelan [8, 9], Datko [3], and Marshall *et al* [15] have written on the extension of this concept to the delay systems area. The main technical challenge for us will be to find those properties of the functional $R(x)$ which carry over to this area while allowing us to focus on the system eigenvalue abscissas. After making definitions and introducing notation, we proceed in Section 2 to give the basic matrix function which will be used in the analysis, focusing on the case in which the delay system is asymptotically stable. There are different ways of defining this matrix function, which in our case will be defined by way of a boundary value functional differential equation. In Section 3 we examine a question involving the existence of this matrix function when the delay system is allowed to be unstable. Here we show that the same defining boundary value functional equation will have a unique solution provided only that a certain determinant is nonzero, and we give a counterexample in the case of zero determinant. In Section 4 we use this matrix function to construct a quadratic functional analogous to $R(x)$, and to give a theorem showing that, in the abscissa parameter, eigenvalues of a delay system generate poles of the endpoint values of this matrix function. This is the main theorem of the chapter, and immediately afterward, in Section 5, we will be able to give a simple numerical procedure for accurately determining eigenvalue abscissas. In Section 5 we also give examples, including one in which we measure the performance improvement obtained over a quadratic optimal controller when one uses induced time delays, with improvement measured in terms of the decay rate. In Section 6 conclusions are given and some directions for future research are suggested, especially as regards the possibility of speeding up the computations associated with this procedure.

2 The Matrix Function

In this section we introduce the matrix function we will use in our analysis of the system stability exponent and other eigenvalue abscissas. Although in the following section we show that asymptotic stability is not required to define this function, it will be easier to motivate the associated boundary value functional equations if we first examine the case of an asymptotically stable system. This is the route usually taken, and in fact the theorems in the next section constitute new information on the functional equations. For comparison, one can consult the works mentioned in Section I.

Consider the delay - differential equation (*) $x'(t) = A_0x(t) + A_1x(t-h)$, where $A_0, A_1 \in \mathbb{R}^{n\times n}$ and $h > 0$. Let $X(t)$ denote the solution to the matrix delay equation $X'(t) = A_0X(t) + A_1X(t-h)$ having initial data $X(u) = 0$ for $-h \leq u < 0$, and $X(0) = I$. Now let $f(s) = |sI - A_0 - A_1e^{-sh}|$, let $M(s)$ be the Laplace transform of $X(t)$, i.e. $M(s) = (sI - A_0 - A_1e^{-sh})^{-1}$, and recall that the condition that $f(s)$ has no zeros in the closed right half - plane $Re(s) \geq 0$ is equivalent to exponential asymptotic stability for the system (*). In this case of exponential asymptotic stability, we introduce the matrix function $\vartheta(\alpha) = \int_0^\infty X^T(t)X(t-\alpha)dt$, defined for each $\alpha \in \mathbb{R}$. Noting that $X(\cdot) \in L^2(-\infty,\infty)$, and that for the Fourier transform of $X(\cdot)$ we have $F\{X(\cdot)\} = \frac{1}{\sqrt{2\pi}}M(i\omega)$, one will find by a use of Parseval's formula that $\vartheta(\alpha) = \frac{1}{2\pi}\int_{-\infty}^{\infty} M^*(i\omega)M(i\omega)e^{-i\omega\alpha}d\omega$. Thus we can immediately see that $\vartheta^T(\alpha) = \vartheta^*(\alpha) = \vartheta(-\alpha)$ for all real α. With $\vartheta(\cdot) = Q(\cdot)$, we have :

$$Q^T(\alpha) = Q(-\alpha). \tag{2.1}$$

It is straightforward to give a type of two point relation satisfied by $\vartheta(\cdot)$. In fact, we begin with the formula $X'(t) = X(t)A_0 + X(t-h)A_1$, valid for $t \geq 0+$. Multiplying both sides on the left by $X^T(t)$, we obtain

$$X^T(t)X'(t) = X^T(t)X(t)A_0 + X^T(t)X(t-h)A_1.$$

Transposing, we find that

$$(X')^T(t)X(t) = A_0^T X^T(t)X(t) + A_1^T X^T(t-h)X(t).$$

Adding, we easily see that

$$(X^TX)'(t) = A_0^T X^T(t)X(t) + X^T(t)X(t)A_0 + A_1^T X^T(t-h)X(t) + X^T(t)X(t-h)A_1$$

for $t \geq 0+$.

Integrating over $[0+,\infty)$, we find that the following is satisfied with $\vartheta(\cdot) = Q(\cdot)$:

$$-I = A_0^T Q(0) + Q(0)A_0 + A_1^T Q^T(h) + Q(h)A_1. \tag{2.2}$$

It is also straightforward to show that $\vartheta(\cdot)$ is differentiable over $\mathbb{R} - \{0\}$, and to give a functional differential equation for $\vartheta(\cdot)$. In fact, since $X(\cdot)$ decays exponentially, and is differentiable over $(0,\infty)$, one can apply a somewhat routine analysis with differentiation under the integral, and show that $\vartheta'(\alpha) = -\int_0^\infty X^T(t)X'(t-\alpha)dt$ for $\alpha < 0$. Noting that $X'(t-\alpha) = X(t-\alpha)A_0 + X(t-\alpha-h)A_1$, we find that $\vartheta'(\alpha) = -\vartheta(\alpha)A_0 - \vartheta(\alpha+h)A_1$ for $\alpha < 0$. Now writing $\vartheta(\alpha) = \vartheta^T(-\alpha)$ for $\alpha > 0$, we have $\vartheta'(\alpha) = (-\vartheta'(-\alpha))^T$, so that $\vartheta'(\alpha) = A_0^T\vartheta^T(-\alpha) + A_1^T\vartheta^T(-\alpha+h) = A_0^T\vartheta(\alpha) + A_1^T\vartheta(\alpha-h)$ for $\alpha > 0$. We have arrived at the following functional differential equation satisfied with $\vartheta(\cdot) = Q(\cdot)$:

$$Q'(\alpha) = -Q(\alpha)A_0 - Q(\alpha+h)A_1 \quad \text{for } \alpha < 0 \tag{2.3}$$
$$Q'(\alpha) = A_0^T Q(\alpha) + A_1^T Q(\alpha-h) \quad \text{for } \alpha > 0. \tag{2.4}$$

Proceeding carefully with differentiation under the integral, one can use the ideas above to show that the matrix function $\vartheta(\alpha)$ is both left differentiable and right differentiable at $\alpha = 0$, with right derivative $\vartheta'(0+) = -I - \vartheta(0)A_0 - \vartheta(h)A_1$, and left derivative $\vartheta'(0-) = -\vartheta(0)A_0 - \vartheta(h)A_1$. From this and the equations (2.3)-(2.4) we see that $\vartheta(\cdot)$ is continuous throughout $\mathbb{R}$.

Starting with these differential equations (2.3)-(2.4), and requiring (2.1), one can write an interesting system of ordinary differential equations for the pair

$Q(\alpha), R(\alpha) = Q(\alpha - h)$. In fact, for $0 < \alpha < h$, we find since $\alpha - h < 0$ that $R'(\alpha) = Q'(\alpha - h) = -Q(\alpha - h)A_0 - Q(\alpha)A_1 = -Q(\alpha)A_1 - R(\alpha)A_0$. Likewise, we can set $W(\alpha) = Q(\alpha + h)$ for $-h < \alpha < 0$, and find that $W'(\alpha) = A_0^T W(\alpha) + A_1^T Q(\alpha)$. Thus, with $R(\cdot) = Q(\cdot - h)$ and $W(\cdot) = Q(\cdot + h)$, we have :

For $-h < \alpha < 0$:

$$\begin{aligned} Q'(\alpha) &= -Q(\alpha)A_0 - W(\alpha)A_1 \\ W'(\alpha) &= A_1^T Q(\alpha) + A_0^T W(\alpha) \end{aligned} \tag{2.5}$$

and for $0 < \alpha < h$:

$$\begin{aligned} Q'(\alpha) &= A_0^T Q(\alpha) + A_1^T R(\alpha) \\ R'(\alpha) &= -Q(\alpha)A_1 - R(\alpha)A_0. \end{aligned} \tag{2.6}$$

For $\alpha \in [-h, h]$, we can determine any continuous $Q(\alpha)$ satisfying (2.1) - (2.4). We begin with $R(\alpha) = Q(\alpha - h) = Q^T(h - \alpha)$, and use $R(0) = Q^T(h)$ as a boundary requirement as well as in the boundary equation (2.2). Provided the associated determinant is nonzero, we will have the values of $Q(0)$, $R(0)$, and so will have the solution to (2.6) for $0 \le \alpha \le h$. Finally, we set $Q^T(\alpha) = Q(-\alpha)$ for $-h \le \alpha \le 0$. We explain these details below, noting that they will have special value in Section 3.

First, it will be convenient to introduce the elementary transformations ξ, ξ^o : $\mathbb{R}^{n\times n} \to \mathbb{R}^{n^2}$, which we define for members $M = \begin{bmatrix} m_1 \\ \vdots \\ m_n \end{bmatrix} = [\mu_1 \ \dots \ \mu_n]$ of $\mathbb{R}^{n\times n}$ by $\xi M = \begin{bmatrix} m_1^T \\ \vdots \\ m_n^T \end{bmatrix}$, $\xi^o M = \begin{bmatrix} \mu_1 \\ \vdots \\ \mu_n \end{bmatrix}$.

We also consider the transformation $\chi : \mathbb{R}^{n\times n} \to \mathbb{R}^{n^2\times n^2}$ by $\chi(M) = \begin{bmatrix} I \otimes \mu_1^T \\ \vdots \\ I \otimes \mu_n^T \end{bmatrix}$, where $I = I_n$ and $\otimes$ denotes the Kronecker product. If we now write $q = \xi Q$, $r = \xi^o R$, then the above system (2.6) of differential equations, taken over $[0, h]$ to emphasize boundary continuity, can be written as

$$\begin{bmatrix} q'(\alpha) \\ r'(\alpha) \end{bmatrix} = J \begin{bmatrix} q(\alpha) \\ r(\alpha) \end{bmatrix}, \quad \text{with } J = \begin{bmatrix} A_0^T \otimes I & \chi(A_1) \\ -\chi(A_1) & -A_0^T \otimes I \end{bmatrix}. \tag{2.7}$$

Thus $\begin{bmatrix} q(\alpha) \\ r(\alpha) \end{bmatrix} = e^{J\alpha} \begin{bmatrix} q(0) \\ r(0) \end{bmatrix}$ for $0 \le \alpha \le h$. From $R(0) = Q^T(h)$, we have $R^T(0) = Q(h)$, so that $\xi^o R(0) = \xi Q(h)$, i.e. $r(0) = q(h)$. For $2k \times 2k$ matrices $N = \begin{bmatrix} n_1 \\ \vdots \\ n_{2k} \end{bmatrix}$, we write $P_+(N) = \begin{bmatrix} n_1 \\ \vdots \\ n_k \end{bmatrix}$, $P_-(N) = \begin{bmatrix} n_{k+1} \\ \vdots \\ n_{2k} \end{bmatrix}$, and we see that

$r(0) = q(h) = P_+ e^{Jh} \begin{bmatrix} q(0) \\ r(0) \end{bmatrix}$, or $(P_+ e^{Jh} - [0\ I]) \begin{bmatrix} q(0) \\ r(0) \end{bmatrix} = 0$ in homogeneous form. Now note that equation (2.2) can be written as

$$[(A_0^T \otimes I) + (I \otimes A_0^T)\quad I \otimes A_1^T + \chi(A_1)] \begin{bmatrix} q(0) \\ r(0) \end{bmatrix} = -\xi(I).$$

Using this form for the boundary data (2.2) and using the homogeneous equation above, we set

$$L = \begin{bmatrix} (A_0^T \otimes I) + (I \otimes A_0^T) & I \otimes A_1^T + \chi(A_1) \\ & P_+ e^{Jh} - [0\ I] \end{bmatrix},$$

and arrive at

$$L \begin{bmatrix} q(0) \\ r(0) \end{bmatrix} = - \begin{bmatrix} \xi(I) \\ 0 \end{bmatrix}. \tag{2.8}$$

We now have the initial values $q(0)$, $r(0)$ for (2.7) provided $|L| \neq 0$, and this determines any continuous $Q(\cdot)$ satisfying (2.1)-(2.4). Particularly, if the delay system (*) is asymptotically stable, then we have determined the original matrix function $\vartheta(\cdot)$, defined at first via an integral.

Theorem 1. *If there exists a continuous solution $Q(\cdot)$ to the system of functional differential equations (2.1)-(2.4), then let $R(\cdot)$, ξ, ξ^o, χ, P_+, J, L be as defined above, and let $q(\alpha) = \xi Q(\alpha)$, $r(\alpha) = \xi^o R(\alpha)$. The entries of the matrix $[Q(0)\ R(0)]$ will satisfy the linear equation $L \begin{bmatrix} q(0) \\ r(0) \end{bmatrix} = - \begin{bmatrix} \xi(I) \\ 0 \end{bmatrix}$. If $|L| \neq 0$, then this equation uniquely determines $[Q(0)\ R(0)]$ from $q(0) = \xi Q(0)$, $r(0) = \xi^o R(0)$, and uniquely determines $[Q(\alpha)\ R(\alpha)]$ over $[0, h]$ from $\begin{bmatrix} q(\alpha) \\ r(\alpha) \end{bmatrix} = e^{J\alpha} \begin{bmatrix} q(0) \\ r(0) \end{bmatrix}$. This in turn determines $[Q(\alpha)\ R(\alpha)]$ over $[-h, h]$ from $Q(-\alpha) = Q^T(\alpha)$. If the delay system (*) $x'(t) = A_0 x(t) + A_1 x(t-h)$ is asymptotically stable, then there does exist such continuous $Q(\cdot)$, it is determined uniquely if $|L| \neq 0$, and we have $Q(\alpha) = \vartheta(\alpha) = \int_0^\infty X^T(t) X(t-\alpha) dt$ for $-h \leq \alpha \leq h$, where $X(\cdot)$ is the fundamental solution of (*).*

3 The Functional Equation

In this section, now without the hypothesis of asymptotic stability for the delay - differential equation (*) $x'(t) = A_0 x(t) + A_1 x(t-h)$, we consider whether the system of functional differential equations (2.1) - (2.4) has a unique continuous solution. Before proceeding, it is worth noting that Castelan and Infante [2] developed an elegant solution to the functional equation $Q'(\alpha) = A_0^T Q(\alpha) + A_1^T Q^T(h-\alpha)$ with the boundary data (2.2) replaced by a condition having considerably more symmetry. They set $R(\alpha) = Q^T(h-\alpha)$, and then solve the ordinary differential equation (2.6) for $Q(\cdot)$, $R(\cdot)$ with initial data $Q(\frac{h}{2}) = K$,

$R(\frac{h}{2}) = K^T$, where K is any member of $\mathbb{R}^{n\times n}$. These initial conditions are somewhat generous in that they are guaranteed to give the ordinary differential equation a unique solution over $(-\infty,\infty)$ for each $K \in \mathbb{R}^{n\times n}$, even if the original delay system (*) has an imaginary axis eigenvalue. Thus much of the link with stability theory is lost. On the other hand, given the hypothesis of asymptotic stability for (*), both Infante and Castelan [9], and Marshall *et al* [15] have examined (2.1), (2.3)-(2.4) with the boundary data (2.2). In the absence of the stability hypothesis for (*), it has not been thought to examine (2.1)-(2.4), nor to make a connection with the eigenvalue behavior of (*). Presently we show, with the matrix L as defined in Section 2, that the system of functional differential equations (2.1)-(2.4) does have a unique continuous solution provided $|L| \neq 0$, and we give a counterexample in a case having $|L| = 0$. Then, in Section 4, we make the link with the eigenvalue behavior of the delay system (*).

Now to begin, we consider $f_d(s) = f(s+d)$, where $f(s) = |sI - A_0 - e^{-sh}A_1|$ is the characteristic function for (*), and $d \in \mathbb{R}$. It is easily seen that $f_d(s)$ is the characteristic function for the delay system $(*_d\ )\ x'(t) = (A_0 - dI)x(t) + (e^{-dh}A_1)x(t-h)$, we can always define the matrices J_d, L_d in Theorem 1 accordingly, and equations such as $(2.2)_d$ have the obvious meaning associated with $(*_d)$. We let $\mathrm{adj}(M)$ denote the adjugate of a square matrix M, we set $\mathbf{c} = -\begin{bmatrix} \xi(I) \\ 0 \end{bmatrix}$, and we will have particular interest in the vector function $d \to v(d) = \begin{bmatrix} \sigma_d \\ \kappa_d \end{bmatrix}$ obtained from $v(d) = \frac{1}{|L_d|}\mathrm{adj}(L_d)\cdot\mathbf{c}$, since of course $L_d v(d) = \mathbf{c}$ if $|L_d| \neq 0$.

In the case that $(*_d)$ is asymptotically stable, we can also define the solutions $Q_d(\alpha)$, $R_d(\alpha)$. In fact, given $A_0, A_1 \in \mathbb{R}^{n\times n}$ and $h > 0$, we define $d_0 = \inf\{d : f(s)$ has no zeros with $\mathrm{Re}(s) \geq d\}$. Naturally, d_0 is the stability exponent of the system (*), and since $d_0 = \inf\{d : f_d(s)$ has no zeros with $\mathrm{Re}(s) \geq 0\}$, we see that the delay system $(*_d)$ is asymptotically stable for all $d > d_0$. For each d having both $d > d_0$, $|L_d| \neq 0$, Theorem 1 tells us that the functional system $(2.1)_d$-$(2.4)_d$ has a unique continuous solution, and also gives an explicit formula for the solution. We let $Q_d(\alpha) = Q(d,\alpha)$, $R_d(\alpha) = R(d,\alpha)$, and similarly for $q_d(\cdot)$, $r_d(\cdot)$, and we look carefully at this formula :

$$\begin{bmatrix} q(d,\alpha) \\ r(d,\alpha) \end{bmatrix} = e^{\alpha J_d}\begin{bmatrix} \sigma_d \\ \kappa_d \end{bmatrix} \quad \text{for } 0 \leq \alpha \leq h, \text{ with } v(d) = \begin{bmatrix} \sigma_d \\ \kappa_d \end{bmatrix} = \frac{1}{|L_d|}\mathrm{adj}(L_d)\cdot\mathbf{c}. \tag{3.1}$$

Now for fixed d with $d > d_0$, $|L_d| \neq 0$, we know that $q(d, h-\alpha) = r(d,\alpha)$ for $0 \leq \alpha \leq h$, and we write this as $[P_+e^{(h-\alpha)J_d} - P_-e^{\alpha J_d}]\begin{bmatrix} \sigma_d \\ \kappa_d \end{bmatrix} = 0$. Recalling the Kronecker product form for the boundary value equation (2.2) given prior to (2.8), we set

$$F(d,\alpha) = \begin{bmatrix} (A_0 - dI)^T \otimes I + I \otimes (A_0 - dI)^T \quad I \otimes e^{-dh}A_1^T + \chi(e^{-dh}A_1) \\ P_+e^{(h-\alpha)J_d} - P_-e^{\alpha J_d} \end{bmatrix},$$

and we have

$$F(d,\alpha)v(d) = \mathbf{c}. \tag{3.2}$$

Multiplying both sides by $\mathrm{adj}(F(d,\alpha))$, we have

$$|F(d,\alpha)|v(d) = \mathrm{adj}(F(d,\alpha)) \cdot \mathbf{c},$$

and multiplying by $|L_d|$, we arrive at

$$|F(d,\alpha)|\mathrm{adj}(L_d) \cdot \mathbf{c} = |L_d|\mathrm{adj}(F(d,\alpha)) \cdot \mathbf{c}. \tag{3.3}$$

Now suppose there merely exists some $d' \in \mathbb{R}$ with $|L_{d'}| \neq 0$. Then $d \to |L_d|$ is analytic, and we conclude that there exists an interval (δ_1, δ_2) such that for each $d \in (\delta_1, \delta_2)$, we have $d > d_0$, $|L_d| \neq 0$. Thus (3.3) will hold for $(d,\alpha) \in (\delta_1,\delta_2) \times (0,h)$, an open subset of $\mathbb{R}^2$. Since both sides of (3.3) have entries analytic in (d,α) throughout $\mathbb{R}^2$, we know that (3.3) will be true for all real d, α. Noting the identity $F(d,0) = L_d$, we see that for each *fixed* real d having $|L_d| \neq 0$, there is a positive α_0 such that for each $\alpha \in [0,\alpha_0)$, one has $|F(d,\alpha)| \neq 0$. For such α we can write (3.3) as (3.2), and since both sides of (3.2) are analytic in α, we see that (3.2) holds for all real α. For our fixed d, again noting $F(d,0) = L_d$, we can define $q(d,\alpha)$, $r(d,\alpha)$ by (3.1), and this now certainly gives us a unique continuous solution for $0 \leq \alpha \leq h$ to the ordinary differential equation $(2.7)_d$ with the boundary value equation $(2.2)_d$ and the functional requirement $q_d(h-\alpha) = r_d(\alpha)$. From this in turn one can easily give the solution to the functional differential equation $(2.1)_d$-$(2.4)_d$ for $-h \leq \alpha \leq h$.

Theorem 2. *Let $A_0, A_1 \in \mathbb{R}^{n\times n}$, and let $h > 0$. Consider the functional differential equation $(2.1)_d$- $(2.4)_d$, with continuity required. If there exists some $d \in \mathbb{R}$ with $|L_d| \neq 0$, then for each $d \in \mathbb{R}$ with $|L_d| \neq 0$, this system has a unique solution over $[-h,h]$. This solution is given by (3.1) for $0 \leq \alpha \leq h$, and by $Q(d,\alpha) = Q^T(d,-\alpha)$ for $-h \leq \alpha \leq 0$.*

Corollary 3. *Let $A_0, A_1 \in \mathbb{R}^{n\times n}$, and let $h > 0$. Consider the functional differential equation (2.1)-(2.4). If $|L| \neq 0$, then this system has a unique continuous solution for $-h \leq \alpha \leq h$. This solution is given as in the theorem above with $d = 0$.*

With the amazing symmetries inherent in (2.7) and (2.8), and in $F(\cdot,\alpha)$, it is wise to give some words of caution. One important point is that even for an asymptotically stable delay system, the solution $Q(\alpha)$ to the functional system (2.1)-(2.4) represents the integral $\vartheta(\alpha)$ for $-h \leq \alpha \leq h$, not for all real α. In fact, it is known [2] that the spectrum for (2.1)-(2.4) has negative symmetry. If the delay system (*) is asymptotically stable, then from the comments opening Section 2, we know $\vartheta(\alpha)$ is the Fourier transform of the square integrable matrix function $\frac{1}{\sqrt{2\pi}}M^*(i\omega)M(i\omega)$, and so also has square integrable entries over $(-\infty,\infty)$. Considering negative symmetry of the spectrum, this is not the case for the solution to (2.1)-(2.4).

Continuing with cautionary remarks, one might note the analyticity of the identities leading to the conclusion in Theorem 2, and then think that the system (2.1)-(2.4) always has a solution, or at least does whenever the equation (2.8) has a solution. This is also not the case, as we presently show with the following counterexample.

Example 1. Consider the delay equation (*) $x'(t) = ax(t) - ax(t-h)$, with $h > 0$, $a \in \mathbb{R} - 0$. Here we have $A_0 = a$, $A_1 = -a$, and for (2.2) we have $-1 = [2a \ \ -2a]\begin{bmatrix} q_0 \\ r_0 \end{bmatrix}$, while $J = \begin{bmatrix} a & -a \\ a & -a \end{bmatrix}$. This gives us $|sI - J| = s^2$, and $e^{tJ} = \begin{bmatrix} 1+at & -at \\ at & 1-at \end{bmatrix}$, so that $L = \begin{bmatrix} 2a & -2a \\ 1+ah & -1-ah \end{bmatrix}$, and $|L| = 0$.

Now $\xi(I) = 1$, and for a solution to (2.1)-(2.4) to exist we must at least have a solution to the equation (2.8) $L\begin{bmatrix} q_0 \\ r_0 \end{bmatrix} = \begin{bmatrix} -1 \\ 0 \end{bmatrix}$. Attempting to solve, we obtain $0 = (1+ah)(-1)$, i.e. $0 = 1 + ah$. Thus there can be no solution to (2.1)-(2.4) in the case that $ah \neq -1$. In the case $ah = -1$, the solution to (2.8) is given as $p = \begin{bmatrix} q_0 \\ r_0 \end{bmatrix} = z\begin{bmatrix} 1 \\ 1 \end{bmatrix} + \begin{bmatrix} 0 \\ .5/a \end{bmatrix}$, $z \in \mathbb{R}$, so that $\begin{bmatrix} q(t) \\ r(t) \end{bmatrix} = e^{tJ}\begin{bmatrix} q_0 \\ r_0 \end{bmatrix} = \begin{bmatrix} z - \frac{t}{2} \\ z - \frac{t}{2} + \frac{1}{2a} \end{bmatrix}$. A routine check shows that $q(h-t) = z - \frac{h}{2} + \frac{t}{2}$, $r(t) = z - \frac{h}{2} - \frac{t}{2}$, so that $q(h-t)$, $r(t)$ are equal only at $t = 0$, and there can be no solution to (2.1)-(2.4) even in the case $ah = -1$.

Astute observers will note in the case $ah = -1$ that we solved the boundary value ordinary differential equation (2.7), (2.8) without solving the functional equation (2.1)-(2.4). In fact, if we consider the system ($*_d$), setting $A_0(d) = a - d$ and $A_1(d) = -e^{-hd}a$ as usual, then using (3.1) we will obtain a solution $v(d)$ which is infinite at $d = 0$, so that $v(d)$ does not converge as $d \to 0$ to the solution p above for any finite value of the solution parameter z. It is noteworthy that the delay system (*) $x'(t) = ax(t) - ax(t-h)$ has an imaginary axis eigenvalue, namely zero, and we are prepared now to look into this matter.

4 The Eigenvalue Abscissas

In this section we explore the link between the solution of the functional differential system (2.1)-(2.4) and the eigenvalue behavior of the delay equation (*) $x'(t) = A_0x(t) + A_1x(t-h)$.

Lemma 4. *If there exists $d' \in \mathbb{C}$ having $|L_{d'}| \neq 0$, then the entries of the vector function $v(d) = \begin{bmatrix} \sigma_d \\ \kappa_d \end{bmatrix}$ are meromorphic throughout $\mathbb{C}$.*

Proof. Noting the definition of $v(d)$ opening Section 3, we see that the entries of $v(d)$ are ratios of entire functions, and are thus meromorphic. □

It might seem appealing to define $\Lambda_d = \frac{1}{2\pi}\int_{-\infty}^{\infty} M_d^*(i\omega)M_d(i\omega)d\omega$ for complex d, where $M_d(s)$ is the Laplace transform of the fundamental solution of $(*_d)$, then note that $\Lambda_d = \vartheta_d(0)$ whenever the system $(*_d)$ is asymptotically stable, and attempt to analyze abscissas of system eigenvalues by analytic continuation. However, this complex matrix function $d \to \Lambda_d$ has a natural boundary at each vertical line which includes an eigenvalue of the system (*). Thus the advantages of analytic continuation can not be directly applied to show that the vector function $v(d)$ has poles at the x-coordinates of the zeros of the characteristic function $f(s)$. Nonetheless, using a quadratic functional constructed with $Q(\cdot)$, it will be shown that $v(d)$ does have poles at these locations, and on this basis we will have a means of accurately determining the abscissas of system eigenvalues.

In fact, noting that $f_d(s) = f(s+d)$ is the characteristic function for the system $(*_d)$ above, we show this by merely showing that $v(d)$ has a pole at $d = 0$ whenever $f(i\omega)$ has a real zero ω. Given any system of the form (*) having $|L| \neq 0$, we now introduce the following quadratic functional $V(\phi)$ [8] defined for each $\phi \in \mathbb{C}[-h,0]$, the space of continuous functions mapping the interval $[-h,0]$ into $\mathbb{C}^n$:

$$\begin{aligned} V(\phi) &= \phi^*(0)Q(0)\phi(0) + \phi^*(0)(\int_{-h}^{0} Q(u+h)A_1\phi(u)du) \\ &\quad +(\int_{-h}^{0} \phi^*(u)A_1^T Q^T(u+h)du)\phi(0) \\ &\quad + \int_{-h}^{0}\int_{-h}^{0} \phi^*(u)A_1^T Q(v-u)A_1\phi(v)dvdu. \end{aligned}$$

For solutions $x(\cdot)$ of (*), we define the member x_t of $\mathbb{C}[-h,0]$ as usual, i.e. by $x_t(u) = x(t+u)$ for $-h \leq u \leq 0$. One can calculate the time derivative of the function $V(x_t)$, obtaining the following after a long, tedious, direct calculation which will use the equations (2.3)-(2.4):

$$\dot{V}(x_t) = x^*(t)(A_0^T Q(0) + Q(0)A_0 + A_1^T Q^T(h) + Q(h)A_1)x(t).$$

Noting the boundary value condition (2.2), we have

$$\dot{V}(x_t) = -x^*(t)x(t), \quad \text{i.e.} \quad \dot{V}(\phi) = -\phi^*(0)\phi(0) \ \ for\ \phi \in \mathbb{C}[-h,0].$$

It is interesting to see why the boundary value functional equation for $Q(\cdot)$ could not have a solution if the system (*) has an imaginary axis eigenvalue. In this case one has $\omega \in \mathbb{R}, z \in \mathbb{C}^n - \{0\}$, and a complex vector function $x(t) = e^{i\omega t}z$ satisfying the delay - differential equation (*) over $(-\infty,\infty)$. By simply writing out the expression for $V(x_t)$ as defined above, one will find for $x(t) = e^{i\omega t}z$ that $V(x_t)$ is not dependent on time, i.e. $\dot{V}(x_t) = 0$. Since we also have $\dot{V}(x_t) = -x^*(t)x(t) = -z^*z$, this is certainly a contradiction, and now we see that the boundary value system (2.1)-(2.4) does not have a solution. Thus $|L| = 0$ if the system (*) has an imaginary axis eigenvalue. This does not in itself mean that $v(d)$ has a pole at $d = 0$. However, in the following theorem we show that this is the case if there exists real d with $|L_d| \neq 0$.

Theorem 5. *If the system (*) $x'(t) = A_0x(t) + A_1x(t-h)$ has an imaginary axis eigenvalue and there exists real d' with $|L_{d'}| \neq 0$, then the vector function $v(d)$ has a pole at $d = 0$.*

Proof. From the immediately above comments, we know that $|L_d| = 0$ at $d = 0$, and since $d \to |L_d|$ is analytic, there is a neighborhood $\mathcal{D}$ of $\mathbb{R} - \{0\}$ in which $|L_d|$ has no zeros. Now since $v(d)$ is meromorphic, we know that either $v(d)$ has a pole at $d = 0$, or $v(0) = \lim_{d\to 0} v(d)$ is finite. If we have continuity, then the following maps are continuous for $(d,\alpha) \in (D \cup \{0\}) \times [0,h]$:

1. $d \to v(d)$
2. $(d,\alpha) \to e^{\alpha J_d}$
3. $(d,\alpha) \to e^{\alpha J_d}v(d)$
4. $(d,\alpha) \to (P_+e^{(h-\alpha)J_d} - P_-e^{\alpha J_d})v(d)$.

Noting that the right side of 4 is equal to $q_d(h-\alpha) - r_d(\alpha)$, which is zero for $d \in \mathcal{D}, \alpha \in [0,h]$, we see from the above that setting $\begin{bmatrix} q_0(\alpha) \\ r_0(\alpha) \end{bmatrix} = e^{\alpha J_0}v(0)$ gives us a solution to the boundary value ordinary differential equation (2.7), (2.8) satisfying $q_0(h-\alpha) = r_0(\alpha)$ for $0 \leq \alpha \leq h$. From this we would immediately have a continuous solution to the functional system (2.1)-(2.4). Since we know this is impossible, we conclude that $v(d)$ has a pole at $d = 0$. □

Noting that $f(s) = |sI - A_0 - e^{-sh}A_1|$ will have a zero with abscissa x_0 if and only if $f_{x_0}(s)$ has an imaginary axis zero, we immediately arrive at the following theorem, the final theorem of the chapter.

Theorem 6. *Consider the delay-differential equation (*) $x'(t) = A_0x(t) + A_1x(t-h)$. For each $d \in \mathbb{R}$, define the system ($*_d$) as in Section 3, with $A_0(d) = A_0 - dI$, $A_1(d) = e^{-dh}A_1$. Suppose there exists some $d \in \mathbb{R}$ with $|L_d| \neq 0$, and set $v(d) = \begin{bmatrix} \sigma_d \\ \kappa_d \end{bmatrix} = \frac{-1}{|L_d|}\mathrm{adj}(L_d)\begin{bmatrix} \xi(I) \\ 0 \end{bmatrix}$ for all such d. Then the components of $v(d)$ are meromorphic functions of d. If the system (*) has an eigenvalue with x-coordinate equal to x_0, then the vector function $v(d)$ has a pole at $d = x_0$, and hence the matrix function $d \to [Q_d(0)R_d(0)]$ also has a pole at $d = x_0$.*

5 Computation

In this section we illustrate the value of Theorem 6 in accurately determining eigenvalue abscissas for the linear delay - differential system (*) $x'(t) = A_0x(t) + A_1x(t-h)$. We begin by noting that for $Y_d(t) = e^{-dt}X(t)$, we have $Y_d'(t) = (A_0 - dI)Y_d(t) + (e^{-dh}A_1)Y_d(t-h)$, so that the system ($*_d$) is related exponentially to the original system (*). Recalling the stability exponent d_0 introduced prior to formula (3.1), we know for each $\alpha > d_0$ that $|X(t)| \leq Ce^{t\alpha}$ for $t \geq 0$, so that $|Y_d(t)| \leq Ce^{t(\alpha-d)}$ for $t \geq 0$. Thus $Y_d(t)$ is exponentially decaying for each $d > d_0$, and $Q_d(0) = \int_0^\infty Y_d^T(t)Y_d(t)dt = \int_0^\infty X^T(t)X(t)e^{-2dt}dt$, which of course means that the matrix function $d \to Q_d(0)$ is the Laplace transform of $X^T(t)X(t)$, evaluated at $s = 2d$. Similarly, we will find that $d \to Q_d(h)$ is again a Laplace transform, i.e. $R_d^T(0) = Q_d(h) = e^{dh} \cdot \int_0^\infty X^T(t)X(t-h)e^{-2dt}dt$ for each $d > d_0$, and the entries of both these matrix functions are analytic for $d > d_0$. With these comments we can present the following simple but necessary lemmas.

Lemma 7. *Consider (*) $x'(t) = A_0x(t)+A_1x(t-h)$. Suppose there exists $d' \in \mathbb{R}$ with $|L_{d'}| \neq 0$. Then the entries of the vector function $v(d)$ are analytic on the interval (d_0, ∞).*

Proof. We know that $v(d) = \begin{bmatrix} q_d(0) \\ r_d(0) \end{bmatrix}$ at each $d > d_0$ having $|L_d| \neq 0$. Recalling Lemma 4, the entries of $v(d)$ are meromorphic throughout $\mathbb{C}$, and now since $q_d(0)$, $r_d(0)$ are both analytic on (d_0, ∞), we know that the only singularities of $v(d)$ in (d_0, ∞) are removable. □

Lemma 8. *For $\beta = \|A_0\| + \|A_1\|$, there exist no complex zeros of $f(s) = |sI - A_0 - A_1e^{-sh}|$ which lie in $Re(s) > \beta$.*

Proof. Let $\lambda \in \mathbb{C}$ have $f(\lambda) = 0$ and $Re(\lambda) \geq 0$. Then λ is an eigenvalue of the matrix $A_0 + A_1e^{-\lambda h}$, and since $|e^{-\lambda h}| \leq 1$ for $Re(\lambda) \geq 0$, we have $|\lambda| \leq \|A_0 + A_1e^{-\lambda h}\| \leq \|A_0\| + \|A_1\|$. Thus $|\lambda| \leq \beta$ if $Re(\lambda) \geq 0$, $f(\lambda) = 0$, and indeed there are no zeros of $f(s)$ which lie in $Re(s) > \beta$. □

We can now explain our approach to the accurate determination of the stability exponent and the eigenvalue abscissas for the delay system (*). From Lemma 8, we know for $d > \|A_0\| + \|A_1\|$ that the system ($*_d$) is asymptotically stable. From Theorem 6, we know that $v(d) \to \infty$ as $d \downarrow d_0$. We thus begin by selecting some $d^+ > \|A_0\| + \|A_1\|$, and we compute $v(d)$ for values of $d < d^+$ until a pole is observed. From Lemma 7, the value at which this singularity occurs is d_0.

To determine the eigenvalue abscissa immediately left of d_0, we let $d_1 = \inf\{d : f(s)$ has no zeros with $d_0 > Re(s) \geq d\}$, and compute $v(d)$ for values of $d < d_0$ until the next pole is observed. Here we still have Theorem 6, but not Lemma 7, and to be certain that the value at which this singularity occurs is d_1, we apply the principle of the argument to see if $f(s)$ does indeed have a zero with $Re(s) = d_1$. This process can be continued.

Now provided only that there exists some d' having $|L_{d'}| \neq 0$, we know since $d \to |L_d|$ is a nonzero analytic function that the zeros of $|L_d|$ and hence also of $\sigma_{\min}(L_d)$ are isolated. Thus, considering points d where $|L_d|$ vanishes, we can expect removable singularities of $v(d)$ to appear as such, and likewise poles. Given these limited comments on numerics, we can now present examples. The computations for these examples are conveniently performed using MATLAB.

Example 2. We consider the system (*) $x'(t) = A_0x(t) + A_1x(t - 1.3)$, where $A_0 = \begin{bmatrix} -1.6 & .8 \\ 2.4 & 2.7 \end{bmatrix}$, $A_1 = \begin{bmatrix} 2.5 & 4.1 \\ 1.5 & -3.2 \end{bmatrix}$. Set $A = A_d = A_0 - dI =$

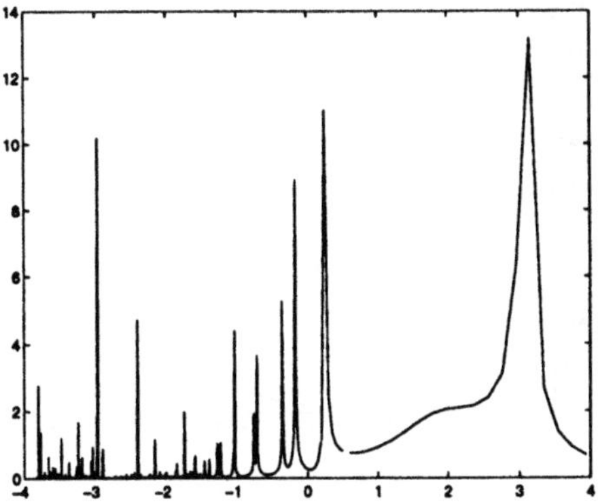

Fig. 1. $y = |v(d)|$

$\begin{bmatrix} -1.6-d & .8 \\ 2.4 & 2.7-d \end{bmatrix}$, $B = B_d = e^{-1.3d}A_1$, and examine the system $(*_d)$. One can easily show that $||A_0|| \leq 5.1$, $||A_1|| \leq 7.3$, and we compute $v(d)$ for $d < 12.5 = 5.1 + 7.3 + .1$, finding from Figure 2 that $0 < d_0 < 5$. From Figure 3, we see that $3.09 < d_0 < 3.10$. If we refined further, we would see that $3.0982 < d_0 < 3.0983$. The growth exponent thus lies in $(3.0982, 3.0983)$. As is well-known [1], the growth exponent for this type of system is also the leading zero of the characteristic function.

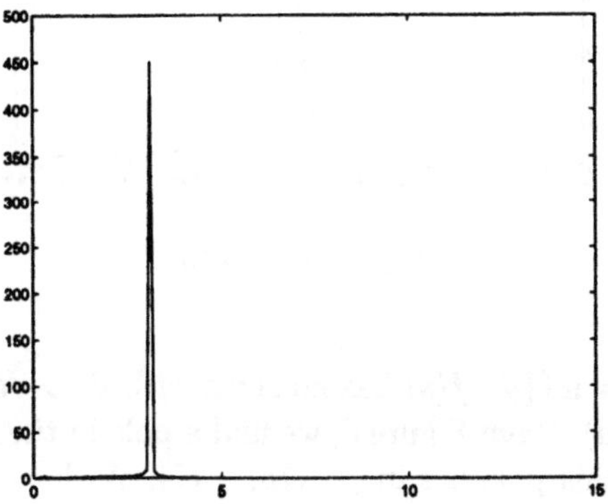

Fig. 2. $y = |v(d)|$

For $d_1 = \inf\{d : f(s)$ has no zeros with $d_0 > Re(s) \geq d\}$, we know that $v(d)$ has a pole at $d = d_1$. Examining Figure 4, we find a pole in the interval $(.25, .3)$. In Figure 5, we see that this pole lies in the interval $(.255, .26)$. If one desires more precision, one will find this pole in the interval $(.2577, .2578)$. Using the principle of the argument, one can find that the characteristic function $f(s)$ has one complex zero to the right of $.2578$, and has three complex zeros to the

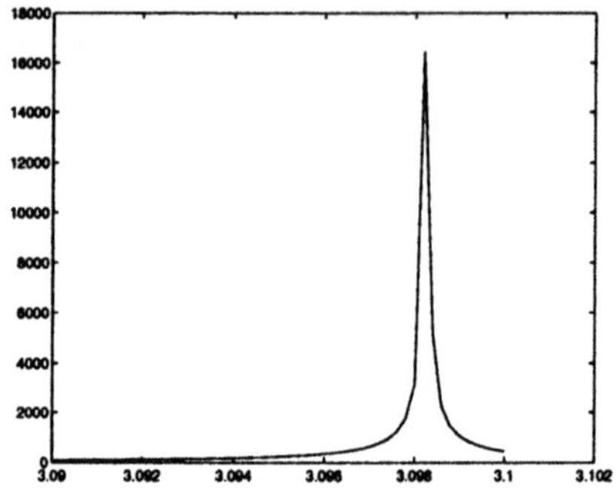

Fig. 3. $y = |v(d)|$

right of .2577. Thus the system (*) has two eigenvalues with x-coordinate in $(.2577, .2578)$, and we have $.2577 < d_1 < .2578$.

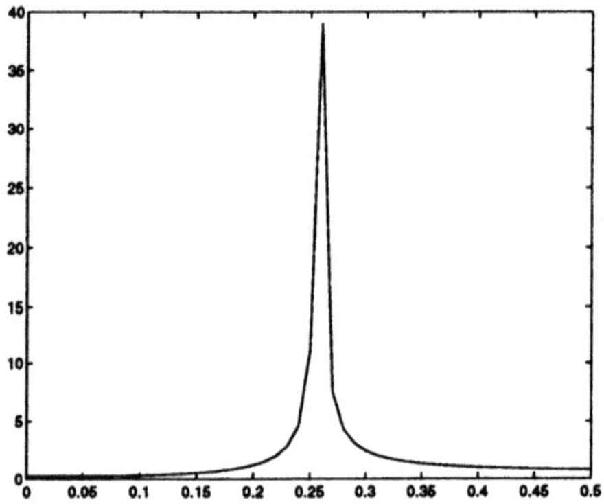

Fig. 4. $y = |v(d)|$

Now to find $d_2 = \inf\{d : f(s) \text{ has no zeros with } d1 > Re(s) \geq d\}$, we again look for a pole of $v(d)$. From Figure 6, we find a pole in the interval $(-.2, -.15)$, and from Figure 7 this pole lies in $(-.175, -.17)$. With even greater precision, the pole would be seen in $(-.1741, -.1740)$. Again using the principle of the argument, we find that the system has three eigenvalues to the right of $-.1740$, and five eigenvalues to the right of $-.1741$. Thus $-.1741 < d_2 < -.1740$.

From the above analysis, the following picture has emerged. The system has a real eigenvalue in the interval $(3.0982, 3.0983)$. The system has a conjugate complex eigenvalue pair with x-coordinate in the interval $(.2577, .2578)$, and also has a conjugate complex eigenvalue pair with abscissa lying in the interval $(-.1741, -.1740)$. This process can be continued to find as many eigenvalue abscissas as desired.

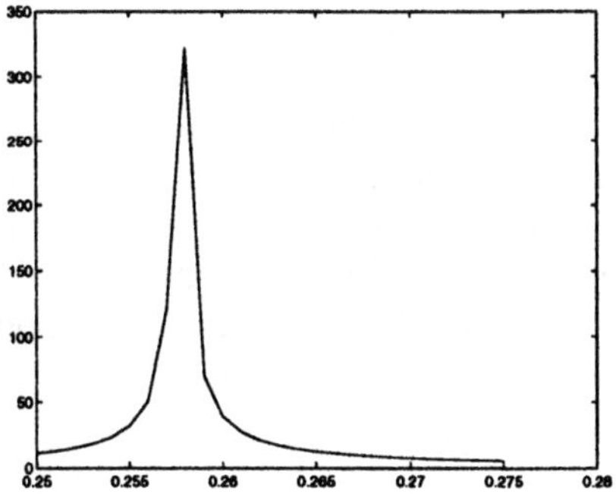

Fig. 5. $y = |v(d)|$

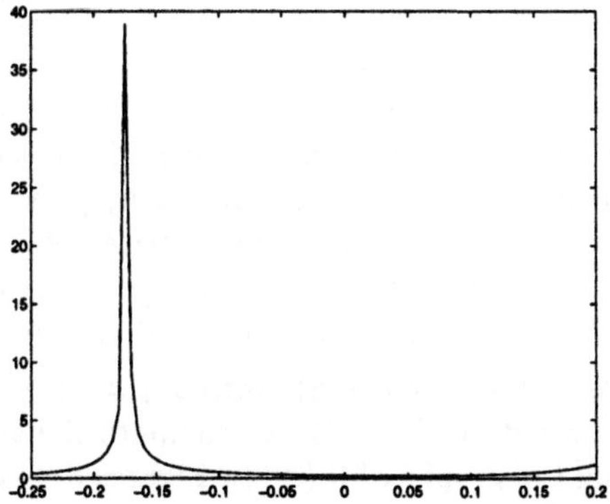

Fig. 6. $y = |v(d)|$

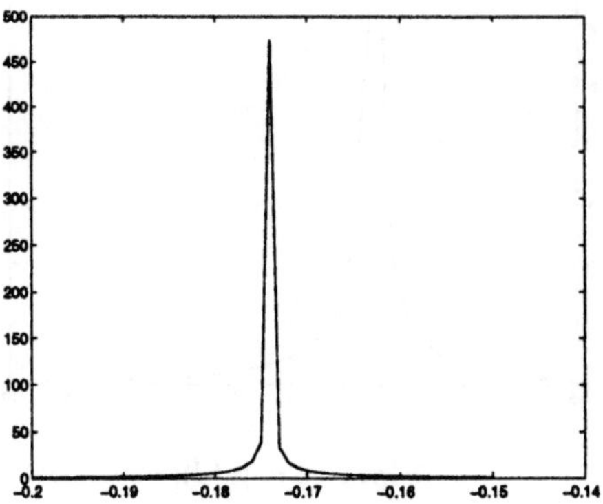

Fig. 7. $y = |v(d)|$

Example 3. The above computational method for determining the system stability exponent is useful in evaluating the benefits of time-delay feedback. In several recent papers [7, 10, 20], it has been shown that time delayed feedback of control systems represented by ordinary differential equations can improve the performance of feedback systems. The improvements come in the form of disturbance rejection, time delay stability margins, and response speed. If this is so, then it is nearly certain that time delay feedback can improve the stability exponent of a feedback system. To show that this is the case, we apply our computational method to an example of Kwon, Lee, and Kim [10].

We consider the controlled system $x'(t) = Ax(t) + Bu(t)$, with free dynamics given by $A = \begin{bmatrix} 2 & 1.2 \\ 1 & 1 \end{bmatrix}$, with control matrix $B = \begin{bmatrix} 1 \\ 3 \end{bmatrix}$, and for simplicity, having free variables directly accessible for feedback. Choosing $Q = 100I$ and $r = 1$ in the standard linear quadratic regulator minimizing $J = \int_0^\infty (x^T(t)Qx(t) + ru(t)^2)dt$, one will obtain optimal state feedback[1] $u = -K_1x(t)$ with $K_1 = [27.27\ 3.07]$. The closed loop system $x'(t) = (A - BK_1)x(t)$ has eigenvalues $\lambda_1 = -31.69$, $\lambda_2 = -1.78$, giving a stability exponent $\lambda = -1.78$ with a corresponding time constant $\tau = 1/1.78 = .56$. In Kwon *et al*, the feedback $u(t) = -K_1x(t) - K_2x(t-h) - \int_{t-h}^{t} K_3(t-v)x(v)dv$ is used, where K_1 is the quadratic optimal feedback matrix, and K_2, $K_3(\cdot)$ are chosen to satisfy performance requirements such as disturbance rejection, robustness against parameter variations, and delay stability margins. The matrices chosen are $K_2 = [-2.6\ \ -6.5]$, $K_3(v) = [5\ \ -1]$ for all v. Frequency domain calculations show that this leads to the closed loop equations $x'(t) = A_0x(t) + A_1x(t-h)$, with $A_0 = A - B(K_1 + \frac{1-e^{-h^2}}{h}K_3)$, $A_1 = -BK_2$. With time delay $h = .04$, we compute the stability exponent for this system, obtaining the graphs of $|v(d)|$ in Figures 8-9 below.

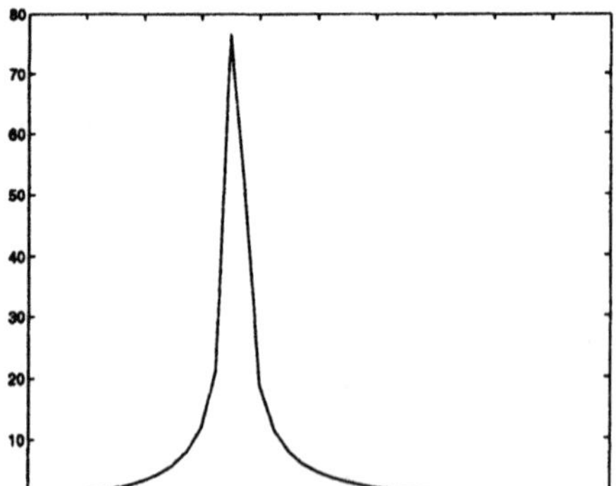

Fig. 8. $y = |v(d)|$

[1] The value given here for K_1 was obtained using MATLAB, and differs slightly from that in Kwon *et al.* This does not change the order of the improvement.

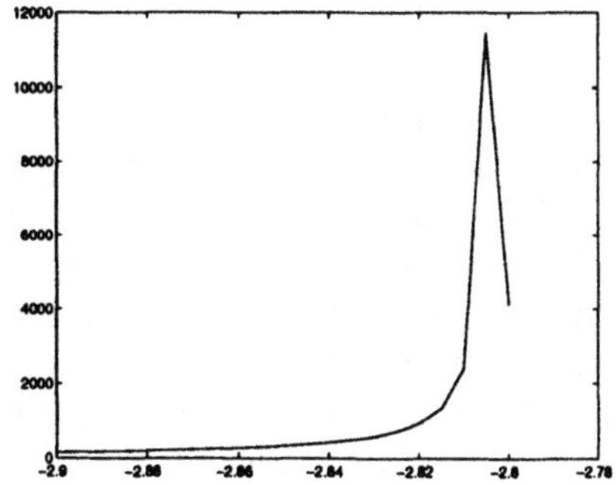

Fig. 9. $y = |v(d)|$

We see here that the stability exponent is no greater than $\lambda_h = -2.80$, giving a closed loop time constant no greater than $\tau_h = 1/2.80 = .36$. Thus this feedback strategy does indeed provide the system with an improved stability exponent and the associated faster response.

Comment : Authors such as Freudenberg, Liang, and Looze [5, 13] have pointed out that time delays in a feedback system are accompanied by bandwidth difficulties, making the system more vulnerable to model uncertainty. Kwon et al address this carefully in their paper giving the above example. In other examples, particularly those involving time delay uncertainty compensation [7, 20], this has become a point of contention, with Liang and Looze [13] arguing that the use of time delays in the feedback loop leads to bandwidth which is too large for engineering practice. The time constant shrinks and response speed improves as bandwidth grows and vulnerability to uncertainty increases. The author's work gives a convenient computational means of determining the improvement in the time constant, thus providing a useful tool for assessing the tradeoff.

6 Conclusion

In this chapter we have given a computational method for determining the stability exponent and the other eigenvalue abscissas in a linear delay - differential system. We have illustrated the theorems with examples, including one in the area of time-delay feedback. For future research, we have given one topic special mention. That is the challenge mentioned above, of obtaining computational procedures for determining stability exponents which can be implemented within system bandwidth constraints, i.e. in terms of constraints on lower and upper bounds of the singular values of system transfer function matrices.

There are other possible directions for research, e.g. efforts at reducing the computational size occurring in constructing $Q(\cdot)$ for systems having multiple commensurate delays, carrying the work in this chapter over to neutral delay systems, the development of asymptotic expressions for the matrix function $d \rightarrow$

$Q_d(0)$, which we know from Section 5 is a Laplace transform, and others. We discuss two interesting possibilities below.

To begin, we note that even given relentless advances in computing speed, there would still be value in a procedure for computing the stability exponent which did not depend on sweeping through d-values. If one could establish the convergence to the stability exponent of a Newton - Raphson method applied to some function of the matrix Q_d, then the computation time for determining the stability exponent would be greatly reduced.

Another point of special interest is the calculation of $Q(\cdot)$ for delay systems having multiple commensurate delays. Although the algebraic techniques necessary for the calculation are rather routine modifications of those occurring in the single delay case, the computational proportions are somewhat daunting. However, when one writes differential equations for $Q(\cdot)$, $R(\cdot) = Q(\cdot - h)$, $S(\cdot) = Q(\cdot - 2h)$, ..., one discovers enough structure to hope for new methods of reducing the computational size, with the possible effect of making it practical to determine the stability exponent and eigenvalue abscissas. This could have relevance to the analysis of delay feedback systems connected in series or parallel, or even, optimistically, in the approximation of $Q(\cdot)$ for systems having incommensurate delays.

References

1. R. Bellman and K. Cooke, *Differential Difference Equations*, Academic Press, New York, 1963.
2. W. B. Castelan and E. F. Infante, "On a Functional Equation Arising in the Stability Theory of Difference - Differential Equations," *Quarterly of Applied Mathematics*, vol. 35, pp. 311–319, 1977.
3. R. Datko, "Lyapunov Functionals for Certain Linear Delay - Differential Equations in a Hilbert Space," *J. Math. Anal. Appl.*, vol. 76, pp. 37–57, 1980.
4. O. Diekman, S. A. van Gils, S. M. V. Lunel and H.- O. Walther, *Delay Equations: Functional, Complex, and Nonlinear Analysis*, Springer - Verlag, New York, 1995.
5. J. S. Freudenberg and D. P. Looze, "A Sensitivity Tradeoff for Plants with Time Delay," *IEEE Transactions on Automatic Control*, vol. 32, pp. 99–104, 1987.
6. J. K. Hale and S. M. V. Lunel, *Introduction to Functional Differential Equations*, Springer-Verlag, New York, 1993.
7. T. C. S. Hsia, T. A. Lasky and Z. Guo, "Robust Independent Joint Controller Design for Industrial Robot Manipulators," *IEEE Transactions on Industrial Electronics*, vol. 38, pp. 21–25, 1991.
8. E. F. Infante, *Some Results on the Lyapunov Stability of Functional Equations*, In *Volterra and Functional Differential Equations*, Lecture Notes in Pure and Applied Mathematics, 81, Marcel Dekker, New York, 1982.
9. E. F. Infante and W. B. Castelan, "A Lyapunov Functional for a Matrix Difference - Differential Equation," *Journal of Differential Equations*, vol. 29, pp. 439–451, 1978.
10. W. H. Kwon, G. W. Lee and S. W. Kim, "Performance Improvement Using Time Delays in Multivariable Controller Design," *International Journal of Control*, vol. 52, pp. 1455–1473, 1990.

11. E. B. Lee and W. S. Lu, *Feedback with Delays : Stabilization of Linear Time - Delay and Two - Dimensional Systems*, In *Signal Processing : Part II*, IMA Vol. Math. Appl., 23, Springer - Verlag, New York, 1990.
12. B. Lehman, J. Bentsman, S. M. V. Lunel and E. I. Verriest, "Vibrational Control of Nonlinear Time Lag Systems with Bounded Delay: Averaging Theory, Stabilizability, and Transient Behavior," *IEEE Transactions on Automatic Control*, vol. 39, pp. 898–912, 1994.
13. Y. J. Liang and D. P. Looze, "Evaluation of Time - Delayed Uncertainty Cancellation Systems," *Proceedings of the 1992 American Control Conference*, pp. 1950–1954, 1992.
14. H. Logemann, "On the Transfer Matrix of a Neutral System: Characterizations of Exponential Stability in Input - Output Terms," *Systems & Control Letters*, vol. 9, pp. 393–400, 1987.
15. J. E. Marshall, H. Gorecki, A. Korytowski and K. Walton, *Time - Delay Systems : Stability and Performance Criteria with Applications*, Ellis Horwood, New York, 1992.
16. T. Mori, N. Fukuma and M. Kuwahara, "On an Estimate of the Decay Rate for Stable Linear Delay Systems," *Int. J. Contr.*, vol. 36, pp. 95–97, 1982.
17. A. Olbrot, "Stabilizability, Detectability, and Spectrum Assignment for Linear Autonomous Systems with General Time Delays," *IEEE Trans. Automat. Contr.*, vol. 23, pp. 887–890, 1978.
18. L. Pontryagin, "On the Zeros of Some Elementary Transcendental Functions," *American Mathematical Society Translations*, vol. 2, pp. 95–110, 1955.
19. G. Tadmor, "Trajectory Stabilizing Controls in Hereditary Linear Systems," *SIAM J. Control and Optim.*, vol. 26, pp. 138–154, 1988.
20. K. Youcef-Toumi, Y. Sasage, J. Ardini and S. Y. Huang, "The Application of Time Delay Control to an Intelligent Cruise Control System," *Proceedings of the 1992 American Control Conference*, pp. 1743–1747, 1992.

Moving Averages for Periodic Delay Differential and Difference Equations

B. Lehman and S. Weibel

Dept. of Electrical and Computer Engineering
Northeastern University, Boston, MA 02115, USA

Abstract. In this chapter we consider the averaging of periodic delay differential and delay difference equations using the method of moving averages. Specifically, we prove formal averaging theorems for both types of systems. This work is based in part on fundamental work in the averaging of delay systems performed in the 1960's by Halanay[11, 12], Hale[13], and Medvedev[23]. The analysis and theorems presented here differ from the earlier works in that our analysis gives greater importance to the delay terms which appear in the averaged system. To illustrate our results, we consider two simple examples of delay systems with periodic excitation - a cart and pendulum stabilization problem in the presence of periodic disturbances and feedback delays, and the adpative identification of chemical concentrations in a pipe mixing problem.

1 Introduction

The development of theory for the stability analysis of dynamical systems represents one of the broadest sustained efforts in mathematics and applied science. At the moment, there exist well-developed tools for studying the stability of systems of the form $\dot{x} = f(x)$, where $f : \mathbb{R}^n \to \mathbb{R}^n$. What has proven more difficult is the development of tools for the stability analysis of explicitly time-dependent systems, i.e. systems of the form $\dot{x} = f(x,t)$, $f : \mathbb{R}^n \times \mathbb{R} \to \mathbb{R}^n$. Complications often arise when the vector field is periodic in t, i.e. $f(x,t+T) = f(x,t)$ for some $T > 0$. What complicates the stability analysis is that the system will exhibit unstable behavior which is purely a function of T. This phenomenon is, of course, known as parametric resonance. Simple physical examples of systems which exhibit parametric resonance include periodically forced spring-mass systems and periodically forced LRC circuits. There are a variety of methods suited to the analysis of such systems, most notably Floquet theory [22, 29] and the topic of the current work, the method of averaging. In particular, this chapter presents new results on the method of averaging for periodic delay differential and delay difference equations.

1.1 A Brief History of Averaging

One method of eliminating explicit time dependence in some periodically excited systems, which ultimately retains global information about the system, is the method of averaging. Motivating the development of a theory of averaging was the interest in predicting the motions of the planets in the 18th and 19th centuries (a brief discussion pertaining to this history is given in [2]), where averaging was seen as a way of coping with small periodic perturbations in models of the solar system. Current theory is largely based on the work of Russian authors[7, 15, 16] from the first half of this century. These authors were principally interested in studying weakly nonlinear second order systems of the form

$$\ddot{x} + \omega_n^2 x = \epsilon f(x, \dot{x}, t). \tag{1.1}$$

where $x \in \mathbb{R}$, f is continuous in its arguments, $f(x, \dot{x}, t+T) = f(x, \dot{x}, t)$ for $T > 0$, and $0 < \epsilon \ll 1$. Through a series of coordinate changes, (1.1) can be written

$$\dot{\mathbf{x}} = \epsilon F(\mathbf{x}, t) + \mathcal{O}(\epsilon^2),$$

which could be averaged to obtain

$$\dot{\mathbf{x}} = \epsilon \overline{F}(\mathbf{x}), \tag{1.2}$$

where $\mathbf{x} \in \mathbb{R}^2$, $\overline{F}(x) = \frac{1}{T}\int_0^T F(\mathbf{x}, t)dt$, and we have ignored $\mathcal{O}(\epsilon^2)$ terms. Equilibrium and stability analysis could then be performed in terms of (1.2). Such systems arise physically as simple nonlinear mechanical and electrical oscillators. The classic example of such a system is the inverted pendulum forced by vertical oscillations of the hinge, for which it has been shown that the inverted equilibrium can be rendered stable for sufficiently large forcing frequencies[3, 7, 21, 28]. Many other examples of systems from this period can be found in [1].

Since then, the concepts used to average weakly nonlinear systems have been extended to the more general class of vector fields

$$\dot{x} = \epsilon f(x, t, \epsilon), \tag{1.3}$$

where it is assumed that $f : \mathbb{R}^n \times \mathbb{R} \times \mathbb{R}^+ \to \mathbb{R}$, $f(t+T) = f(t)$ for $T > 0$, f is bounded on bounded sets, f is at least twice differentiable in its arguments, and $0 < \epsilon \ll 1$. As described in standard texts[10, 14, 34] there exists a change of coordinates $x = y + \epsilon w(y, t, \epsilon)$ such that (1.3) can be written

$$\dot{y} = \epsilon \overline{f}(y) + \epsilon^2 \widehat{f}(y, t, \epsilon), \tag{1.4}$$

where $\overline{f}(y) = \frac{1}{T}\int_0^T f(y, t, \epsilon)dt$, and $\widehat{f}(y, t+T, \epsilon) = \widehat{f}(y, t, \epsilon)$. The effect of this coordinate change is to push the nonautonomous terms to $\mathcal{O}(\epsilon^2)$. Thus, to $\mathcal{O}(\epsilon)$, (1.3) can approximated by

$$\dot{y} = \epsilon \overline{f}(y). \tag{1.5}$$

The closeness of solutions of (1.3) and (1.5) are given by averaging theorems proven in [10, 14, 34]: specifically if $x(t)$ is a solution of (1.3) and $y(t)$ of is a solution of (1.5), then $|x(t) - y(t)| = \mathcal{O}(\epsilon)$ on the interval $t \in [0, t_1]$, where $t_1 = \mathcal{O}(1/\epsilon)$. In addition, these same averaging theorems associate hyperbolic behaviors of (1.3) and (1.5) (hyperbolic fixed points of (1.5) are associated with hyperbolic periodic orbits of (1.3), asymptotic behavior of solutions on the stable/unstable manifolds of fixed points/periodic orbits in forward/backward time). The theory used for systems with periodic excitation has been extended in [14] for *almost-periodic* excitation. Arnol'd[2] gives an alternate description of the method of averaging from the point of view of *action-angle* coordinates, and discusses conditions under which averaging results for single and multifrequency systems break down.

1.2 Applications of Averaging Theory in Controls Engineering

One field which has benefitted from fundamental advances in averaging theory is automatic control theory. The need for averaging theory arises in automatic control because physical processes often either possess some form of periodic excitation as part of their natural dynamics (external sources of oscillation) or may be controlled by periodic excitation. Once again, the inverted pendulum problem described previously has served as a paradigm for controls-oriented applications of periodic excitation and the method of averaging.

Averaging techniques have also proven useful in the synthesis of so-called open-loop periodic controls, sometimes referred to as *vibrational controllers* (see [6] for a tutorial). *Open-loop* control laws are those which do not take into account the state of the system being controlled. The control is merely some physical input (force, torque, voltage, etc.) chosen in such a way to excite some desired system behavior. Recent literature has focused on time-periodic functions as control inputs, making the method of averaging a desirable analytical tool. Motivated by the stabilization of the inverted pendulum by high frequency vertical forcing, recent work[6] has focused on the use of open-loop periodic inputs in the control and stabilization of a variety of systems. One class of systems where periodic controls have been employed and have been observed to be robust is a class of *superarticulated mechanical systems*[26, 32, 33], of which the inverted pendulum is a somewhat trivial member. Concepts developed in the study of superarticulated mechanical systems are currently being extended to permit the systematic synthesis and analysis of open-loop control laws for more general systems[5, 6].

1.3 Motivation for the Averaging of Delay Systems

While the averaging of delay equations is interesting for its own sake, the importance of the research is immediately apparent in the context of control. One invariant in the synthesis of closed-loop control laws is that any conventional feedback loop will possess some delay or latency. In "analog" control systems, delays may arise as the result of unmodelled internal capacitances in the controlled system or the controller. The effect of these capacitances is that a phase lag may be induced in the feedback signal which will also be manifested in the control. In "digital" control systems, delays are often the result of the computational overhead in performing the control task and the finite clock cycle of digital components. While the increase in the allowable clock rates for digital control hardware is making clock precision less of an issue, there is current interest in the control of complex systems where the computational overhead is a significant source of control latency. Examples of such complicated systems where computation-induced delays might arise are infinite dimensional systems such as fluid/chemical systems and elastic structures. Another significant source of delays in control systems is transmission delays resulting from tele-operation. Examples of systems where transmission delays are significant include the Viking Mars lander and the robots which will be deployed as part of the upcoming Mars exploration.

The need for developing extensive theoretical results for averaging of delay systems should also be apparent. Such theory is needed to design high-frequency open-loop control laws for the above mentioned applications. Interesting examples of periodic forcing in delay systems can be found in [17, 18, 19]. As has been mentioned in [17], a limiting factor in the development of open-loop oscillatory controllers has been the lack of mathematical theory that can be used to describe the behavior of periodic delay systems.

The current work in this chapter focuses on developing a comprehensive theory of computing averages for periodic delay differential and delay difference equations. In Section 2, we present theorems for the averaging of delay differential equations. This work has been motivated by the earlier work of Halanay[11, 12], Hale[13], and Medvedev[23], and in fact extends these earlier results to yield averaged systems which more accurately approximate the original nonautonomous dynamics in the presence of significant delays. In Section 3, we modify the results of Section 2 for delay discrete equations, and obtain completely analogous results. In Section 4, we apply the theory of Sections 2 and 3 to two examples from controls engineering. First, we study the closed-loop feedback control of an inverted pendulum in the presence of external oscillations. Next, we study the adaptive identification and control of a discrete pipe mixing problem. We summarize and conclude in Section 5.

2 Averaging of Continuous-Time Delay Systems

This section considers continuous-time delay differential equations given by

$$\dot{x}(t) = \epsilon f\left(t, x(t), x(t-r)\right); \qquad x(t) = \phi(t), t \in [t_0 - r, t_0] \tag{2.1}$$

where f is continuous with respect to all its argments, $f : \mathbb{R} \times D \times D \to \mathbb{R}^n$, $D \subset \mathbb{R}^n$, and ϵ and r are positive real parameters. Furthermore, assume that f is T- periodic, i.e., $f(t+T, x_1, x_2) = f(t, x_1, x_2)$ for all $(t, x_1, x_2) \in \mathbb{R} \times D \times D$. The function ϕ is continuous on $t \in [t_0 - r, t_0]$. Let the solution of (2.1) be denoted by $x(t) = x(t; t_0, \phi)$.

Numerous papers[11, 12, 13, 23] have studied (2.1) and shown when, for sufficiently small ϵ, the solution to (2.1) can be approximated by the solution to the corresponding averaged ODE

$$\dot{y}(t) = \epsilon f_0\left(y(t), y(t)\right); \qquad y(t_0) = \phi(t_0) \tag{2.2}$$

where

$$f_0\left(c_1, c_2\right) = \frac{1}{T}\int_0^T f\left(s, c_1, c_2\right) ds. \tag{2.3}$$

In (1.2), the explicit time dependence of f has been averaged out. As a result the complexity of analysis has been reduced. Additionally, all information on the delay has been ignored, greatly simplifying the problem even more. Unfortunately, though, it is precisely this oversimplification of ignoring the delay that causes inaccuracies in the approximation when ϵ is not infinitesimely small [18]. In fact, the work of [18] indicates that there are two separate, important values of ϵ. First, there exists some ϵ_1, sufficiently small, such that, for $0 < \epsilon \leq \epsilon_1$, the time dependence can be averaged out in (1.1). Next, there exists an upper bound on ϵ, denoted by ϵ_2, such that, for $0 < \epsilon \leq \epsilon_2$, the delay can be ignored. As [18] has noted, it is common that $\epsilon_2 \ll \epsilon_1$.

In general averaging results are stated for 'sufficiently small ϵ.' Therefore, there is no need to distinguish between ϵ_1 and ϵ_2 in any of the proofs of averaging. However, in practice, a physical system may not admit an ϵ infinitesimely small, and therefore, the classical averaging results of [11, 12, 13, 23] may not be applicable. It is then of interest to develop averaging theory which permits ϵ to have a larger upper bound.

The approach taken in this chapter is to perform moving averages on the solution, $x(t)$, of (2.1). That is, we simply take the forward moving average value of $x(t)$ on the time interval $[t, t+T]$, denoted by $\overline{x}(t)$. Since $x(t)$ is continuous and bounded on finite time intervals, its moving average will always remain reasonably close. In particular, when ϵ is small, the rate of change on $x(t)$ is small, suggesting that $|x(t) - \overline{x}(t)|$ will also be small.

Because the higher harmonics of $x(t)$ have been averaged out to obtain $\overline{x}(t)$, it is then possible to approximate $\overline{x}(t)$ by the solution of an autonomous differential equation - obtained by taking the (newly proposed) average value of the vector-field of the right-hand-side of (2.1). In this way, it is possible to relate $x(t)$ with the solution to an averaged equation. Let us formally define the notion of a moving average.

Definition 1. Suppose that $x(t) = x(t; t_0, \phi)$ is the solution to (2.1) with continous initial function $\phi \in \mathcal{C}$. The **moving average** of $x(t)$ is denoted by $\overline{x}(t)$ and is defined as

$$\overline{x}(t) \equiv \begin{cases} \phi(t), & \text{for } t \in [t_0 - r, t_0) \\ \frac{1}{T}\int_t^{t+T} x(s) ds, & \text{for } t \geq t_0, \end{cases}$$

where $T > 0$ is the period of f.

Theorem 2. *Assume that the solution to (2.1) satisfies $x(t) \in D$ for $t \in [t_0 - r, t_0 + L_1 + T]$ where $L_1 > 0$ and $T > 0$. Assume further that f is T-periodic and satisfies $|f(t, x_1, x_2)| \leq M$ for all (t, x_1, x_2) on $([t_0 - r, t_0 + L_1] \times D \times D)$. Then $|x(t) - \overline{x}(t)| = \mathcal{O}(\epsilon T)$ for all $t \in [t_0 - r, t_0 + L_1]$.*

Proof. For $t \in [t_0 - r, t_0)$, we have by Definition 1 that $x(t) - \overline{x}(t) = 0$. On $t \in [t_0, L_1]$,

$$\begin{aligned}
|x(t) - \overline{x}(t)| &= \left| x(t) - \frac{1}{T}\int_t^{t+T} x(s)ds \right| \\
&= \left| \frac{1}{T}\int_t^{t+T} [x(t) - x(s)]ds \right| \\
&= \left| \frac{\epsilon}{T}\int_t^{t+T} \int_s^t f\left(\tau, x(\tau), x(\tau - r)\right) d\tau ds \right| \\
&\leq \left| \frac{\epsilon}{T}\int_t^{t+T} M(t - s)ds \right| \\
&= \left| \frac{\epsilon M T}{2} \right| = \mathcal{O}(\epsilon T),
\end{aligned}$$

where we have used the fact that f is uniformly bounded by M whenever $x(t) \in D$. □

In this theorem, uniform convergence of the average value of $f(t, \cdot)$ was never used. This implies that the above theorem is valid even when f is not periodic. Indeed, fewer restrictions are needed on vector-fields when taking moving averages, in comparison to when performing classical averaging (as in Theorem 3 below). On the other hand, in order to take a moving average in (2.1), the solution to time-varying delay differential equation (2.1) is needed. Hence, analysis will only be simplified if it is possible to approximate the moving average of a system by another trajectory that is easier to obtain.

As in [18], introduce the alternate averaged model to (2.2) given by

$$\dot{z}(t) = \epsilon f_0(z(t), z(t - r)); \qquad z(t) = \phi(t), t \in [t_0 - r, t_0] \tag{2.4}$$

where f_0 is defined in (2.3). It will be shown that (2.4) is a more natural representation of an averaged approximate model of (2.1).

Remark 1. In (2.4), information on the delay has been retained, and hence, the proposed new averaged model is more complicated than (2.2). However, as this chapter demonstrates, it is often important to retain the delay in order for the averaged approximation to remain valid. The results of [9, 11, 12, 13, 23, 30] were developed before the advent of advanced computer technology. In the 1960's, qualitatively analyzing the behavior of (2.4) was considered to be difficult. However, advances in computers and numerical algorithms have made the simulation of delay systems relatively easy. (Although analytically, the problem of analyzing the behavior of nonlinear delay differential equations in the form of (2.4) remains an active area of research, the problem has become more tractable in recent times.)

To this extent, in no way are we suggesting that the results of [9, 11, 12, 13, 23, 30] are incorrect in claiming that (2.2) is an averaged approximation of (2.1). For ϵ 'sufficiently small,' the approximation is valid. In fact, these earlier averaging results might have been criticized if they proposed (2.4) to be the simplified averaged approximation of (2.1), since it remains infinite dimensional. On the other hand, with the capabilities of modern computing hardware, the increased accuracy of (2.4) should more than warrant any additional added complexity. ◇

We will now relate the solutions of (2.4) to $\overline{x}(t)$. Assuming right-hand derivatives, the function $\overline{x}(t)$ satisfies the differential equation

$$\begin{aligned} \dot{\overline{x}}(t) &= \frac{1}{T}[x(t+T;t_0,\phi) - x(t;t_0,\phi)] \\ &= \frac{\epsilon}{T}\int_t^{t+T} f\left(s, x(s;t_0,\phi), x(s-r;t_0,\phi)\right) ds \end{aligned} \tag{2.5}$$

on $t \geq t_0$, with $\overline{x}(t) = \phi(t)$ for $t \in [t_0 - r, t_0)$.

The right hand side of the above differential equation depends on $x(t)$ and not on $\overline{x}(t)$. Hence, $\overline{x}(t)$ can be interpreted as the solution to the differential equation whose vector field is equal to the 'local average' of $\epsilon f(\cdot)$ along the solution to (2.1). (The work of [24] examines local averaging techniques for ODE's.)

By rewriting (2.5) as

$$\begin{aligned} \dot{\overline{x}}(t) = \frac{\epsilon}{T}\int_t^{t+T} f\left(s, \overline{x}(t), \overline{x}(t-r)\right) ds \\ + \frac{\epsilon}{T}\int_t^{t+T} \left[f\left(s, x(s;\phi), x(s-r;\phi)\right) - f\left(s, \overline{x}(t), \overline{x}(t-r)\right)\right] ds \end{aligned} \tag{2.6}$$

perturbation theory can be applied to relate $\overline{x}(t)$ to the solution of (2.4). That is, due to Theorem 2 and Lipschitz arguments, the terms inside the square brackets in the second integral in (2.6) can be shown to be $\mathcal{O}(\epsilon)$. Hence, when f is periodic, the limit (2.3) exists uniformly in t, and the following result naturally follows.

Theorem 3. *Let the assumptions of Theorem 2 be true for $L_1 = L/\epsilon$ and assume that $|f(t,x_1,x_2) - f(t,\tilde{x}_1,\tilde{x}_2)| \leq K(|x_1 - \tilde{x}_1| + |x_2 - \tilde{x}_2|)$ for all $(t, x_i, \tilde{x}_i) \in$*

$([t_0 - r, t_0 + L/\epsilon] \times D \times D)$. Suppose, also, that both $\overline{x}(t)$ and $z(t)$ remain in D for all $t \in [t_0 - r, t_0 + L/\epsilon]$, with $x(t) = z(t) = \phi(t)$ on $t \in [t_0 - r, t_0]$. Then

$$|\overline{x}(t) - z(t)| = \mathcal{O}(\epsilon T) + \mathcal{O}(\epsilon r)$$

for all $t \in [t_0 - r, t_0 + L/\epsilon]$.

Proof. Let $z(t) = z(t; t_0, \phi)$ denote the solution to (2.4). By assumption, on $t \in [t_0 - r, t_0)$, $|\overline{x}(t) - z(t)| = 0$. We note that $\overline{x}(t)$ usually has a discontinuity at $t = t_0$, and therefore, $\overline{x}(t - r)$ has a discontinuity at $t = t_0 + r$. This requires us to be especially careful at $t = t_0$ and at $t = t_0 + r$. Let $\delta > 0$ be an arbitrarily small constant, and consider $|\overline{x}(t) - z(t)|$ on $t \in [t_0, t_0 + r + \delta]$, for which we can always write

$$\begin{aligned}|\overline{x}(t) - z(t)| = \Bigg|&\overline{x}(t_0) - z(t_0) \\ &+\epsilon \int_{t_0}^{t} \left[\frac{1}{T}\int_{s}^{s+T} f(\tau, x(\tau), x(\tau - r))d\tau - f_0(z(s), z(s - r))\right] ds\Bigg|.\end{aligned}$$

However, we know that on $t \in [t_0, t_0 + r + \delta]$ and $z \in D$

$$|f_0(z(s), z(s - r))| \leq \frac{1}{T}\int_0^T |f(\tau, z(s), z(s - r))|\, d\tau \leq \frac{1}{T}\int_0^T M d\tau = M.$$

Likewise, $|f(\tau, x(\tau), x(\tau - r))| \leq M$ on this interval since it has been assumed that $x \in D$. Therefore, for $t \in [t_0, t_0 + r + \delta]$

$$|\overline{x}(t) - z(t)| \leq |\overline{x}(t_0) - z(t_0)| + \epsilon \int_{t_0}^{t_0+r+\delta} \left(\frac{1}{T}\int_{s}^{s+T} M d\tau + M\right) ds.$$

From the assumption that $z(t_0) = x(t_0)$, Theorem 1 guarantees that $|z(t_0) - \overline{x}(t_0)| \leq \epsilon MT/2$. Therefore, for $t \in [t_0, t_0 + r + \delta]$, we have $|\overline{x}(t) - z(t)| \leq \epsilon M(T/2 + 2r + 2\delta)$. Next, assume $L/\epsilon \geq r + \delta$. Then for $t \in [t_0 + r + \delta, t_0 + L/\epsilon]$, we write

$$\begin{aligned}\overline{x}(t) - z(t) &= \overline{x}(t_0 + r + \delta) - z(t_0 + r + \delta) \\ &+\epsilon \int_{t_0+r+\delta}^{t} \left[\frac{1}{T}\int_{s}^{s+T} f\left(\tau, x(\tau), x(\tau - r)\right) d\tau - f_0\left(z(s), z(s - r)\right)\right] ds.\end{aligned}$$

Using (2.6), this leads to

$$\begin{aligned}|\overline{x}(t) - z(t)| &\leq |\overline{x}(t_0 + r + \delta) - z(t_0 + r + \delta)| \\ &+\epsilon \int_{t_0+r+\delta}^{t} |f_0\left(\overline{x}(s), \overline{x}(s - r)\right) - f_0(z(s), z(s - r))|\, ds \\ &+\epsilon \int_{t_0+r+\delta}^{t} \left|\frac{1}{T}\int_0^T f(s + \tau, x(s + \tau), x(s + \tau - r))d\tau - f_0(x(s), x(s - r))\right| ds \\ &+\epsilon \int_{t_0+r+\delta}^{t} |f_0(x(s), x(s - r)) - f_0(\overline{x}(s), \overline{x}(s - r))|\, ds. \qquad (2.7)\end{aligned}$$

From above, we know that $|\overline{x}(t_0+r+\delta) - z(t_0+r+\delta)| \le \epsilon M(T/2+2r+2\delta)$. Furthermore, since f is Lipschitz with constant K and since it has been assumed that x, $\overline{x}$ and z remain in D, this implies that $|f_0(\overline{x}(s),\overline{x}(s-r)) - f_0(z(s),z(s-r))| \le K\left(|\overline{x}(s)-z(s)| + |\overline{x}(s-r)-z(s-r)|\right)$ for $t \in [t_0, t_0+L/\epsilon]$. (For a proof of this statement, see the Appendix of [18]).

Similarly, $|f_0(x(s),x(s-r)) - f_0(\overline{x}(s),\overline{x}(s-r))| \le K\left(|x(s)-\overline{x}(s)| + |x(s-r)-\overline{x}(s-r)|\right)$. By the proof of Theorem 1, this implies $|f_0(x(s),x(s-r)) - f_0(\overline{x}(s),\overline{x}(s-r))| \le \epsilon KMT$ for all $s \in [t_0-r, t_0+L/\epsilon]$.

Likewise, for $s \in [t_0+r+\delta, t_0+L/\epsilon]$, $L/\epsilon \ge r+\delta$, and $\tau \in [0,T]$,

$$\left|\frac{1}{T}\int_0^T f(s+\tau, x(s+\tau), x(s+\tau-r))d\tau - f_0(x(s),x(s-r))\right|$$

$$= \frac{1}{T}\left|\int_0^T [f(s+\tau,x(s+\tau),x(s+\tau-r)) - f(s+\tau,x(s),x(s-r)]d\tau\right|$$

$$\le \frac{1}{T}\int_0^T K\left(|x(s+\tau)-x(s)| + |x(s+\tau-r)-x(s-r)|\right)d\tau.$$

For $t_0 \le t_1 \le t_2$, it is known that $x(t_2) - x(t_1) = \epsilon\int_{t_1}^{t_2} f(\lambda, x(\lambda), x(\lambda-r))d\lambda$. This implies for $t_0 \le t_1 \le t_2$

$$|x(t_2)-x(t_1)| \le \epsilon\int_{t_1}^{t_2} |f(s,x(s),x(s-r))|\,ds \le \epsilon M(t_2-t_1).$$

Therefore, for $s \in [t_0+r+\delta, t_0+L/\epsilon]$, $\tau \in [0,T]$, and any constant $d \in [0, r+\delta]$, we have $|x(s+\tau-d) - x(s-d)| \le \epsilon M\tau$.

Using the above inequalities, for $t \in [t_0+r+\delta, t_0+L/\epsilon]$ and $L/\epsilon \ge r+\delta$, (2.7) becomes

$$\begin{aligned}|\overline{x}(t)-z(t)| &\le \epsilon M\left(\frac{T}{2}+2r+2\delta\right)\\ &+\epsilon K\int_{t_0+r+\delta}^{t}\left(|\overline{x}(s)-z(s)| + |\overline{x}(s-r)-z(s-r)|\right)ds\\ &+\frac{2\epsilon^2}{T}\int_{t_0+r+\delta}^{t}\int_0^T KM\tau d\tau ds + \epsilon^2\int_{t_0+r+\delta}^{t} KMTds.\end{aligned}$$

Since each of the integrands is positive, for any $t \in [t_0, t_0+L/\epsilon]$ we can write

$$\begin{aligned}|\overline{x}(t)-z(t)| &\le \epsilon M\left(\frac{T}{2}+2r+2\delta\right)\\ &+\epsilon K\int_{t_0}^{t}\left(|\overline{x}(s)-z(s)| + |\overline{x}(s-r)-z(s-r)|\right)ds\\ &+\frac{2\epsilon^2}{T}\int_{t_0}^{t}\int_0^T KM\tau d\tau ds + \epsilon^2\int_{t_0}^{t} KMTds.\end{aligned}$$

(The discontinuity of $\overline{x}$ at $t = t_0{}^-$ is Lebesgue integrable.) Notice that the above inequality is increasing and that $\overline{x}(s-r) = z(s-r)$ for $s \in [t_0, t_0 - r]$. Therefore for $t \in [t_0, t_0 + L/\epsilon]$

$$\begin{aligned}\sup_{s\in[t_0,t]} |\overline{x}(s) - z(s)| &\leq \epsilon M\left(\frac{T}{2} + 2r + 2\delta\right) + 2\epsilon KMTL \\ &\quad + 2\epsilon K \int_{t_0}^{t} \sup_{\sigma\in[t_0,s]} |\overline{x}(\sigma) - z(\sigma)| ds.\end{aligned}$$

Since $\sup_{s\in[t_0,t]} |\overline{x}(s) - z(s)|$ is a continuous function, we can apply Gronwall's inequality for $t \in [t_0, t_0 + L/\epsilon]$ to obtain

$$\sup_{s\in[t_0,t]} |\overline{x}(s) - z(s)| \leq \left[\epsilon M\left(\frac{T}{2} + 2r + 2\delta\right) + 2\epsilon KMTL\right] e^{2\epsilon K(t-t_0)}.$$

The constant δ is arbitrarily small (e.g. select $\delta = \epsilon r$), and therefore, the above inequality implies that $|\overline{x}(t) - z(t)| = \mathcal{O}(\epsilon T) + \mathcal{O}(\epsilon r)$ on $t \in [t_0, t_0 + L/\epsilon]$. □

Combining the Theorems 2 and 3 leads to the following averaging theorem.

Theorem 4. *Let the assumptions of Theorems 2 and 3 be true. Then* $|x(t) - z(t)| = \mathcal{O}(\epsilon T) + \mathcal{O}(\epsilon r)$ *for all* $t \in [t_0 - r, t_0 + L/\epsilon]$.

Proof. Write $|x(t) - z(t)| \leq |x(t) - \overline{x}(t)| + |\overline{x}(t) - z(t)|$. The result now follows from Theorems 1 and 2. □

Remark 2. The fact that we considered only one delay is for pure convenience. The results can easily be extended to systems with multiple constant delays. Generalization of the results to non-periodic functional differential equations with time-varying delays is more complicated and is the subject of a present research effort by the authors.

3 Moving Averages of Discrete-Time Systems with Delays

In this section, we demonstrate how the techniques of taking moving averages can be used to develop averaging theory for discrete-time delay difference equations. Let discrete-time be denoted by n, and consider the periodic time-varying delay difference equation given by

$$x(n+1) = x(n) + \epsilon f\left(n, x(n), x(n-p)\right); \qquad x(n) = \phi(n), \;\; n \in [n_0 - p, n_0] \quad (3.1)$$

where $x \in \mathbb{R}^m$, $p \geq 0$ is the integer delay, ϵ is a non-negative constant and f is sufficiently continuous so that a solution to (3.1) exists. For simplicity, let $J[a, b]$ denote the set of all integers between a and b. Then, using this notation, $n \in J[a, b]$ denotes the set of integers satisfying $a \leq n \leq b$. In this chapter. it will always be assumed that f is N-periodic, i.e., there exists an integer $N > 0$ such that $f(n, c_1, c_2) = f(n+N, c_1, c_2)$ for any integer N and any $c_i \in \mathbb{R}^m$.

Once again, our goal will be to relate the solutions of (3.1) to the solutions of a corresponding averaged autonomous delay difference equation. In particular, let the averaged equation be described by

$$z(n+1) = z(n) + \epsilon f_0\left(z(n), z(n-p)\right); \qquad z(n) = \phi(n), \;\; n \in J[n_0 - p, n_0] \quad (3.2)$$

where $z \in \mathbb{R}^m$ and $f_0\left(z(n),\; z(n-p)\right) \equiv \frac{1}{N}\sum_{j=0}^{j=N-1} f\left(j+n, z(n), z(n-p)\right)$. The fact that we consider only one delay is purely arbitrary and for ease of presentation. That is, we could also consider $x(n+1) = x(n) + \epsilon f\left(n,\; x(n), x(n-p_1), \ldots,\; x(n-p_j)\right)$ and then relate solutions to $z(n+1) = z(n) + \epsilon f_0\left(z(n),\right.$ $\left.z(n-p_1), \ldots, z(n-p_j)\right)$.

In comparison with the results in the previous section for averaging of continuous-time delay *differential* equations, the results of this section are simpler. The discrete equations of motion given in (3.1) are finite dimensional. In fact, any delay difference equation can be rewritten as a higher order difference equation without delay. For example, in (3.1) let $x^0(n) = x(n)$, $x^1(n) = x(n-1)$, ... , $x^p(n) = x(n-p)$. Then (3.1) becomes

$$\begin{aligned} x^0(n+1) &= x^0(n) + \epsilon f\left(n, x^0(n), x^1(n), \ldots, x^p(n)\right) \\ x^i(n+1) &= x^i(n) + g\left(x^{i-1}(n), x^i(n)\right); \quad i = 1, 2, \ldots, p \end{aligned}$$

where $g\left(x^{i-1}(n), x^i(n)\right) = x^{i-1}(n) - x^i(n)$.

Now it is possible to attempt to use mixed time scale averaging results for difference equations with no delays to prove averaging theorems, such as those found in Chapter 8 of [27]. As a result, averaging of periodic delay difference equations might be derived from known averaging techniques of periodic difference equations with no delays. Additionally, for periodic difference equations, it is possible to perform 'lifting' and eliminate the time dependence altogether (see [8]).

The approach of this section is somewhat different than the above outlined approaches and is consistent with the approach used in the previous section for delay differential equations in continuous-time. No attempt will be made to increase the size of the vector space in order to eliminate the delay. Instead, we take the moving average of the solution of (3.1) and show that this moving average is 'close' to the solution of (3.1) when ϵ is sufficiently small. Next, the moving average is related to the solution of (3.2). Many of the necessary procedures to prove discrete time averaging results are, therefore, similar to those previously presented for continuous-time systems.

Definition 5. Suppose that $x(n) = x(n; n_0, \phi)$ is the solution to (3.1) with initial function ϕ. The **moving average** of $x(n)$ is denoted by $\overline{x}(n)$ and is defined as

$$\overline{x}(n) \equiv \begin{cases} \phi(n), & \text{for } n \in J[n_0 - p, n_0] \\ \frac{1}{N}\sum_{k=0}^{N-1} x(n+k), & \text{for } n > n_0, \end{cases}$$

where n_0 is an integer starting time and $N > 0$ is the period of f.

As before, this leads to the following theorem.

Theorem 6. *Assume that f in (3.1) is a continuous N-periodic function satisfying $|f(n, c_1, c_2)| \leq M$ for all $(n, c_1, c_2) \in (J[n_0 - p, n_0 + N + L_1] \times D \times D)$, where L_1 is a positive integer and $D \subset \mathbb{R}^m$. Assume further that $x \in D$ for all $n \in J[n_0 - p, n_0 + N + L_1]$. Then $|x(n) - \overline{x}(n)| = \mathcal{O}(\epsilon(N-1))$ for all $n \in J[n_0 - p, n_0 + L_1]$.*

Proof. For $n \in J[n_0 - p, n_0]$, we have that $x(n) - \overline{x}(n) = 0$. By definition, for $n \geq n_0 + 1$

$$\begin{aligned} |x(n) - \overline{x}(n)| &= \left| x(n) - \frac{1}{N}\sum_{k=0}^{N-1} x(n+k) \right| \\ &= \frac{1}{N}\sum_{k=1}^{N-1} |x(n) - x(n+k)| . \end{aligned}$$

However, we know, for $k \geq 1$ and $n > n_0$, that $x(n) - x(n+k) = -\epsilon \sum_{i=n}^{k+n} f(i, x(i), x(i-p))$. Therefore, $|x(n) - \overline{x}(n)| \leq \frac{\epsilon}{N}\sum_{k=1}^{N-1}\sum_{i=n}^{k+n} |f(i, x(i), x(i-p))|$. Since it has been assumed that $x \in D$, this leads to $|x(n) - \overline{x}(n)| \leq \frac{\epsilon}{N}\sum_{k=1}^{N-1}\sum_{i=n}^{k+n} M = \frac{\epsilon}{N}\sum_{k=1}^{N-1} Mk = \frac{\epsilon M(N-1)}{2}$. □

The next step will be to show that $|\overline{x}(n) - z(n)|$ remain close to each other for sufficiently small ϵ. First we note that for $n \geq n_0 + 1$

$$\overline{x}(n+1) - \overline{x}(n) = \frac{1}{N}\sum_{k=0}^{N-1}[x(n+k+1) - x(n+k)] = \frac{1}{N}[x(n+N) - x(n)].$$

Since $x(n+N) = x(n) + \epsilon \sum_{j=0}^{N-1} f(j+n, x(j+n), x(j+n-p))$, the above equation becomes

$$\overline{x}(n+1) = \overline{x}(n) + \epsilon \sum_{j=0}^{N-1} f(j+n, x(j+n), x(j+n-p)) . \tag{3.3}$$

Qualitatively, the idea of averaging delay difference equations becomes more clear by analyzing (3.3). From the previous theorem, it is known that $\overline{x}(n) \approx x(n)$ for sufficiently small ϵ. Therefore, this implies in (3.3) that $\overline{x}(n+1) \approx \overline{x}(n) + \epsilon \sum_{j=0}^{N-1} f(j+n, \overline{x}(j+n), \overline{x}(j+n-p))$. Additionally, when ϵ is sufficiently small, $\overline{x}(n+j)$ changes little over $j \in J[0, N-1]$ and therefore, $\overline{x}(n+j) \approx \overline{x}(n)$. This leads to the approximation of (3.3) by the delay difference equation $\overline{x}(n+1) \approx \overline{x}(n) + \frac{\epsilon}{N}\sum_{j=0}^{N-1} f(j+n, \overline{x}(n), \overline{x}(n-p))$, which is precisely the same equation as (3.2). In summary, these arguments show that for sufficiently small ϵ, the averaged equation (3.2) is a small perturbation of (3.3). Therefore, it is now possible for us to relate solutions of (3.1) to (3.2).

To help formalize these ideas, note that by (3.2) and (3.3)

$$\begin{aligned} z(n) &= z(n_0) + \epsilon \sum_{s=n_0}^{n-1} f_0\,(z(s), z(s-p)) \\ \overline{x}(n) &= \overline{x}(n_0) + \frac{\epsilon}{N} \sum_{s=n_0}^{n-1} \sum_{j=0}^{N-1} f\,(j+s, x(j+s), x(j+s-p)) \end{aligned}$$

for all $n \geq n_0$.

Theorem 7. *Let the assumptions of Theorem 6 be true for $L_1 = L/\epsilon$ and assume that $|f(n, c_1, c_2) - f(n, \tilde{c_1}, \tilde{c_2})| \leq K\,(|c_1 - \tilde{c_1}| + |c_2 - \tilde{c_2}|)$ for all $(n, c_1, c_2) \in (J[n_0 - p, n_0 + L/\epsilon] \times D \times D)$. Suppose, also, that both $\overline{x}(n)$ and $z(n)$ remain in D for all $n \in J[n_0 - p, n_0 + L/\epsilon]$, with $\overline{x}(n) = x(n) = z(n) = \phi(n)$ on $n \in J[n_0 - p, n_0]$. Then $|\overline{x}(n) - z(n)| = \mathcal{O}\,(\epsilon(N-1)) + \mathcal{O}\,(\epsilon p)$.*

Proof. For $n \in J[n_0 - p, n_0]$, it has been assumed that $\overline{x}(n) = z(n)$. For $n \geq n_0 + 1$

$$\begin{aligned} \overline{x}(n) - z(n) &= \overline{x}(n_0) - z(n_0) \\ &\quad + \epsilon \sum_{s=n_0}^{n-1} \frac{1}{N} \sum_{j=0}^{N-1} [f\,(j+s, x(j+s), x(j+s-p)) \\ &\quad - f_0\,(z(s), z(s-p))]. \end{aligned} \tag{3.4}$$

As in the continuous time case, we will break the proof into two separate time intervals. First, consider $n \in J[n_0 + 1, n_0 + p + 1]$. Since $\overline{x}(n_0) = z(n_0)$, we have

$$\begin{aligned} |\overline{x}(n) - z(n)| &\leq \epsilon \sum_{s=n_0}^{n-1} \frac{1}{N} \sum_{j=0}^{N-1} [|f\,(j+s, x(j+s), x(j+s-p))| \\ &\quad + |f_0\,(z(s), z(s-p))|] \\ &\leq \frac{\epsilon}{N} \sum_{s=n_0}^{n_0+p} \frac{1}{N} \sum_{j=0}^{N-1} 2M = 2\epsilon M p, \end{aligned}$$

where we used the fact that f_0 will be bounded by M when f is bounded by M (as we showed in the beginning of the proof of Theorem 3 in continuous-time).

Now consider the interval $n \in J[n_0 + p + 2, n_0 + L/\epsilon]$, $L/\epsilon > p + 2$. We rewrite (3.4) on this interval to become

$$\begin{aligned} &\overline{x}(n) - z(n) = \overline{x}(n_0 + p + 1) - z(n_0 + p + 1) \\ &+ \frac{\epsilon}{N} \sum_{s=n_0+p+1}^{n-1} \sum_{j=0}^{N-1} [f\,(j+s, x(j+s), x(j+s-p)) - f_0\,(z(s), z(s-p))] \end{aligned}$$

which implies that on $n \in J[n_0+p+2, n_0+L/\epsilon]$

$$
\begin{aligned}
|\overline{x}(n) - z(n)| \le & \; |\overline{x}(n_0+p+1) - z(n_0+p+1)| \\
& + \epsilon \sum_{s=n_0+p+1}^{n-1} \left[\frac{1}{N} \left| \sum_{j=0}^{N-1} [f(j+s, \overline{x}(s), \overline{x}(s-p)) - f_0(z(s), z(s-p))] \right| \right. \\
& + \left| \frac{1}{N} \sum_{j=0}^{N-1} [f(j+s, x(s), x(s-p)) - f(j+s, \overline{x}(s), \overline{x}(s-p))] \right| \\
& + \left| \frac{1}{N} \sum_{j=0}^{N-1} [f(j+s, x(j+s), x(j+s-p)) \right. \\
& \left. \left. - f(j+s, x(s), x(s-p))] \right| \right]. \qquad (3.5)
\end{aligned}
$$

From above, we know that $|\overline{x}(n_0+p+1) - z(n_0+p+1)| \le 2\epsilon Mp$. Analyzing the first inner summation yields

$$
\begin{aligned}
& \left| \frac{1}{N} \sum_{j=0}^{N-1} [f(j+s, \overline{x}(s), \overline{x}(s-p)) - f_0(z(s), z(s-p))] \right| \\
& \qquad = |f_0(\overline{x}(s), \overline{x}(s-p)) - f_0(z(s), z(s-p))| \\
& \qquad \le K(|\overline{x}(s) - z(s)| + | \quad s-p) - z(s-p)|)
\end{aligned}
$$

where we have used the fact, without proof, that f_0 is Lipschitz with constant K when f is Lipschitz with constant K (see Appendix of [18] for the continuous time proof). By the proof of Theorem 6 and the assumption that x and $\overline{x}$ remain in D, the second inner summation in (3.5) yields

$$
\begin{aligned}
& \left| \frac{1}{N} \sum_{j=0}^{N-1} [f(j+s, x(s), x(s-p)) - f(j+s, \overline{x}(s), \overline{x}(s-p))] \right| \\
& \qquad \le \frac{K}{N} \sum_{j=0}^{N-1} (|x(s) - \overline{x}(s)| + |x(s-p) - \overline{x}(s-p)|) \\
& \qquad \le \frac{K}{N} \sum_{j=0}^{N-1} \epsilon M(N-1) = \epsilon KM(N-1).
\end{aligned}
$$

Finally, the last inner summation in (3.5)

$$
\begin{aligned}
& \left| \frac{1}{N} \sum_{j=0}^{N-1} [f(j+s, x(j+s), x(j+s-p)) - f(j+s, x(s), x(s-p))] \right| \\
& \qquad \le \frac{K}{N} \sum_{j=0}^{N-1} (|x(j+s) - x(s)| + |x(j+s-p) - x(s-p)|)
\end{aligned}
$$

$$+\frac{K}{N}\sum_{j=1}^{N-1}\left(|x(j+s)-x(s)|+|x(j+s-p)-x(s-p)|\right)$$

$$+\frac{K}{N}\sum_{j=1}^{N-1}\left|\epsilon\sum_{i=s}^{j+s} f\left(i,x(i),x(i-p)\right)\right|+\left|\epsilon\sum_{i=s-1}^{j+s-1} f\left(i,x(i),x(i-p)\right)\right|$$

$$\leq\frac{\epsilon K}{N}\sum_{j=1}^{N-1}2Mj=2\epsilon KM(N-1)$$

for $s\in J[n_0+p+2,n_0+L/\epsilon]$, $L/\epsilon>p+2$. Combining the above bounds for $n\in J[n_0+p+2,n_0+L/\epsilon]$ and $L/\epsilon>p+2$, (3.5) becomes

$$\begin{aligned}|\overline{x}(n)-z(n)| &\leq 2\epsilon Mp+\epsilon\sum_{s=n_0+p+1}^{n-1}\left[3\epsilon KM(N-1)+K\left(|\overline{x}(s)-z(s)|\right.\right.\\ &\quad\left.\left.+|\overline{x}(s-p)-z(s-p)|\right)\right]. \qquad (3.6)\end{aligned}$$

Since the right hand side of (3.6) increases with n and since $\overline{x}(n)=z(n)$ on $n\in J[n_0-p,n_0]$, we can rewrite (3.6) as

$$|\overline{x}(n)-z(n)|\leq 3\epsilon^2KM(N-1)(n-n_0)+\epsilon\sum_{s=n_0}^{n-1}2K\,|\overline{x}(s)-z(s)|$$

which is valid for all $n\in J[n_0,n_0+L/\epsilon]$. This implies that for all $n\in J[n_0,n_0+L/\epsilon]$

$$|\overline{x}(n)-z(n)|\leq 2\epsilon Mp+3\epsilon KM(N-1)L+\epsilon\sum_{s=n_0}^{n-1}2K\,|\overline{x}(s)-z(s)|\,.$$

By Gronwall's inequality for discrete systems (see Appendix C2 of [27]), this implies

$$\begin{aligned}|\overline{x}(n)-z(n)| &\leq [2\epsilon Mp+3\epsilon KM(N-1)L]\,(1+2\epsilon K)^{n-1-n_0}\\ &\leq [2\epsilon Mp+3\epsilon KM(N-1)L]\,e^{2\epsilon K(n-n_0)}\\ &\leq [2\epsilon Mp+3\epsilon KM(N-1)L]\,e^{2KL}\end{aligned}$$

for all $n\in J[n_0,n_0+L/\epsilon]$. □

Combining the previous Theorems leads to the following averaging theorem for delay difference equations.

Theorem 8. *Let the assumptions of Theorems 6 and 7 be true. Then* $|x(n)-z(n)|=\mathcal{O}\left(\epsilon(N-1)\right)+\mathcal{O}\left(\epsilon p\right)$ *for all* $n\in J[n_0-p,n_0+L/\epsilon]$.

Proof. Write $|x(n)-z(n)|\leq|x(n)-\overline{x}(n)|+|\overline{x}(n)-z(n)|$. By Theorems 6 and 7, this implies that $|x(n)-z(n)|=\mathcal{O}\left(\epsilon(N-1)\right)+\mathcal{O}\left(\epsilon p\right)$. □

4 Applications of Averaging to Delay Systems in Controls Engineering

4.1 Cart and Pendulum Control in the Presence of External Vibrations and Feedback Delays

We now present a simple application to a variation of cart and pendulum stabilization by proportional feedback. As illustrated in Figure 1, the system consists of a cart and planar pendulum apparatus in a reference frame which is being

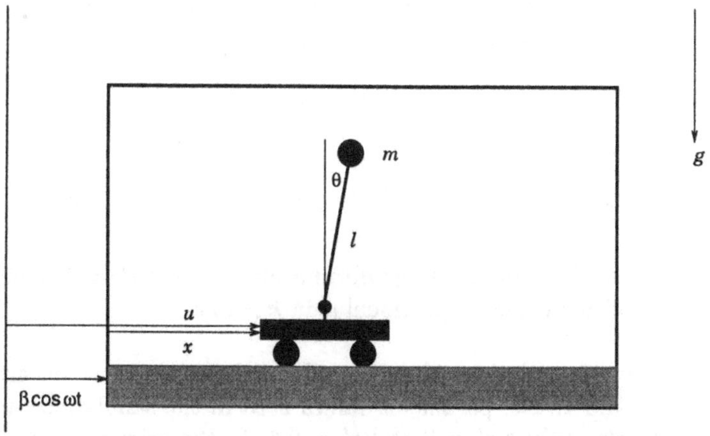

Fig. 1. Cart and Pendulum subjected to periodic disturbances.

subjected to a periodic disturbance of amplitude β and frequency ω along the horizontal axis. This disturbance might be due to an unsteady platform or some external periodic noise signal.

The net motion of the cart is equal to the sum of the disturbance and the cart's motion in the local frame of reference. The pendulum is modelled as a rigid, massless link of length ℓ and a bob of mass m, and its displacement is referenced to the inverted equilibrium. Suppose that the cart position may be precisely controlled. As discussed in [20], this assumption is fairly common. Then the differential equation of motion for the pendulum is

$$m\ell^2\ddot{\theta} + c_d\dot{\theta} - m\ell\left(\ddot{u}\cos\theta + g\sin\theta\right) = 0, \tag{4.1}$$

where $\ddot{u} = -\omega^2\beta\cos\omega t + \ddot{x}$, $\ddot{x}$ is the acceleration of the cart, and c_d is the damping coefficient for the hinge.

To stabilize the inverted equilibrium, we prescribe the proportional control $\ddot{x} = -K_p\theta(t - r)$, where $r > 0$ is a control delay perhaps due to sampling, computation, or even tele-remote operation. Then $\ddot{u}(t) = -\omega^2\beta\cos\omega t - K_p\theta(t - r)$. (4.1) can be written as a dimensionless delay differential equation

$$\theta'' + \epsilon^2 c\theta' - \epsilon\cos\tau + \epsilon^2\left(k\theta(\tau - r')\cos\theta - \sin\theta\right) = 0,$$

where $\tau = \omega t$, $(\cdot)' = d/d\tau$, $\epsilon = \beta/\ell = \sqrt{g/\ell}/\omega$, $c = \ell c_d/g$, $k = \ell K_p/g$, and $r' = \omega r$. Prescribing the coordinate change $\theta = y_1 - \epsilon\cos\tau\cos y_1$, $\theta' = \epsilon y_2 + \epsilon\sin\tau\cos y_1$ and proceeding as in [7], we eventually have the system of first order equations

$$\begin{aligned} y_1' &= \epsilon y_2 + \mathcal{O}(\epsilon^2) \\ y_2' &= \epsilon\left[-cy_2 + \sin y_1\cos y_1\cos^2\tau + y_2\sin y_1\sin\tau\right. \\ &\quad \left.-ky_1(\tau - r')\cos y_1 + \sin y_1\right] + \mathcal{O}(\epsilon^2). \end{aligned} \tag{4.2}$$

By Theorem 2, the average of (4.2) is given as

$$\begin{aligned} z_1' &= \epsilon z_2 \\ z_2' &= \epsilon\left[-cz_2 + \frac{\sin z_1\cos z_1}{2} - kz_1(\tau - r')\cos z_1 + \sin z_1\right]. \end{aligned} \tag{4.3}$$

Linear analysis of the inverted equilibrium for the case $r' = 0$ shows that the system is stabilized if the proportional gain $k > 3/2$.

The results of simulations of the periodic and averaged systems are shown in Figures 2, 3, and 4. The parameter values used in the simulations are $\epsilon = 0.1$, $k = 3$, and delay values $r' = 0$ and $r' = 0.5$. In the original time scale, these delay values scale back to $r = r'/\omega = \epsilon r'\sqrt{g/\ell}$. Initial condition and function data is given by $(y_1(\tau), y_2(\tau)) = (z_1(\tau), z_2(\tau)) = (0.5, 0)$ for $\tau \in [-r', 0]$.

In Figures 2 and 3, we see phase portraits for the averaged and periodic systems both without (Fig. 2) and with (Fig. 3) delays. In both figures, the averaged phase portrait approximates the periodic system's trajectory. The significance of this result is that it shows that the appropriate averaged equations retain the delay term, as opposed to earlier results which suggest that the delay term is not important. In Figure 4, we see a comparison of the averaged trajectories for the system without and with delays. It is clear from the figure that these trajectories are distinct, and that trajectory with delay is not a small perturbation of the trajectory without delay. This is true in spite of the fact that the delay is $\mathcal{O}(\epsilon)$ in the original time scale.

4.2 Adaptive Identification of Pipe Mixing

To demonstrate the applicability of the discrete time averaging results presented in this chapter, we will now discuss an application of adaptive identification in process control. We first note that the identification algorithm that we propose is, perhaps, slower and less robust than other known techniques. However, the algorithm is based on classical Least Mean Squares adaptive identification and

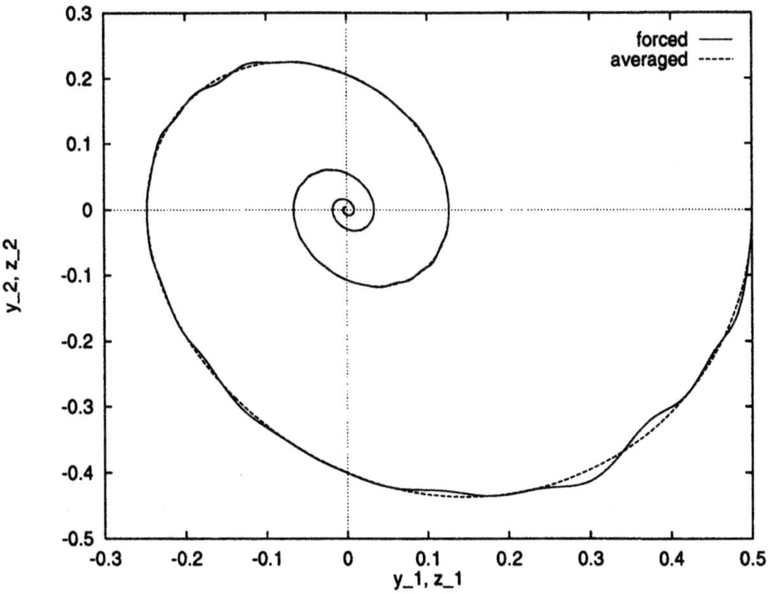

Fig. 2. Periodic system (solid) and averaged system (dashed) trajectories for the controlled cart and pendulum with no feedback delay.

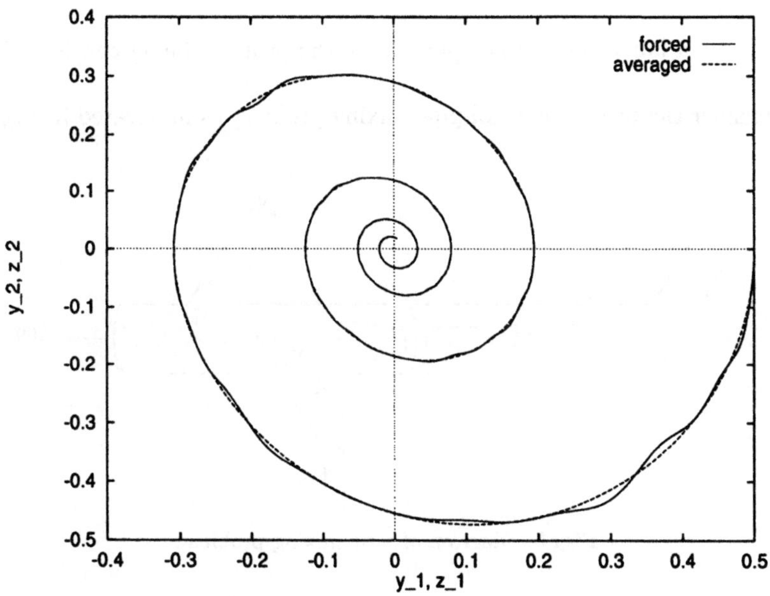

Fig. 3. Periodic system (solid) and averaged system (dashed) trajectories or the controlled cart and pendulum with feedback delay.

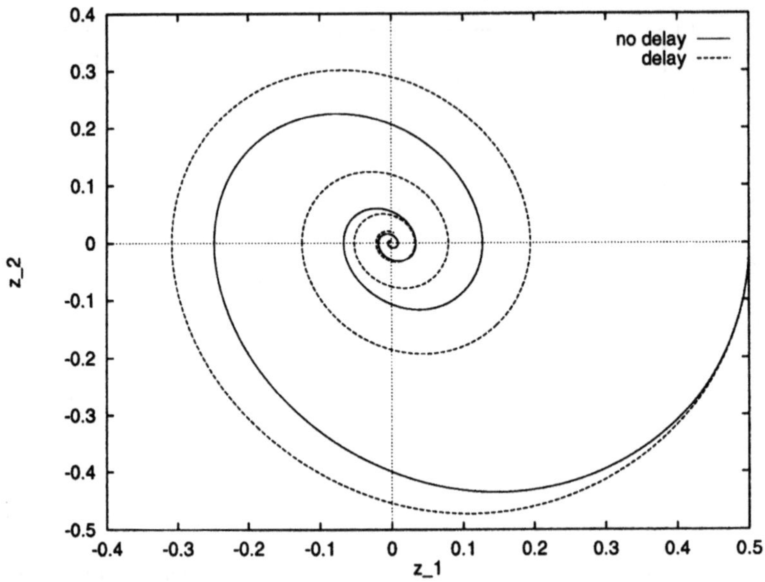

Fig. 4. Averaged trajectories with and without feedback delay. Note from this figure and Figures 2 and 3 that the method of averaging presented in this chapter accurately approximates the periodic system's dynamics in the presence of a significant feedback delay.

seems to be a fairly reasonable approach to the problem being considered.

Consider the process control pipe mixing problem as illustrated in Figure 5.

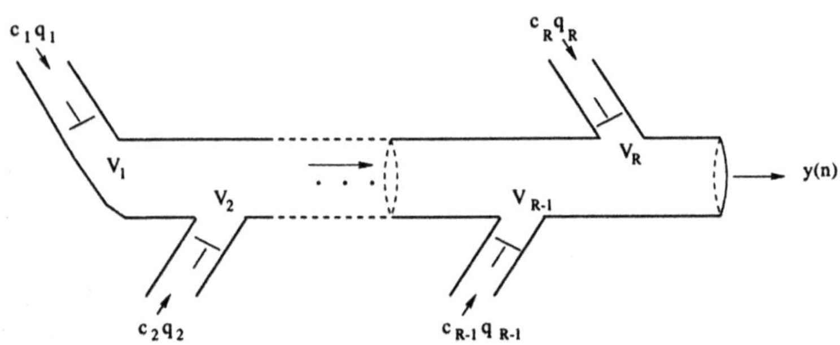

Fig. 5. Diagram of pipe mixing problem.

A large pipe is being fed liquid chemical by R number of smaller pipes. At the exit of each of the feeding smaller pipes there is a valve to regulate the flow. Denote valve i by V_i and the corresponding flow rate of liquid through it at time

n, measured in *meters*^{3}*/second*, by $q_i(n)$. The values of $q_i(n)$ are measureable using the flow sensors in each valve and are also regulatable by V_i. Each fresh feed input pipe carries the identical type of chemical, but with different concentration c_i, measured in *moles/meters*^{3}, into the main feed pipe. The product $c_i q_i(n)$ represents the number of moles flowing out of V_i each second.

Our goal is to develop a recursive identification algorithm that can identify the concentrations, c_i, by measuring the concentration of the traveling liquid somewhere after V_R. Since the feed pipes are each separated by finite distances, there will be transportation lags in the system. Suppose that we are able to measure the transportation lag from V_i to the location where we are reading the chemical concentration, and denote the lag time by p_i. Furthermore, assume that each p_i is a constant value (this is, perhaps, an unrealistic assumption).

The output $y(n)$ represents the number of moles per second at the measurement location, i.e., $y(n) = \sum_{i=1}^{R} c_i q_i(n - p_i)$. In order to estimate the concentrations c_i, consider an adaptive estimate of the output, given by $s(n) = \sum_{i=1}^{R} \lambda_i(n - p_i) q_i(n - p_i)$. Here, $\lambda_i(n)$ represent the concentration estimate at V_i and should be designed to converge to c_i. The error in our estimated output is given by $e(n) = y(n) - s(n)$, or, more precisely,

$$e(n) = \sum_{i=1}^{R} [c_i - \lambda_i(n - p_i)]\, q_i(n - p_i).$$

If we let $e_i(n)$ denote the estimated error V_i, then we can symbolically model each $e_i(n)$ as in Figure 6, implying that $e(n) = \sum_{i=1}^{R} e_i(n)$.

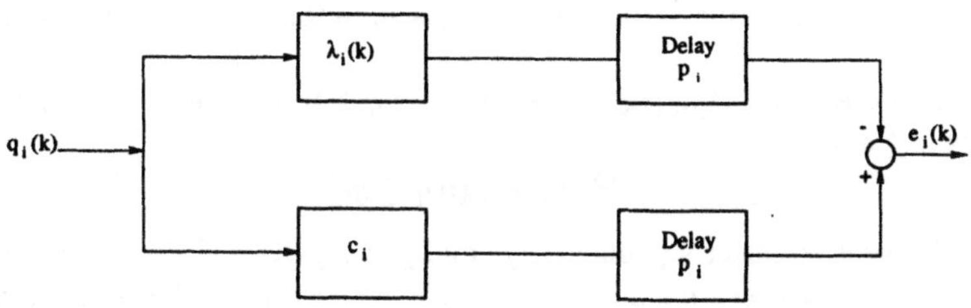

Fig. 6. Block diagram describing the adaptive identification algorithm for the pipe mixing problem.

Define the functions $\theta_i(n) = c_i - \lambda_i(n)$, from which we may define the vector $\theta(n) = \{\theta_1(n), \theta_2(n), \ldots, \theta_R(n)\}$. Then each estimated valve error can be written as $e_i(n) = \theta_i(n - p_i) q_i(n - p_i)$. Now, suppose that we update $\theta(n)$ according to the least means squares algorithm (see [27] for an overview)

$$\theta(n+1) = \theta(n) - \frac{\epsilon}{2}\sum_{i=1}^{R} \frac{\partial e^2(n)}{\partial \theta_i(n-p_i)},$$

which leads to

$$\begin{aligned}\theta_i(n+1) &= \theta_i(n) - \epsilon e(n) q_i(n-p_i) \\ &= \theta_i(n) - \epsilon \sum_{j=1}^{R} q_i(n-p_i) q_j(n-p_j)\theta_j(n-p_j), \end{aligned} \tag{4.4}$$

where $i = 1, 2, \ldots, R$. Successive estimates of λ_i are obtained from the definition of $\theta_i(n)$ and (4.4): we have

$$\begin{aligned}\lambda_i(n+1) &= \lambda_i(n) + \epsilon e(n) q_i(n-p_i) \\ &= \lambda_i(n) + \epsilon \sum_{j=1}^{R} q_i(n-p_i) q_j(n-p_j)\theta_j(n-p_j). \end{aligned} \tag{4.5}$$

Since we are able to regulate each $q_i(n)$ by adjusting V_i, let us implement N-periodic flow rates in each of the valves, implying that $q_i(n+N) = q_i(n)$. Then (4.4) is exactly in the form that averaging can take place. The corresponding average of (4.4) is given by

$$z_i(n+1) = z_i(n) - \epsilon \sum_{j=1}^{R} \overline{q_i(n-p_i) q_j(n-p_j)} z_i(k-p_i) \tag{4.6}$$

for $i = 1, 2, \ldots, R$, and where $\overline{q_i(n-p_i) q_j(n-p_j)}$ represents the (constant) moving average of $q_i(n-p_i)q_j(n-p_j)$, as previously defined. (It is constant because $q_i(n-p_i)q_j(n-p_j)$ will also be periodic.) Assume that $\theta_i(n) = z_i(n)$ on $n \le n_0$.

Hence, for any $n \in J[n_0, n_0 + L/\epsilon]$ and any $\eta > 0$, there exists an $\epsilon_0 > 0$ such that for $0 \le \epsilon \le \epsilon_0$

$$|\theta_i(n) - z_i(n)| \le \eta.$$

The goal now is to select ϵ (sufficiently small) and $q_i(n)$ such that $z_i(n)$ tends to sufficiently close to zero for all $n = \tilde{n} < L/\epsilon$. Then this implies that $\theta_i(\tilde{n}) \approx 0$, i.e, $\lambda_i(\tilde{n}) \approx c_i$. As a result, we would have adaptively identified concentrations c_i, $i = 1, 2, ..R$. We remark that since we can arbitrarily choose q_i, to simplify the algorithm we choose q_i such that $\overline{q_i(n)q_j(n)} = 0$ for $i \neq j$. For example, choosing $q_i(n) = \sin\frac{n\pi}{i}$ will satisfy this criteria. In this case, (4.6) becomes

$$z_i(n+1) = z_i(n) - \epsilon\overline{q_i^2(n-p_i)} z_i(n-p_i),$$

which has a stable trivial solution independent of the delay p_i.

For the purpose of illustration, let $R = 1$ and suppose that $q_1(n) = Q + \alpha sin\frac{n\pi}{2}$, where $0 < \alpha < Q$ are constants. In this case, (4.6) becomes

$$z_1(n+1) = z_1(n) - \epsilon\left[Q^2 + \frac{\alpha^2}{2}\right] z_1(n-p_1),$$

and successive estimates of λ_1 are likewise obtained from (4.5) as

$$\overline{\lambda_i}(n+1) = \overline{\lambda_i}(n) + \epsilon\left[Q^2 + \frac{\alpha^2}{2}\right] z_1(n-p_1),$$

where the overline indicates that we have obtained the estimate from the average z_1.

The algorithms using θ_1 and z_1 have been simulated, with the results given in Figures 7 and 8. In these simulations, we have chosen $c_1 = 1$ and $\lambda_1(0) = 0$, implying that $\theta_1(n) = z_1(n) = 1$ for $n \in J[-p_1, 0]$. The parameter values used were $Q = 1$, $\alpha = 0.3$, $\epsilon = 0.01$, and a delay $p_1 = 3$. The plots in Figure 7 show that the average (z_1) algorithm approximates the periodic (θ_1) algorithm so well that the plots of θ_1 and z_1 vs. n and λ_1 and $\overline{\lambda_1}$ vs. n cannot be distinguished. In evaluating $\lambda_1 - \overline{\lambda_1}$, we see in Figure 8 that there is indeed some error, but it is small and tends to zero as $n \to \infty$. This actually results from both θ_1 and z_1 tending to zero as $n \to \infty$, as shown in the top plot of Figure 7. In addition, from the lower plot of Figure 7, we note that the adaptive identification algorithm we derived correctly identified c_1.

5 Conclusion

In this chapter, we have extended the earlier results of [11, 12, 13, 23] for the averaging of periodic delay differential equations and delay difference equations. Specifically, we have proven for both classes of systems that the delay is not negligible and must be retained for the averaged system to accurately reflect the dynamics of the periodic system. This fundamental result forms a basis for the continued research in the development and application of averaging methods for delay systems in applied mathematics and control theory.

We considered two simple applications in this chapter: we studied the closed-loop feedback control of the cart and pendulum in the presence of high-frequency external oscillations and a feedback delay, and the adaptive identification of chemical concentrations in a discrete-time pipe mixing problem. In the first application, we saw that the averaging method we proposed produced an averaged system which accurately approximated the periodic system. Although the delay, in the original time scale, was small, the phase portraits for the system with delay and the system without delay were distinct. This observation reaffirms that the delay must not be neglected in the averaged system. In the second application, we likewise saw that the averaged adaptive identification algorithm accurately

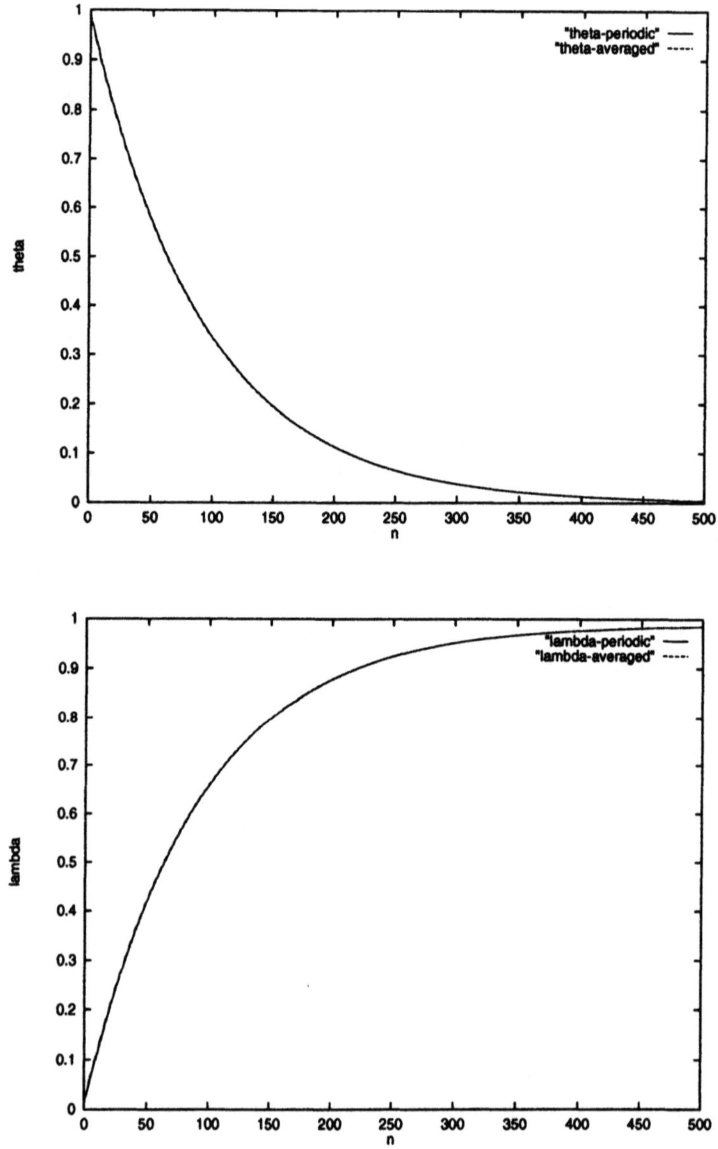

Fig. 7. Plots of θ_1, z_1 vs. n (top), and $\lambda_1, \overline{\lambda_1}$ vs. n (bottom). From the figures, the averaged system approximates the periodic system to high accuracy. In addition, the adaptive identification algorithm correctly identifies the chemical concentration $c_1 = 1$.

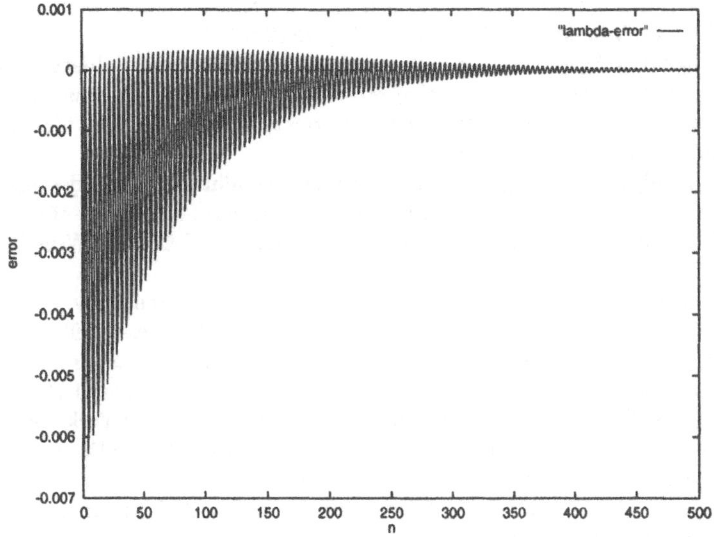

Fig. 8. Difference of λ_1 and $\overline{\lambda_1}$ vs. n from Figure 4.7. Note that the error tends to zero as $n \to \infty$.

captured the average behavior of the periodic algorithm.

The two simple examples presented in Section 4 suggest possible extensions to the main theoretical results of this chapter. First, future research will attempt to extend the averaging theorems presented in this chapter to infinite time intervals. Such theorems will allow us to understand the asymptotic behavior of solutions to nonlinear periodic delay systems which limit on periodic points or periodic orbits. This knowledge is necessary to understand how hyperbolic invariant structures organize the dynamics of the periodically excited system. Another interesting extension is to systems with almost-periodic excitation. This is, of course, of importance in systems where the excitation is composed of at least two irrationally related frequencies. In terms of practical applications, such a result will be useful in the averaging of systems where one of the excitatory inputs is noise. Finally, we hope to extend the results presented here to systems with time-varying delays. These new results and additional examples shall be presented in forthcoming journal publications.

References

1. Andronov, A.A., A.A. Vitt, and S.E. Khaikin. *Theory of Oscillators*, volume 4 of *The International Series of Monographs in Physics*. Pergamon Press, Oxford, 1966.
2. Arnol'd, V.I. *Geometrical Methods in the Theory of Ordinary Differential Equations, 2nd Ed.*, volume 250 of *A Series of Comprehensive Studies in Mathematics*. Springer-Verlag, Berlin, 1988.

3. Arnol'd, V.I. *Mathematical Methods of Classical Mechanics*, volume 60 of *Graduate Texts in Mathematics*. Springer-Verlag, Berlin, 1989.
4. Baillieul, J. Stable average motions of mechanical systems subject to periodic forcing. In *Dynamics and Control of Mechanical Systems: The Falling Cat and Related Problems: Fields Institute Communications*, pages 1–23, Providence, R.I., 1993. American Mathematical Society.
5. Baillieul, J., S. Dahlgren, and B. Lehman. Nonlinear control designs for systems with bifurcations and applications to stabilization and control of compressors. In *Proceedings of the 1995 Conference on Decision and Control, New Orleans*, pages 3062–3067, 1995.
6. Baillieul, J. and B. Lehman. Open-loop control using oscillatory inputs. To appear in *The Control Handbook*, W.S. Levine, Editor. CRC Press, 1995.
7. Bogoliubov, N.N, and Y.A. Mitropolsky. *Asymptotic Methods in the Theory of Nonlinear Oscillators*. International Monographs on Advanced Mathematics and Physics. Gordon and Breach Science Publishers, Inc., New York, 1961.
8. Davis, J.H. Stability conditions derived from spectral theory: Discrete systems with periodic feedback, *SIAM J. Control*, 10:1–13, Feb. 1972.
9. Fodcuk, V.I. The method of averaging for differential equiations of the neutral type. *Ukraine Mat. Z.*, 20:203–209, 1968.
10. Guckenheimer, J. and P. Holmes. *Nonlinear Oscillations, Dynamical Systems, and Bifurcations of Vector Fields*, volume 42 of *Applied Mathematical Sciences*. Springer-Verlag, Berlin, 1983.
11. Halanay, A. The method of averaging in equations with retardation. *Rev. Mat. Pur. Appl. Acad. R.P.R.*, 4:467–483, 1959.
12. Halanay, A. On the method of averaging for differential equations with retarded arguments. *Journal of Mathematical Analysis and Applications*, 14:70–76, 1966.
13. Hale, J.K. Averaging methods for differential equations with retarded arguments with a small parameter. *Journal of Differential Equations*, 2:57–73, 1966.
14. Hale, J.K. *Ordinary Differential Equations*. Texts and Monographs in Pure and Applied Mathematics. Robert E. Krieger Publishing, Malabar, FL, 1969.
15. Krylov, N.M. and N.N. Bogoliubov. *New Methods of Nonlinear Mechanics in their Application to the Investigation of the Operation of Electronic Generators, I.* United Scientific and Technical Press, Moscow, 1934.
16. Krylov, N.M. and N.N. Bogoliubov. *Introduction to Nonlinear Mechanics*. Princeton University Press, Princeton, 1937.
17. Lehman, B., S.V. Lunel, J. Bentsman, and E.I. Verriest. Vibrational control of nonlinear time lag systems with bounded delay: Averaging theory, stabilizability, and transient behavior. *IEEE Transactions on Automatic Control*, AC-39:898–912, 1994.
18. Lehman, B. and V.B. Kolmanovskii. Extensions of classical averaging techniques to delay differential equations. In *Proceedings of the 33rd IEEE Conference on Decision and Control*, pages 411–416, 1994.
19. Lehman, B., I. Widjaya, and K. Shujaee. Vibrational control of chemical reactions in a cstr with delayed recycle. *Journal of Mathematical Analysis and Applications*, 193:28–59, 1995.
20. Lehman, B. and S. Weibel. Fundamental theorems of averaging for functional differential equations. To appear in *Proceedings of the 1997 American Control Conference*, 1997.
21. Levi, M. and W. Weckesser. Stabilization of the inverted pendulum by high frequency vibrations. *SIAM Review*, 27(2):219–223, June 1995.

22. Magnus, W. and S. Winkler. *Hill's Equation*, volume 20 of *Tracts in Mathematics*. Interscience Publishers, New York, 1966.
23. Medvedev, G.N. Asymptotic solutions of some systems of differential equations with deviating argument. *Soviet Math Dokl.*, 9:85–87, 1968.
24. Sanders, J.A. On the fundamental theory of averaging. *SIAM J. Math. Anal.*, 14:1–10, 1983.
25. Sanders, J.A. and F. Verhulst. *Averaging Methods in Nonlinear Dynamical Systems*, volume 59 of *Applied Mathematical Sciences*. Springer-Verlag, Berlin, 1985.
26. Seto, D. and J. Baillieul. Control problems in superarticulated mechanical systems. *IEEE Transactions on Automatic Control*, 39(12), 1994.
27. Solo, V. and X. Kong. *Adaptive Signal Processing - Algorithms, Stability, and Performance*. In *Prentice Hall Information and System Sciences Series*, Thomas Kailath, series editor. Prentice Hall, Englewood Cliffs, NJ, 1995.
28. Stephenson, A. On induced stability. *Phil. Mag.*, 17:765–766, 1909.
29. Stoker, J.J. *Nonlinear Vibrations in Mechanical and Electrical Systems*. Interscience Publishers, New York, 1950.
30. Volosov, V.M., G.N. Medvedev and B.I. Morgunov. Mr 32 #7904. *Vestnik Moskov: Univ. Ser. III. Fiz. Astronom.*, 6:89, 1965.
31. Weibel, S., J. Baillieul, and T.J. Kaper. Small-amplitude periodic motions of rapidly forced mechanical systems. In *Proceedings of the 34th IEEE Conference on Decision and Control*, New Orleans, pages 533–539, 1995.
32. Weibel, S. and J. Baillieul. Averaging and energy methods for robust open-loop control of mechanical systems. To appear in the *Proceedings of the 1993 IMA Workshop on Robotics*, 1997.
33. Weibel, S., T.J. Kaper, and J. Baillieul. Global dynamics of a rapidly forced cart and pendulum. To appear in *Nonlinear Dynamics*, 1997.
34. Wiggins, S. *Introduction to Applied Nonlinear Dynamical Systems and Chaos*, volume 2 of *Texts in Applied Mathematics*. Springer-Verlag, Berlin, 1990.

On Rational Stabilizing Controllers for Interval Delay Systems.*

Leonid Naimark[1], Jacob Kogan[2]**, Arie Leizarowitz[3], Ezra Zeheb[1]

[1] Department of Electrical Engineering, Technion - Israel Institute of Technology, Haifa 32000, Israel
[2] Department of Mathematics and Statistics, University of Maryland Baltimore County, Baltimore, Maryland 21228, USA
[3] Department of Mathematics, Technion - Israel Institute of Technology, Haifa 32000, Israel

Abstract. A control system designer usually prefers to use rational controllers. The question of when such a controller exists is considered in this chapter, for the class of systems composed of a delay element e^{-hs} with interval (finite or infinite) uncertainty in h, followed by a plant characterized by a rational transfer function. Explicit conditions for the existence of such controllers, are given. Also, a computationally tractable design method, which explicitly yields the entire set of all constant gain controllers which robustly stabilize a family of systems with uncertainty, is described. A desired "optimal" controller may then be selected from the feasible set. The method is extended to the case when the rational part of the plant has uncertainties too, and is represented by a transfer function with independent interval coefficients. Illustrative numerical examples are provided.

1 Introduction

This chapter concerns the problem of designing rational robust stabilizing controllers for a system with delay, where interval uncertainty of the delay, as well as interval uncertainties of the coefficients of the rational transfer function (pertaining to the linear part), are assumed.

Although a parameterization of all controllers that stabilize a given plant has been obtained almost two decades ago [23],[4], there is little knowledge to date about the above design problem when uncertainties are assumed. The first question which arises is to what extent are plants with parameter uncertainties stabilizable by (possibly dynamic) output feedback. Using the Nevanlinna-Pick interpolation, an elegant solution [20] is proposed to the special case where the interval uncertainty is only in the gain factor of a linear system with no delay.

* This work was supported by grant no. 94-00010 from the United States-Israel Binational Science Foundation (BSF), Jerusalem, Israel, and by the Fund for the Promotion of Research at the Technion.

** Research of this author is partially supported by NSF grant ECS-9418709

This is referred to as the "blending problem". An extension was shown subsequently in [11] and [9], where the Nevanlinna-Pick interpolation can be applied to a specific choice of a *one parameter* family of plants.

If one suffices with stabilizing only the two end points of the one parameter family, than the problem reduces to simultaneous stabilization of two fixed coefficients plants, studied by [19]. The problem of simultaneous stabilization of a finite number of (fixed coefficients) plants has been studied in [22]. The problem of stabilizing a family of interval coefficients plants characterized by rational transfer functions with no delay, has been studied in [8], where it was shown that a *constant gain* controller K for the above is any (and only) a constant gain controller K which simultaneously stabilizes a certain set of eight fixed coefficients plants. Extensions of these results to first order controllers are derived in [10] and in [2].

A design technique, based on the zero-set approach, which provides the *complete set* of constant gain controllers for multi-input multi-output plants under uncertainty conditions, has been derived in [5]-[6]. However, the computational complexity in applying this technique in the general case, may limit its use.

Some efforts have been made in stability *analysis* of time-delay systems. The Edge Theorem to quasipolynomial families with *constant* delays and coefficients depending affinely on parameters was proposed in [7], a graphical test for quasipolynomial families with one interval delay was proposed in [21], and necessary and sufficient Hurwitz stability conditions for quasipolynomials with interval coefficients and interval delays was proposed in [14] (for a comprehensive list of relevant bibliography consult, for example, [13]).

Hence, it is clear that the problem (in its full capacitance) of stabilizing a system with delay, where there is uncertainty in the delay as well as in a number of coefficients of the rational part, is a difficult one and far from being completely solved.

The first goal of this chapter is to resolve the question of stabilizability of systems with an interval delay and fixed coefficients, by rational controllers and to explain how to design stabilizing controllers using these results. This part of the chapter is based on [16].

The rest of the chapter is devoted to design of constant gain stabilizing output controllers. In this context, the matter is resolved completely. A computationally tractable technique to derive the *entire set* of all constant gain controllers which robustly stabilize a family of interval coefficients plants preceded by an interval delay element, is described. This part of the chapter is based on [17].

The *idea* on which the present design technique is based is not related to previous results. It is based on continuity considerations in addition to a simple observation on what we term the "delay condition", to be explained further. However, the *implementation* of the basic idea becomes possible only by applying some recent results with regard to the frequency response envelopes of interval coefficients transfer functions [15]. Other forms of envelope results are listed in books [1] and [3].

The structure of the chapter is as follows:

In Section 2 we state the problem for infinite interval delay and give some definitions. Section 3 deals with the problem of existence of constant gain and rational dynamic stabilizing controllers and consideration of simple design algorithms. In Section 4 we treat the problem of constant gain controller design for the case where the coefficients of the rational transfer function may be interval ones and the delay is completely unknown (its interval is infinite). The resulting set of controllers robustly stabilize the system independently of delay (IOD). In Section 5 we treat the case where the coefficients of the rational transfer function are fixed, but we now assume partial knowledge of the delay, namely a *finite* interval for the delay. In Section 6 we treat the most general (and most difficult) case where both the delay and the coefficients of the rational transfer function are prescribed in finite intervals, and still obtain the entire set of all (and only) robustly stabilizing constant gain controllers. We provide numerical examples where appropriate, and conclude in Section 7.

2 Statement of the problem

Assume that a system transfer function $P(s)$ is given as a ratio of two relatively prime polynomials with fixed real coefficients, namely

$$P(s) = \frac{n_p(s)}{d_p(s)}, \tag{2.1}$$

where

$$\begin{aligned} n_p(s) &= n_k s^k + n_{k-1} s^{k-1} + \cdots + n_1 s + n_o \\ d_p(s) &= d_l s^l + d_{l-1} s^{l-1} + \cdots + d_1 s + d_o \end{aligned} \tag{2.2}$$

In the sequel we assume that

$$k \leq l. \tag{2.3}$$

Moreover, suppose that the system includes a delay function e^{-hs}, where the delay parameter can vary in the interval $[0, \infty)$. Finally, let $C(s)$ be any rational controller, given by

$$C(s) = \frac{n_c(s)}{d_c(s)}. \tag{2.4}$$

where $n_c(s)$ and $d_c(s)$ are relatively prime polynomials in s. See Fig. 1 for the closed-loop configuration.

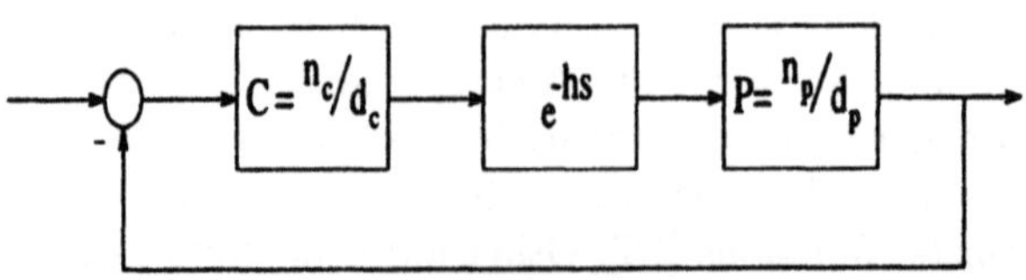

Fig. 1. Closed-loop system.

We pose the following two questions:
Problem 1: When does a constant gain or rational dynamic stabilizing controller C exist.
Problem 2: Find the set of all constant gain or rational dynamic stabilizing controllers.

Let $R(s) = n_r(s)/d_r(s)$ be any rational function.

Definition 1. *A rational function R is strictly proper if $deg(n_r) < deg(d_r)$.*

Definition 2. *A rational function R is proper if $deg(n_r) = deg(d_r)$.*

Note that Definition 2 is different from the common one for a proper function.

3 When does a rational stabilizing controller exist

Mathematically, this problem can be stated as follows: detect if there exists at least one rational $C(s)$ such that

$$d_c(s)d_p(s) + n_c(s)n_p(s)e^{-hs} \neq 0 \text{ in } \mathrm{Re}s \geq 0 \ \forall \ h \in [0, \infty) \tag{3.1}$$

Let H be the set of all Hurwitz polynomials i.e., the set of all polynomials whose zeros are in the open left half complex plane.

Theorem 3. *A rational stabilizing controller exists if and only if $d_p \in H$.*

For the proof of this theorem we need the following result:

Lemma 4. *Let U be an open set in R^n such that $f(\cdot)$ is defined and differentiable on the closure of U and $f(U)$ contains a ball of radius R around the origin. Let $g(\cdot)$ be differentiable on the closure of U and such that $|g(s)| < r$ for every $s \in U$, where $r < R$. Then $f - g$ has a zero in U.*

An exact proof of this lemma is follows from degree theory and will be omitted for the sake of brevity. Instead, we provide some intuitive explanations. Consider the image of the boundary ∂U under f, which is far away from the origin by more that R, and since g perturb by less than r and $r < R$ the image of the boundary under $f - g$ will surround the origin. Since the image is simply connected then the result will follow.
Denote

$$f_h := d_p(s) + Ce^{-hs}n_p(s) \tag{3.2}$$

Then we have the following theorem:

Theorem 5. *If f_h is stable for $h \in [0, \infty)$ then $d_p \in H$.*

Proof. Assume to the contrary that $d_p \notin H$ and distinguish between two cases:

Case 1. There exists s_0 such that $\text{Re}s_0 > 0$ with $d_p(s_0) = 0$. Then there is a neighborhood U of s_0 such that $q(U)$ contains the ball of radius R around the origin, and $\text{Re}s > 0$ for every s in the closure of U. Since C and n_p are fixed then there is sufficiently large $h_0 > 0$ such that $|Ce^{-h_0 s}n_p(s)| < R/2$ for every $s \in U$. By the Lemma there exists a zero of f_{h_0} in U, which is a contradiction.

Case 2. This part of the proof is based on the argument principle. Assume that $d_p(j\omega) = 0$ for some real ω. Then consider a contour Γ, which is constructed as follows: let $\rho > 0$ be a small radius, we take half a circle of radius ρ going through $(\omega - \rho)j$ and $(\omega + \rho)j$ in the right half plane, and then go along the imaginary axis from $(\omega + \rho)j$ down to $(\omega - \rho)j$. Taking h very large, we obtain that along the straight line segment there will be many encircling of the origin for the image of $d_p + Ce^{-hs}n_p$, while there will be fewer encirclings along the half circle. The last point follows from the following argument. Denote the half circle by Γ_0, let C and n_p be fixed and choose some small $\epsilon > 0$. We then denote by Γ_ϵ the arc $\Gamma_0 \cap \{s : \text{Re}s > \epsilon\}$. If h is chosen sufficiently large then $|d_p(s)|$ dominates $|Ce^{-hs}n_p(s)|$ on Γ_ϵ and consequently there will be no encircling of the origin for the image of f_h on Γ_ϵ.

The complement of Γ_ϵ in Γ_0 is composed of two little arcs, the lower arc is denoted γ_1 and the upper one is denoted γ_2. The number of encirclings of the origin by the image of f_h on both γ_1 and γ_2 is much smaller than the number of encircling by the image of the straight line segment $\{tj : \omega - \rho \leq t \leq \omega + \rho\}$, which we denote by l. This claim follows from a straightforward estimate of the number of encirclings of the origin by the images of f_h on l, γ_1 and γ_2. This number of encircling for, e.g., the line segment l is estimated by $\rho h/\pi$ (recalling that $|Cn_p(s)|$ is much larger than $|d_p(s)|$ on l). The above assertion follows from a similar estimate for γ_1 and γ_2, and the observation that the lengths of γ_1 and γ_2 is much smaller than 2ρ, the length of l, if ϵ is chosen sufficiently small. □

Now we are ready to prove Theorem 3.

Proof of Theorem 3. The delay system can be stabilized for the interval $h \in [0, \infty)$ if and only if there exist two polynomials $n'_c = Cn_c$ and d_c such that

$$f_c = d_c(s)d_p(s) + e^{-hs}n'_c(s)n_p(s) \tag{3.3}$$

is stable for every $h \in [0, \infty)$. If $d_p \in H$ then we can choose n_c and d_c such that $d_c \in H$, and $\text{deg}d_c d_p \geq \text{deg}n_c n_p$ and then we can choose C to make f_c stable for every $h \in [0, \infty)$.

On the other hand, suppose that $d_p \notin H$. Then for every choice of polynomials n_c and d_c whatsoever, $d_c d_p \notin H$ and hence by Theorem 5 there cannot exist a constant C such that f_c in (3.3) is stable for every $h \in [0, \infty)$. This estblishes the other part of the Theorem. □

Theorem 6. *When there exists a rational stabilizing controller, there also exists a constant gain stabilizing controller.*

Proof. The proof follows immediately because of assumption $k \leq l$. □

4 Stabilizing controllers for IOD systems.

Now we consider how to find the set of all constant gain stabilizing controllers if such exists. Additionally, we assume interval uncertainties in the coefficients of the rational part of the system, i.e

$$\underline{n}_i \le n_i \le \overline{n_i}\,, \qquad \underline{d}_i \le d_i \le \overline{d}_i \tag{4.1}$$

The design procedure yields *all* stabilizing constant gain controllers, independent of delay.

In the case of a proper plant $k = l$, we need the following assumption

$$d_l + Ce^{-hs} n_l \neq 0 \tag{4.2}$$

for each $h \in [0, \infty)$ and s : Re$s \ge 0$. This assumption guarantees that the coefficient of s to the highest degree does not vanish. Obviously, if $k < l$, the avoidance of "degree reduction" is guaranteed in any case, since $d_l \neq 0$ by definition.

The assumption (4.2) is readily seen to be equivalent to the requirement that a constant gain controller for a proper plant may only take on values in the open interval

$$C \in (-|d_l/n_l|, |d_l/n_l|) \tag{4.3}$$

By continuity (zero exclusion principle), one can see that, assuming (4.2) for the case $k = l$, (3.1) is equivalent for constant gain C to the following conditions (4.4)+(4.5)

$$d_p(s) + Cn_p(s) \neq 0 \quad \text{in} \quad \text{Re}s \ge 0 \tag{4.4}$$

$$d_p(j\omega) + Cn_p(j\omega)\, e^{j\Theta} \neq 0, \;\; \omega \in \mathbb{R}, \;\; 0 \le \Theta \le 2\pi \tag{4.5}$$

The explanation of this observation is clear: condition (4.4) corresponds to stability of the closed-loop system without delay, $h = 0$, (and thus we term it "stability condition without delay", or simply, "stability condition"). Increasing h from zero to infinity, while keeping the zeros of (3.1) from crossing the imaginary axis ensures (3.1), provided there is no "degree reduction". Assuming (4.2) for the case $k = l$ ensures that there is no "degree reduction", and condition (4.5) with $0 \le \Theta \le 2\pi$ ensures the "no crossing" of the imaginary axis.

In order to interpret condition (4.5), which we term "delay condition", we rewrite it in the following form

$$C \neq -\frac{d_p(j\omega)}{e^{j\Theta} n_p(j\omega)}\,, \quad \omega \in \mathbb{R}\,, \quad 0 \le \Theta \le 2\pi \tag{4.6}$$

Since for $h \in [0, \infty)$, Θ may take on *any* value in $[0, 2\pi]$, and C is a real number, then (4.6) is equivalent to

$$|C| \neq \left| \frac{d_p(j\omega)}{n_p(j\omega)} \right| = |P^{-1}(j\omega)|\,, \qquad \omega \in \mathbb{R}\,. \tag{4.7}$$

Note that the absolute value in the right hand side of (4.7) refers to the complex value of $d_p(j\omega)/n_p(j\omega)$, whereas the absolute value in the left hand side of (4.7) refers to the sign of C. Hence, from the delay condition as expressed in (4.7), we obtain the following constraints

$$|C| > \max_{\omega} \left|P^{-1}(j\omega)\right| \ , \quad \text{or} \tag{4.8}$$

$$|C| < \min_{\omega} \left|P^{-1}(j\omega)\right| \tag{4.9}$$

However, the constraint (4.8) is not valid for a strictly proper plant because $|P^{-1}(j\omega)|$ is unbounded and not valid for a proper plant because of Eqn. (4.3). On the other hand, the constraint (4.9) *includes* the requirement (4.3), since $|d_l/n_l|$ (in the case $k = l$) is $|P^{-1}(j\infty)|$. Hence, the only requirements for a constant gain stabilizing controller are (4.4) and (4.9).

The problem of finding the set of all real values of C satisfying the "stability condition" (4.4) can be overcome with the aid of the results in [8], where it was shown that an interval rational transfer function is stabilizable by a positive constant gain controller if and only if certain four rational transfer functions with fixed coefficients are simultaneously stabilizable by that same controller. Similarly, for a negative constant gain controller, but with another set of four rational transfer functions with fixed coefficients. In the context of our problem, define

$$R(s,C) = d_p(s) + Cn_p(s) = \sum_{i=0}^{n} r_i s^i \tag{4.10}$$

with

$$\underline{r}_i \le r_i \le \overline{r}_i \qquad i = 0, 1, \ldots, n \tag{4.11}$$

where

$$\underline{r}_i = \underline{d}_i + C\underline{n}_i \ , \quad \overline{r}_i = \overline{d}_i + C\overline{n}_i \ ; \quad \text{if} \quad C > 0 \tag{4.12}$$

and

$$\underline{r}_i = \underline{d}_i + C\overline{n}_i \ , \quad \overline{r}_i = \overline{d}_i + C\underline{n}_i \ ; \quad \text{if} \quad C < 0 \tag{4.13}$$

Denote the four Kharitonov [12] polynomials associated with the interval polynomial $R(s,C)$ for positive C (Eqn. (4.12)), by $R_i^+(s,C)$, $i = 1,\ldots,4$, and denote the four Kharitonov polynomials associated with $R(s,C)$ for negative C (Eqn. (4.13)), by $R_i^-(s,C)$, $i = 1,\ldots,4$. The coefficients of each of these 8 polynomials depend linearly on only one parameter C. To find the intervals of C for such a polynomial to be stable, we propose the following theorem, which is simpler and more explicit than just to use a Routh table parametrically in C and solve the inequalities for the first column of the table to be positive.

Theorem 7. *Let $R(s,C)$ as in (4.10), represent one of the pertinent Kharitonov polynomials. Let $d_e(\omega)$ and $d_o(\omega)$ be, respectively, the even and odd parts of $d_p(s)$, where $s = j\omega$. Let $n_e(\omega)$ and $n_o(\omega)$ be, respectively, the even and odd parts of $n_p(s)$, where $s = j\omega$. Then, the values of C which are the end points of the* *intervals of C for which $R(s,C)$ is stable (satisfies (4.4)), are from the set given by:*

$$C = - \; d_k/n_k \qquad (\textit{if } k=l) \tag{4.14}$$

and

$$C = - \; d_o(\omega_i)/n_o(\omega_i) = -d_e(\omega_i)/n_e(\omega_i) \tag{4.15}$$

where ω_i are the real zeros of the following equation

$$d_o(\omega_i)n_e(\omega_i) - d_e(\omega_i)n_o(\omega_i) = 0 \tag{4.16}$$

Remark 1. If $d_e(\omega_i) = n_e(\omega_i) = 0$ or $d_o(\omega_i) = n_o(\omega_i) = 0$, the meaningful expression in the right hand side of (4.15) should be used. We assume that $n_p(s)$ and $d_p(s)$ have no common factor, thus the case $d_e(\omega_i) = n_e(\omega_i) = d_o(\omega_i) = n_o(\omega_i) = 0$ is void.

Proof of Theorem 7. By continuity considerations, a transition from a stability interval to an instability interval, or vice versa, can only occur when a zero of $R(s,C)$ crosses the imaginary axis of the s plane at a finite point, or there is a degree reduction in $R(s,C)$. For the latter to occur, (4.14) must be satisfied. A finite zero crossing implies

$$d_p(j\omega) + Cn_p(j\omega) = 0 \tag{4.17}$$

which is equivalent to the two simultaneous equations

$$d_e(\omega) + Cn_e(\omega) = 0\,, \qquad d_o(\omega) + Cn_o(\omega) = 0\,. \tag{4.18}$$

It is readily verified that the solution of (4.18) is given by the pairs (ω_i, C) defined in (4.15) and (4.16). □

Turn now to the "delay condition" (4.5). Firstly, it is obvious that since the interval coefficients of the numerator polynomial are independent of those of the denominator polynomial, the ratio of the maximum (minimum) of the numerator amplitude and the minimum (maximum) of the denominator amplitude yields the exact envelope of the amplitude of the family of rational functions. Thus, we only need to be able to treat the amplitude of a family of interval polynomials:

$$F(s) = \sum_{i=0}^{n} \alpha_i s^i\,, \qquad \underline{\alpha}_i \le \alpha_i \le \overline{\alpha}_i\,. \tag{4.19}$$

We are interested in

$$F_{\max}(\omega) \triangleq \max_{\underline{\alpha}_i \le \alpha_i \le \overline{\alpha}_i} \{|F(j\omega)|\} \text{ and } F_{\min}(\omega) \triangleq \min_{\underline{\alpha}_i \le \alpha_i \le \overline{\alpha}_i} \{|F(j\omega)|\} \tag{4.20}$$

Denote

$$\begin{aligned} F_{e\max} &\triangleq \overline{\alpha}_o - \underline{\alpha}_2\omega^2 + \overline{\alpha}_4\omega^4 - \underline{\alpha}_6\omega^6 + \cdots \\ F_{o\max} &\triangleq \omega(\overline{\alpha}_1 - \underline{\alpha}_3\omega^2 + \overline{\alpha}_5\omega^4 - \underline{\alpha}_7\omega^6 + \cdots) \\ F_{e\min} &\triangleq \underline{\alpha}_o - \overline{\alpha}_2\omega^2 + \underline{\alpha}_4\omega^4 - \overline{\alpha}_6\omega^6 + \cdots \\ F_{o\min} &\triangleq \omega(\underline{\alpha}_1 - \overline{\alpha}_3\omega^2 + \underline{\alpha}_5\omega^4 - \overline{\alpha}_7\omega^6 + \cdots) \end{aligned} \tag{4.21}$$

and

$$\alpha_i^o \triangleq \underline{\alpha}_i + \overline{\alpha}_i\,, \qquad i = 0, \ldots, n \tag{4.22}$$

Then, it is shown in [15] that $F_{\max}(\omega)$ coincide with one of the following four fixed coefficients expressions

$$|F_{e\,\max} + jF_{o\,\max}|,\ |F_{e\,\max} + jF_{o\,\min}|,\ |F_{e\,\min} + jF_{o\,\max}|,\ |F_{e\,\min} + jF_{o\,\min}|, \tag{4.23}$$

and $F_{\max}(\omega)$ can change from one of the fixed coefficients expressions in (4.23) to another of the fixed coefficients expressions in (4.23) only at a finite number of frequencies, which can be computed. Concerning $F_{\min}(\omega)$, it may coincide with *nine* fixed coefficients expressions: the four expressions listed in the right hand side of (4.23), in addition to $|F_{e\,\max}|$, $|F_{e\,\min}|$, $|F_{o\,\max}|$, $|F_{o\,\min}|$ and zero. The "change frequencies" for $F_{\min}(\omega)$ can also be computed.

Then, we can summarize the design method which yields *all* stabilizing constant gain controllers for this case, as follows:

Algorithm 1

A. Stability condition

1. Determine the values of positive C yielded by Theorem 7 (Eqns. (4.14) and (4.15)) for each of the polynomials $R_i^+(s, C)$, $i = 1, \ldots, 4$.
2. Choose an arbitrary value of C in each of the intervals created by the values computed in step 1, and determine whether the corresponding $R_i^+(s, C)$, $i = 1, \ldots, 4$, is stable.
3. Intersect the results in step 2, to determine the intervals of positive C for which all four $R_i^+(s, C)$, $i = 1, \ldots, 4$, are simultaneously stable.
4. Repeat steps 1, 2 and 3 for negative C, with the polynomials $R_i^-(s, C)$, $i = 1, \ldots, 4$.

B. Delay condition

1. Determine $n_{p\,\max}(\omega)$ by (4.23).
2. Determine $d_{p\,\min}(\omega)$ by one of the nine expressions indicated in the text.
3. Compute $|P_{\min}^{-1}(j\omega)|$ using $d_{p\,\min}(\omega)/n_{p\,\max}(\omega)$.
4. The set of all stabilizing C according to the "delay condition" is

$$C \in (-\min_{\omega} |P_{\min}^{-1}(j\omega)|,\ \min_{\omega} |P_{\min}^{-1}(j\omega)|). \tag{4.24}$$

C. Intersect results of A and B.

Note that the case of fixed coefficients rational transfer function with unknown delay, is a special case of the above.

Example 1. Consider a second order fixed coefficients transfer function

$$P(s) = \frac{s+1}{s^2 + 5s + 6} = \frac{(s+1)}{(s+2)(s+3)} \tag{4.25}$$

From the "stability condition" we obtain that the pertinent interval is $C \in (-5, \infty)$. The "delay condition" renders $|C| < 4.56$. Combining these two results we readily obtain a final answer: The all and only interval of constant gain controllers which stabilize (4.25) cascaded with an unknown delay is

$$C \in (-4.56,\ 4.56) \tag{4.26}$$

In the design algorithm we limited ourselves to the case of constant gain stabilizing controllers. One can easily extend the results to the case of rational dynamic stabilizing controllers. As shown previously, if the plant is not stabilizible with a constant gain controller, then it is not stabilizible with any rational controller. A parameterization of all rational stabilizing controllers is the following:

1. Choose an arbitrary Hurwitz denominator $d_c \in H$.
2. Choose an arbitrary numerator n_c, but $\deg(n_c n_p) \leq \deg(d_c d_p)$ and there is no unstable zero-pole cancellation between n_c and d_p.
3. For each pair (n_c, d_c) determine by algorithm 1 the set of constants such that Cn_c/d_c is a stabilizing controller.

Remark 2. If we remove the assumption $k \leq l$ (Eqn. (3)) and consider the case $k > l$, then a constant gain stabilizing controller does not exist. Instead, it is always possible to divide the transfer function of the plant by any *Hurwitz* polynomial, say $t(s)$, such that the extended transfer function $n_p(s)/(d_p(s)t(s))$ is now a proper or strictly proper function with a stable denominator. This extended plant is evidently stabilizible by a constant gain controller, say K. So, a controller $K/t(s)$ is a pertinent dynamic stabilizing controller of the original plant in the case $k > l$.

5 Stabilizing controllers for finite interval delay systems

In this section we discuss a design method for constant gain controllers for delay systems, where the delay is partially known. It is more reasonable to assume that the delay is known to be bounded by a certain finite bound H than to assume that the delay is not known at all and may take on any value without bound. For methodical purposes (clearness of presentation) we consider first systems with fixed coefficients, and then add the assumption of interval coefficients, treated in Section 6.

This case of a finite delay interval is much more complex than the corresponding case of stability independent of delay, treated in Section 4. However, we are still able to provide a design method which yields *all* stabilizing constant gain controllers, where the plant is preceded (or cascaded) by a delay element (see Fig. 1), with

$$0 \leq h \leq H \tag{5.1}$$

and H is a *given* real number.

Assume that the polynomials $n_p(s)$ and $d_p(s)$ are polynomials with constant coefficients. Consider a vector in the complex plane $e^{j\omega h}$ for a fixed value of the frequency $\omega = \omega_1$, and h in the interval $[0, H]$. The magnitude of this vector is equal to 1 and the phase is equal to $\omega_1 h$. So, the set of all possible phases for $\omega = \omega_1$ and h in (5.1) is the interval $[0, \omega_1 H]$. Let us define a critical frequency ω_c by

$$\omega_c = \frac{2\pi}{H} \tag{5.2}$$

If $\omega_1 \geq \omega_c$ then by a proper choice of h (not necessarily unique) we can obtain any arbitrary phase of $e^{j\omega h}$ from $[0, 2\pi]$. If $\omega_1 < \omega_c$ then by a proper choice of h we can obtain any phase in the interval $[0, \omega_1 H]$ (see Fig. 2a).

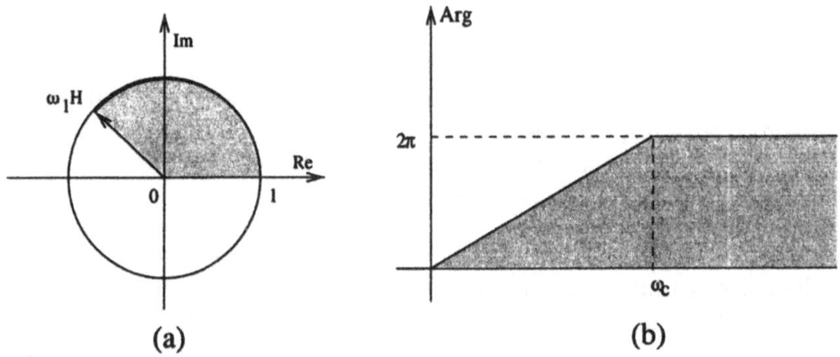

Fig. 2. (a) Possible phases of $e^{j\omega_1 h}$ for various h. (b) The family of all possible phases for $e^{j\omega h}$.

Fig. 2b illustrates the family of possible phases for different frequencies and h as in (5.1). Recall the delay condition in form (4.6)

$$C \neq -\frac{d_p(j\omega)}{n_p(j\omega)} e^{jh\omega} \tag{5.3}$$

Define $G(\omega) := d_p(j\omega)/n_p(j\omega) = P^{-1}(j\omega)$.
For ω_1 fixed, $G(\omega_1)$ can be interpreted as a vector in the complex plane (see Fig. 3a).

Observation:

1. If there exists any h_1 in (5.1) such that

$$Arg\,[G(\omega_1)] + \omega_1 h_1 = 0\,(mod\,2\pi) \tag{5.4}$$

then the right hand side of (5.3) becomes real and negative. The corresponding $C < 0$ does not satisfy the "delay condition".

2. If there exists any h_2 in (5.1) such that

$$Arg\,[G(\omega_1)] + \pi + \omega_1 h_2 = 0\,(mod\,2\pi) \tag{5.5}$$

then the right hand side of (5.3) becomes real and positive. The corresponding $C > 0$ does not satisfy the "delay condition".

3. If there exist both h_1 and h_2 such that (5.4), (5.5) are satisfied, then the corresponding $\pm C$ do not satisfy the "delay condition".

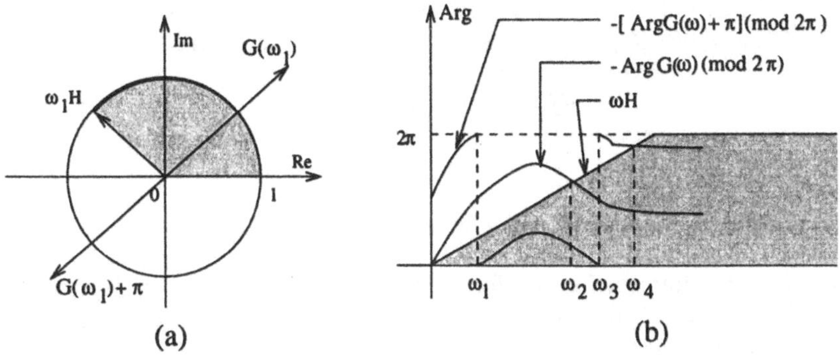

Fig. 3. (a) Vectors $G(\omega_1)$ and $G(\omega_1) + \pi$. (b) Different kinds of frequency intervals.

In order to find all possible values of C satisfying the "delay condition", we should plot $-Arg[G(\omega)](mod2\pi)$ and $-(Arg[G(\omega)] + \pi)(mod2\pi)$ and check for each ω whether the plots are "inside" or "outside" the family ωh. If for some ω_o, a phase plot is inside the family, then the value of $C = |G(\omega_o)|$ (or $C = -|G(\omega_o)|$) does not satisfy the "delay condition". Thus, we make a search of C over all frequencies inside the family ωh. Fig. 3b illustrates all possible different situations: for the frequency interval $[0, \omega_1]$ both $\pm C$ satisfy the "delay condition"; for the interval $[\omega_1, \omega_2]$, $C < 0$ satisfies the "delay condition" but $C > 0$ does not satisfy the "delay condition"; for the intervals $[\omega_2, \omega_3]$ and $[\omega_4, \infty]$ both $\pm C$ do not satisfy the "delay condition"; and, finally, for the interval $[\omega_3, \omega_4]$, $C > 0$ satisfies the "delay condition" but $C < 0$ does not satisfy the "delay condition". We summarize the "delay condition" in this case by the following algorithm which replaces the one in Algorithm 1 (Evidently, the "stability condition" is a special case of the one in Algorithm 1).

Algorithm 2 *(Delay condition)*

1. Solve the equation

$$Arg\,[G(\omega)] + \omega H = 0\,(mod\,2\pi) \tag{5.6}$$

   Let the real positive solutions of (5.6), in addition to the frequencies ω_i for which $-Arg\,[G(\omega_i)] = 2\pi$, be denoted by the ordered sequence $\omega_1 < \omega_2 < \cdots < \omega_\ell$.

2. If ℓ is odd, we construct $m := (\ell + 1)/2$ intervals

$$I_1 = [\omega_1, \omega_2],\ I_2 = [\omega_3, \omega_4], \ldots, I_m = [\omega_\ell, \infty]\,.$$

   If ℓ is even, we construct $m := \ell/2 + 1$ intervals

$$I_1 = [0, \omega_1],\ I_2 = [\omega_2, \omega_3], \ldots, I_m = [\omega_\ell, \infty]\,.$$

3. Denote by

$$\underline{G}_i = \min_{\omega \in I_i} |G(\omega)|\, , \; \overline{G}_i = \max_{\omega \in I_i} |G(\omega)|\, , \; i = 1, \ldots, m \tag{5.7}$$

Then, all and only values of $C < 0$ which satisfy the "delay condition" are those C for which $|C|$ do not belong to any of the intervals

$$[\underline{G}_i\, , \, \overline{G}_i] \quad , \quad i = 1, \ldots, m \tag{5.8}$$

4. Replace (5.6) with

$$Arg\,[G(\omega)] + \pi + \omega H = 0\,(mod\, 2\pi) \tag{5.9}$$

and add the frequencies ω_i for which $-Arg\,[G(\omega_i)] = \pi$ (instead of $-Arg\,[G(\omega_i)] = 2\pi$), to the ordered sequence $\omega_1, \omega_2, \ldots, \omega_\ell$. Repeat steps 2-3 for the positive values of C which satisfy the "delay condition".

5. If $k = l$, intersect obtained result with interval $(-|d_l/n_l|, |d_l/n_l|)$ (assumption (4.3) is still valid for $h \in [0, H]$).
Note that, if $k < l$ then $G(\infty) = \infty$ and hence interval $[\underline{G}_m, \overline{G}_m] = [\underline{G}_m, \infty]$.

Remark 3. An extended version of this algorithm for bounded interval delay system $h \in [H_1,\ H_2]$ can be found in [18].

Example 2. Consider the following transfer function with fixed coefficients

$$P(s) = \frac{s^2 + 5s + 6}{s^2 + 4s - 5} = \frac{(s+2)(s+3)}{(s-1)(s+5)} \tag{5.10}$$

Since this function has an unstable pole, it is impossible to stabilize it IOD by a rational controller. However, assume now a finite interval of uncertainty in the delay, say, $h \in [0, 0.5]$. In this case Fig. 4 illustrates the procedure:

An intersection of $-Arg[G(\omega)]$ and ωH (solid and dashdot curves) takes place at only one frequency $\omega = 12.50$. Namely, ℓ is odd and we should consider the frequency interval $[12.50, \infty)$ in order to check the delay condition for negative C. From Fig. 4a we obtain $C \in (-\infty, -1.04) \cup (-1, 0)$. In the same manner, an intersection of $-Arg[G(\omega)] - \pi$ and ωH (dashed and dashdot curves) occurs at $\omega = 5.73$ defining the frequency interval $[5.73, \infty)$ for positive C. From Fig. 4a we obtain $C \in (0, 1) \cup (1.13, \infty)$. So, from the delay condition we obtain: $C \in (-\infty, -1.04) \cup (-1, 1) \cup (1.13, \infty)$. Taking into account the fact that $P(s)$ is proper, condition (4.3) renders $C \in (-1, 1)$.

The stability condition renders the following intervals of C:
$C \in (-\infty, -1) \cup (0.83, \infty)$.

Intersecting the intervals obtained from the "stability condition" with the intervals obtained from the "delay condition" and condition (4.3), we obtain the final result of all and only intervals of constant gain controllers which stabilize (5.10) cascaded with a delay element in the interval [0, 0.5],

$$C \in (0.83, 1.0) \tag{5.11}$$

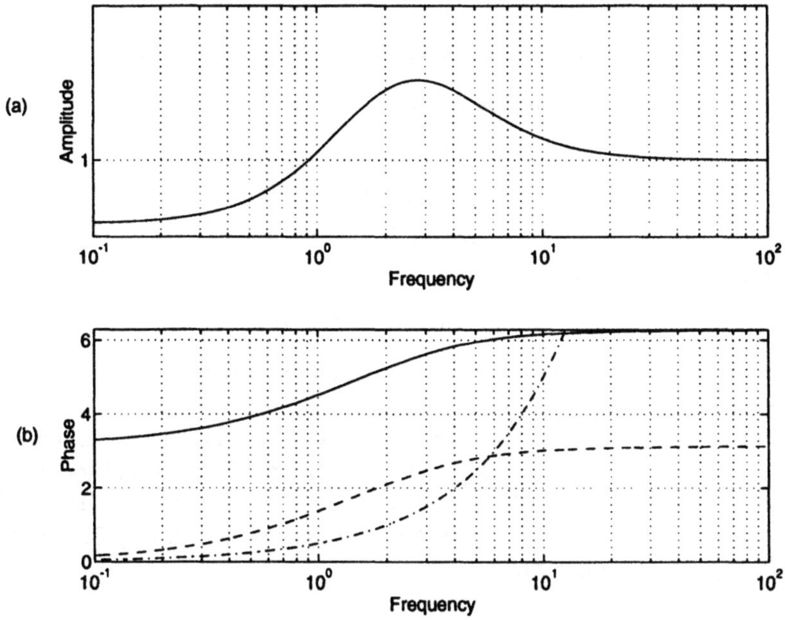

Fig. 4. (a) Amplitude of $G(j\omega)$. (b) $-Arg[G(\omega)]$ (solid curve), $-Arg[G(\omega)]-\pi$ (dashed curve) and ωH (dashdot curve).

As expected, although it is not possible to stabilize this system IOD by a constant gain controller, it *is* stabilizible by a constant gain controller for a finite delay interval $[0, 0.5]$. So, Example 2 gives immediately the positive answer to the following question: do there exist such cases for $d_p \notin H$ which *are* stabilizable by a rational controller? Moreover, our approach allows to solve the following problem: we are given a rational part of the plant and we are limitited in use of low-order (for example, first order) controller, say, with positive constant gain. Estimate the upper value H of the delay parameter h such that the closed-loop system still remains stable.

Example 3. Let the transfer function of a plant be

$$P(s) = \frac{s-2}{s-1} \tag{5.12}$$

If the interval of the unknown delay is infinite, this plant cannot be stabilized by a rational controller, as implied by Theorem 3. Assume that $h \in [0,\ H]$, where H is finite. Clearly a positive constant gain stabilizer does not exist, no matter what the value of H is. However, consider a first order rational controller

$$C(s) = C\frac{s+a}{s+b} \tag{5.13}$$

where a, b and C are real numbers. The closed-loop characteristic polynomial for the "stability condition" is

$$(C+1)s^2 + (Ca - 2C + b - 1)s - (2ac + b) = 0 \tag{5.14}$$

To have a closed-loop stability we demand

$$\begin{aligned} C + 1 &> 0 \\ Ca - 2C + b &> 1 \\ 2aC + b &< 0 \end{aligned} \tag{5.15}$$

For example a satisfactory choice is

$$a = -6; \quad b = 10; \quad C = 0.9. \tag{5.16}$$

One can check that assumption (4.3) is satisfied for this controller, i.e becomes $0.9 \in [-1, 1]$.

In order to find the maximal value of H for which the "delay condition" is also satisfied, we use algorithm 2. To this end, we need the amplitude and the phase of $P(j\omega)^{-1}C(j\omega)^{-1}$, which are shown in Fig. 5. The solid line in Fig. 5b describes $-Arg[P^{-1}(j\omega)C^{-1}(j\omega)]$ and the dashed line in Fig. 5b describes $-[Arg[P^{-1}(j\omega)C^{-1}(j\omega)] + \pi]$.

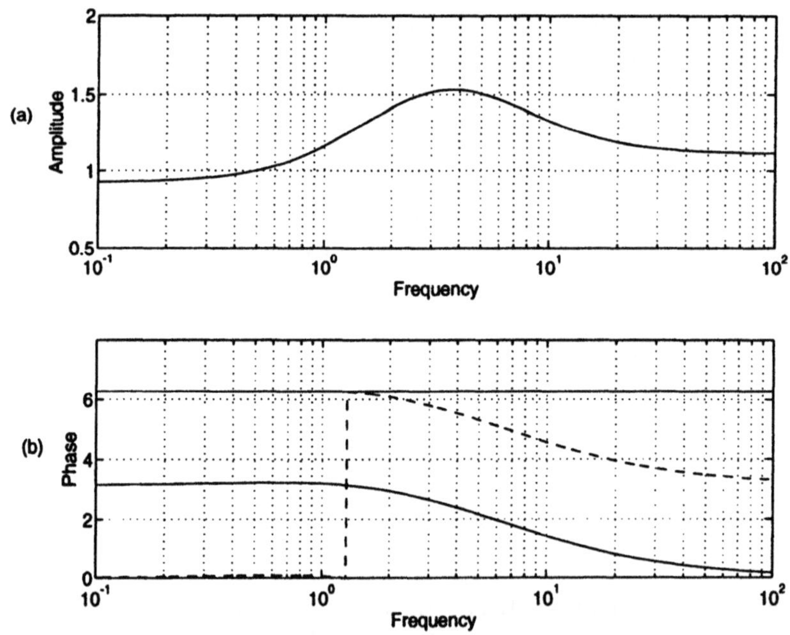

Fig. 5. Amplitude (a) and phase (b) of the open-loop transfer function.

Adding the plot of ωH versus ω to Fig. 5b, it can be deduced, that for any $H < 0.18$ this plant is stabilizable by the first order controller (5.13), (5.16).

6 Systems with interval coefficients

Suppose now that the delay parameter h can vary in the finite interval (5.1) in addition to the assumption that $n_p(s)$ and $d_p(s)$ are polynomials with interval coefficients (4.1). This case is the most realistic of all previous cases, from the practical point of view. As expected, this problem is also the most difficult from the computational and algorithmical points of view. The algorithm for the "stability condition" (part A) is absolutely identical to the one in Algorithm 1. However, the algorithm related to the "delay condition" (part B), becomes more complicated. We have to modify the one described in Section 5 in order to incorporate the fact that in the present case both the amplitude *and* phase of $G(\omega) = P^{-1}(j\omega)$ are not fixed.

To this end we use again the results in [15]. In addition to the possibility of computing the bounds (envelope) of

$$G_{\max}(\omega) \triangleq \max |G(\omega)|, \quad G_{\min}(\omega) \triangleq \min |G(\omega)|. \tag{6.1}$$

where maximum and minimum are taken over the set of all possible coefficients, discussed above, it is also shown in [15] how to compute the bounds (envelope) of the phase of an interval rational function. Let $\Phi_a(\omega)$ and $\Phi_b(\omega)$ denote the two bounds of $Arg[G(\omega)]$, for each ω. Let $\Phi_b(\omega)$ be the lower bound of the phase and $\Phi_a(\omega)$ the upper bound of the phase. Then, it is shown in [15] how to explicitly determine a finite number of intervals on the frequency axis ω, in each of which $\Phi_a(\omega)$ and $\Phi_b(\omega)$ take on the values of the phases of certain explicit *fixed* coefficients rational functions.

Having the ability to compute $G_{\max}(\omega)$, $G_{\min}(\omega)$, $\Phi_a(\omega)$ and $\Phi_b(\omega)$ in a tractable way, it is now clear how to modify the algorithm described in Section 5. Since this case is the most realistic one (and the most complex one), the design method which yields *all* stabilizing constant gain controllers for this case, is formulated explicitly as follows:

Algorithm 3

A. **Stability condition** (exactly as in Algorithm 1)

B. **Delay condition**

1. Solve equation
$$\Phi_b(\omega) + \omega H = 0 \,(mod\, 2\pi) \tag{6.2}$$
Let the real positive solutions of (6.2), in addition to the frequencies ω_i for which $-\Phi_a(\omega_i) = 2\pi$, be denoted by the ordered sequence $\omega_1 < \omega_2 < \cdots < \omega_\ell$.
2. If ℓ is odd, construct $m := (\ell + 1)/2$ intervals
$$I_1 = [\omega_1, \omega_2],\ I_2 = [\omega_3, \omega_4], \ldots, I_m = [\omega_\ell, \infty].$$
If ℓ is even, construct $m := \ell/2 + 1$ intervals
$$I_1 = [0, \omega_1],\ I_2 = [\omega_2, \omega_3], \ldots, I_m = [\omega_\ell, \infty].$$

3. Denote by

$$\underline{G}_i = \min_{\omega \in I_i} G_{\min}(\omega)\ ,\ \overline{G}_i = \max_{\omega \in I_i} G_{\max}(\omega)\ ,\ i = 1, \ldots, m \tag{6.3}$$

Then, all and only values of $C < 0$ which satisfy the delay condition are those C for which $|C|$ do not belong to any of the intervals

$$[\underline{G}_i\,,\,\overline{G}_i]\quad ,\quad i = 1, \ldots, m \tag{6.4}$$

4. Replace (6.2) with

$$\Phi_b(\omega) + \omega H + \pi = 0\,(mod\,2\pi) \tag{6.5}$$

and add the frequencies ω_i for which $-\Phi_a(\omega_i) = \pi$ (instead of $-\Phi_a(\omega_i) = 2\pi$) to the corresponding ordered sequence. Repeat steps 2-3 for the positive values of C which satisfy the "delay condition".
5. If $k = l$, intersect obtained result with interval $(-|\hat{d}_l/\hat{n}_l|, |\hat{d}_l/\hat{n}_l|)$, where $\hat{n}_l = \max\{|\underline{n}_l|, |\overline{n}_l|\}$ and $\hat{d}_l = \min\{|\underline{d}_l|, |\overline{d}_l|\}$. Note again that, if $k < l$ then $G(\infty) = \infty$ and hence interval $[\underline{G}_m, \overline{G}_m] = [\underline{G}_m, \infty]$.

C. Intersect results of A and B.

Example 4. Consider again the same transfer function $H(s)$ as in example 2, but with 5% uncertainty in the coefficients of s^1 and s^0 and a finite delay interval $h \in [0, 0.5]$. In this case, Fig. 6 illustrates the procedure of Algorithm 3.

An intersection of $-\Phi_b(\omega)$ and ωH (lower solid and dashdot curves) takes place at only one frequency $\omega = 12.30$. The solution of equation $-\Phi_a(\omega) = 2\pi$ is empty. Namely, ℓ is odd ($\ell = 1$) and we should consider the frequency interval $[12.30, \infty)$ in order to check the delay condition for negative C. From Fig. 6a we obtain

$$C \in (-\infty,\ -1.07) \cup (-1,\ 0) \tag{6.6}$$

In the same manner an intersection of $-(\Phi_b(\omega) + \pi)$ and ωH (lower dashed and dashdot curves) occurs at $\omega = 5.60$ defining the frequency interval $[5.60, \infty)$ for positive C. (Again, the solution of $-\Phi_a(\omega) = \pi$ is empty).

From Fig. 6a we obtain

$$C \in (0,\ 1) \cup (1.19,\ \infty) \tag{6.7}$$

Taking into account condition (4.3) we have

$$C \in (-1, 1). \tag{6.8}$$

The pertinent intervals obtained from the "stability condition" for each of the polynomials
$R_i^+(s, C)$, $R_i^-(s, C)$, $(i = 1, \ldots, 4)$ are

$$R_1^+ : (0.754, \infty)\ R_2^+ : (0.754, \infty)\ R_3^+ : (0.921, \infty)\ R_4^+ : (0.921, \infty)$$
$$R_1^- : (-\infty, -1)\ R_2^- : (-\infty, -1)\ R_3^- : (-\infty, -1)\ R_4^- : (-\infty, -1)$$

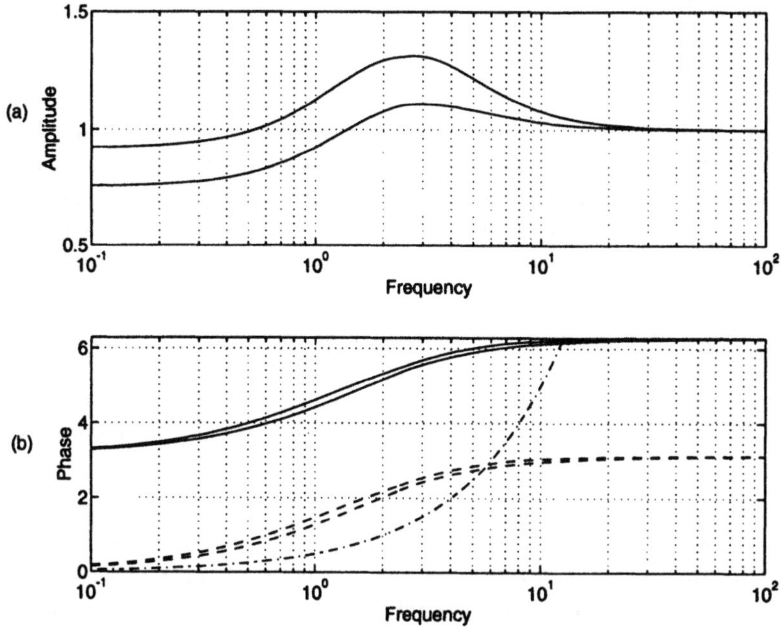

Fig. 6. (a) Amplitude envelope of the family $G(\omega)$ (minimal and maximal curves). (b) $-\Phi_a(\omega)$, $-\Phi_b(\omega)$ (solid curves), $-\Phi_a(\omega) - \pi$, $-\Phi_b(\omega) - \pi$ (dashed curves), and ωH (dashdot curve).

The intersection of the above intervals renders, as a result of the "stability condition", the interval

$$C \in (-\infty, \quad -1) \cup (0.921, \quad \infty) \tag{6.9}$$

Intersecting the intervals obtained from the "stability condition" (6.9) with the intervals obtained from the "delay condition" intersected with condition (4.3), namely (6.8), we obtain the final result. All and only intervals of constant gain controllers which stabilize (5.10) with 5% uncertainty in coefficients and cascaded with a delay element in the interval [0, 0.5] are:

$$C \in (0.921, 1) \tag{6.10}$$

Note that, as expected,

$$(6.10) \subset (5.11) \tag{6.11}$$

Taking 10% uncertainty in coefficients instead of 5%, we obtain that the stability condition renders the following interval: $C \in (-\infty, -1) \cup (1.019, \infty)$. Intersection of this interval with (6.8) yields an empty interval. We conclude that there is no constant gain controller which can stabilize the system with 10% uncertainty.

7 Conclusion

Assuming that an engineering system has an unknown (or partially known) delay, in addition to uncertainties in the coefficients of the rational part of its transfer function, is on one hand very realistic from the standpoint of real life systems, and on the other hand almost has not been considered in the literature. The reason for this stems from the mathematical complexity of this difficult problem.

First, this chapter deals with the problem of stabilization of a delay system by a rational controller. We focus on delay systems independently of delay. The necessary and sufficient conditions for existence of a constant gain controller are derived. These conditions are easily extended to the case of stabilization by any rational controller.

Next we discuss the problem of designing stabilizing controllers for such systems. We treat only the simplest ones namely, static constant gain controllers. However, for this class of controllers, we are able to derive a tractable systematic design method which yields the *entire* set of such feasible stabilizing controllers. The designer is now in an excellent position to choose the optimal controller (in this class) according to whatever criteria is best for the case in hand. Evidently, the method also algorithmically answers the question of existence of a constant gain controller. An empty set result would mean that no constant gain controller can stabilize the system.

The derivation of the design method is carried out in an increasing complexity order. Firstly, it is assumed that the delay is completely unknown (infinite interval). Then, we treat the case of fixed coefficients, but assume that the delay is known to take on a value in a given finite interval. Finally, we derive the method in the case where the delay is in a finite interval and the coefficients of the rational function are also interval ones. By working out the same example for all cases, we can see the change in the interval of the feasible controllers, which is *consistent* with the various assumptions. For example, assuming an infinite interval for the delay yields an empty set of feasible controllers whereas assuming a finite interval for the delay (partial knowledge), yields a non-empty set. Also, assuming uncertainty in the coefficients of the rational transfer function yields a smaller (included) interval of feasible controllers than assuming fixed (known) coefficients. If the uncertainty in the coefficients increase, we again find an empty set of feasible controllers, as in the case of infinite daley interval, although the finite delay interval remains as previously.

These results allow to obtain design algorithms for stabilizing controllers, when appropriate. Conditions for existence of rational controllers for finite interval delay systems and design algorithms require future research. Other extensions of the approach derived in this paper may be studied in the following directions: dependency of the coefficients of the rational transfer function on physical interval parameters and multi-input multi-output systems.

References

1. Barmish, B.R., New tools for robustness of linear systems, Macmillan, 1994.
2. Barmish, B.R., Hollot, C.V., Kraus, F.J. and R. Tempo, "Extreme point results for robust stabilization of interval plants with first order compensators," *IEEE Trans. Aut. Contr.*, **AC-37**, pp. 707–714, 1992.
3. Bhattacharyya, S.P., Chapellat, H. and H.Keel, Robust Control: the parametric approach, Prentice Hall, 1995.
4. Desoer, C.A., Liu, R.W., Murray J. and R. Saeks, "Feedback system design: The fractional representation approach to analysis and synthesis", *IEEE Trans. Aut. Contr.*, **AC-25**, pp. 399–412, 1980.
5. Fruchter, G., Srebro, U. and E. Zeheb, "On several variable zero sets and application to MIMO robust feedback stabilization," *IEEE Trans. Circ. and Syst.*, **CAS-34**, pp. 1208–1220, 1987.
6. Fruchter, G., Srebro, U. and E. Zeheb, "Conditions on the boundary of the zero set and applications to stabilization of systems with uncertainty," *Jour. of Math. Anal. and Appl.*, **161**, pp. 148–175, 1991.
7. Fu, M., Olbrot, A.V. and M.P. Polis, "Robust stability for time-delay systems: the edge theorem and graphical tests", *IEEE Trans. on AC*, **34**, pp. 813–820, 1989.
8. Ghosh, B.K., "Some new results on the simultaneous stabilizability of a family of single input single output systems," *Syst. and Contr. Lett.*, **6**, pp. 39–45, 1985.
9. Ghosh, B.K., "An Approach to Simultaneous System Design. Part II: Nonswitching Gain and Dynamic Feedback Compensation by Algebraic Geometric Methods," *SIAM J. Control and Optim.*, **26**, pp. 919–963, 1988.
10. Hollot, C.V. and F. Yang, "Robust stabilization of interval plants using lead or lag compensators," *Syst. and Contr. Lett.*, **14**, pp. 9–12, 1990.
11. Khargonekar, P.P. and A. Tannenbaum, "Non-Euclidean metrics and the robust stabilization of systems with parameter uncertainty," *IEEE Trans. Aut. Contr.*, **AC-30**, pp. 1005–1015, 1985.
12. Kharitonov, V.L, "Asymptotic stability of an equilibrium position of a family of systems of linear differential equations", *Diff. Eqn.*, **14**, pp. 1483–1485, 1979.
13. Kogan, J., Robust Stability and Convexity: An Introduction, Springer, London, 1995.
14. Kogan, J. and A. Leizarowitz, "Frequency domain criterion for robust stability of interval time-delay systems", *Automatica*, **31**, pp. 463–469, 1995.
15. Levkovich, A., Zeheb, E. and N. Cohen, "Frequency Response Envelopes of a Family of Uncertain Continuous-Time Systems", *IEEE Trans. on CAS*, Vol. 42, pp. 156–165, 1995.
16. Naimark, L., Kogan, J., Leizarowitz, A. and E. Zeheb, "When is a delay system stabilizable by a rational controller", (submitted to 4th ECC, Brussel), 1997.
17. Naimark, L. and E. Zeheb, "All Constant Gain Stabilizing Controllers for an Interval Delay System with Uncertain Parameters", Technical report EE-981, Technion, IIT, 1995, (also submitted (revised version) to *Automatica*).
18. Naimark, L. and E. Zeheb, "Constant Gain Stabilizing Controller Design for Bounded Interval Delay System", 19th convention of Electrical and Electronics Engineers in Israel, IEEE, Jerusalem, Nov 5-6, pp. 519–522, 1996.
19. Saeks, R. ans J. Murray, "Fractional representation, algebraic geometry and the simultaneous stabilization problem," *IEEE Trans. Aut. Contr.*, **AC-27**, pp. 895–903, 1982.

20. Tannenbaum, A., "Feedback stabilization of linear dynamical plants with uncertainty in the gain factor," *Int. J. Cont.*, **32**, pp. 1–16, 1980.
21. Tsypkin, Ya.Z. and M. Fu, "Robust stability of time-delay systems with an uncertain time-delay constant", *Int. J. Contr.*, **59**, pp. 865–879, 1993.
22. Vidyasagar, M. and N. Viswanadham, "Algebraic design techniques for reliable stabilization," *IEEE Trans. Aut. Contr.*, **AC-27**, pp. 1085–1095, 1982.
23. Youla, D.C., Bongiorno, J.J. and H.A. Jabr, "Modern Wiener-Hopf design of optimal controllers", Part I, *IEEE Trans. Aut. Contr.*, **AC-21**, pp. 3–13, 1976.

Stabilization of Linear and Nonlinear Systems with Time Delay

Wassim M. Haddad[1], Vikram Kapila[2] and Chaouki T. Abdallah[3]

[1] School of Aerospace Engineering, Georgia Institute of Technology, Atlanta, GA 30332-0150
[2] Department of Mechanical, Aerospace, and Manufacturing Engineering, Polytechnic University, Brooklyn, NY 11201
[3] Department of Electrical and Computer Engineering, University of New Mexico, Albuquerque, NM 87131

Abstract. This chapter considers the problem of stabilizing linear and nonlinear continuous-time systems with state and measurement delay. For linear systems we address stabilization via fixed-order dynamic output feedback compensators and present sufficient conditions for stabilization involving a system of modified Riccati equations. For nonlinear systems we provide sufficient conditions for the design of static full-state feedback stabilizing controllers. The controllers obtained are delay-independent and hence apply to systems with infinite delay.

1 Introduction

In dynamical systems such as the control of flexible structures with noncollocated sensors and actuators, teleoperators, biological systems [1], and electrical networks [2], time delay arises frequently and can severely degrade closed-loop system performance and in some cases drive the system to instability. Since controllers designed with the assumption of instantaneous information and power transfer may fail to stabilize dynamic systems with time delay [3] it is of paramount importance that delay system dynamics be accounted for in the control-system design process. There exists an extensive literature devoted to the control of dynamical systems with time delay (see, for example, [4, 5, 6, 7, 8, 9, 10, 11, 12] and the numerous references therein). Three main approaches can be distinguished for designing stabilizing controllers for delay systems. Namely:

i) Stabilization independent of delay amount [13, 14]: In this approach the delay can be large (even infinite) without destabilizing the closed-loop system. However, the conditions for stabilization are often conservative.
ii) Stabilization dependent on delay amount [15, 16, 17]: Such approaches rely on Razumikin-like theorems [18] and provide stabilization conditions if the delay is less than a given amount.
iii) Stabilization based on delay amount [19, 20]: In this approach there exist delay windows which allow a stabilizing compensator to exist, while no

stabilizing compensators are possible outside these windows. This approach however applies to a restricted class of systems.

In this chapter we design feedback controllers which are independent of the delay amount. Furthermore, we address both linear and nonlinear dynamical systems. Specifically, we present a rigorous development of sufficient conditions via fixed-order dynamic compensation and static full-state feedback controllers for stabilization of systems with state and measurement delay. For linear plants these sufficient conditions are in the form of a coupled system of algebraic Riccati equations that explicitly characterize dynamic controllers of fixed dimension while for nonlinear plants our sufficient condition is given by a modified Riccati equation for characterizing static full-state feedback controllers. We emphasize that our approach is constructive in nature rather than existential. In particular, as opposed to the results of [6] which are based on the total stability theorem [21] our sufficient conditions provide explicit formulae for controller gains that guarantee stabilization of systems with time delay. For the linear plant case, in order to account for closed-loop system performance our framework also includes minimization of a given performance functional. Finally, even though for simplicity of exposition we do not address system parametric uncertainty as in [7, 22, 23] the proposed approach can be merged with the guaranteed cost control approach [24] to provide robust stability and performance in the face of system uncertainty and system delay.

The contents of the chapter are as follows. In Section 2 we state the problem of fixed-order dynamic compensation for systems with state and measurement delay. Sufficient conditions for stabilization of systems with time delay are given in Section 3 Section 4 provides design equations for characterizing fixed-order dynamic controllers for linear systems with time delay. In Section 5 we state the full-state feedback control problem for nonlinear systems with time delay and provide design equations for full-state feedback controllers. Section 6 provides two illustrative numerical examples. Finally, Section 7 gives conclusions.

Nomenclature

$\mathbb{R}, \mathbb{R}^{r\times s}, \mathbb{R}^r$	$-$real numbers, $r \times s$ real matrices, $\mathbb{R}^{r\times 1}$
$()^T$, $()^{-1}$, $\mathrm{tr}\,()$	$-$transpose, inverse, trace
$I_r, 0_r$	$-r \times r$ identity matrix, $r \times r$ zero matrix
$\Vert \cdot \Vert_2$	$-$Euclidean vector norm
$\lambda_{\min}(Z)$	$-$minimum eigenvalue of the symmetric matrix Z
$\alpha, \gamma, \epsilon, \sigma$	$-$real positive scalars
$n, l, m, n_c, \tilde{n}$	$-$positive integers; $1 \le n_c \le n$; $\tilde{n} = n + n_c$
$x, u, y, x_c, \tilde{x}$	$-n-, m-, l-, n_c-, \tilde{n}-$ dimensional vectors
A, B, C	$-n \times n, n \times m, l \times n$ matrices
$A_{\mathrm{d}}, C_{\mathrm{d}}$	$-n \times n, l \times n$ matrices
A_c, B_c, C_c, K	$-n_c \times n_c, n_c \times l, m \times n_c, m \times n$ matrices
V_1, V_2	$-n \times n, l \times l$ matrices
R_1, R_2	$-n \times n, m \times m$ matrices

2 Fixed-Order Controller Synthesis for Systems with Time Delay

In this section we introduce the fixed-order dynamic compensation problem for linear systems with state and measurement delays. Specifically, given the nth-order stabilizable and detectable dynamical system, where stabilizability and detectability are defined in the sense of [25], with state and measurement delay

$$\begin{aligned}\dot{x}(t) &= Ax(t) + A_{\mathrm{d}}x(t-\tau_{\mathrm{d}}) + Bu(t), \quad t \in [0,\infty), \quad \tau_{\mathrm{d}} > 0, \\ &\quad x(t) = \phi(t), \quad t \in [-\tau_{\mathrm{d}}, 0], \quad x(0) = \phi(0) = x_0, \end{aligned} \tag{2.1}$$

$$y(t) = Cx(t) + C_{\mathrm{d}}x(t-\tau_{\mathrm{d}}), \tag{2.2}$$

where $u(t) \in \mathbb{R}^m$, $y(t) \in \mathbb{R}^l$, and $\phi : \mathbb{R}^+ \to \mathbb{R}^n$ is a continuous vector valued function specifying the initial state of the system, determine an n_c^{th}-order $(1 \leq n_c \leq n)$ dynamic compensator

$$\dot{x}_c(t) = A_c x_c(t) + B_c y(t), \qquad x_c(0) = x_{c0}, \tag{2.3}$$

$$u(t) = C_c x_c(t), \tag{2.4}$$

which satisfies the following design criteria:

i) the closed-loop system (2.1)–(2.4) is asymptotically stable; and

ii) the performance functional

$$J(x(t), x_c(t), x(t-\tau_{\mathrm{d}})) \triangleq \int_0^\infty L(x(t), x_c(t), x(t-\tau_{\mathrm{d}}))dt, \tag{2.5}$$

where $L : \mathbb{R}^n \times \mathbb{R}^{n_c} \times \mathbb{R}^n \to \mathbb{R}$, is minimized. An explicit characterization of $L(x(t), x_c(t), x(t-\tau_{\mathrm{d}}))$, $t \geq 0, \tau_{\mathrm{d}} > 0$, is given in Section 3

3 Sufficient Conditions for Stabilization of Systems with Time Delay

In this section we provide a Riccati equation that guarantees that the closed-loop system (2.1)–(2.4) consisting of the nth-order time-delayed system (2.1), (2.2) and the n_c^{th}-order dynamic compensator (2.3), (2.4) is asymptotically stable. First note that for a given fixed-order controller (A_c, B_c, C_c) the closed-loop system (2.1)–(2.4) can be written as

$$\dot{\tilde{x}}(t) = \tilde{A}\tilde{x}(t) + \tilde{A}_{\mathrm{d}}\tilde{x}(t-\tau_{\mathrm{d}}), \quad \tilde{x}(0) = \tilde{x}_0, \quad t \in [0,\infty), \quad \tau_{\mathrm{d}} > 0, \tag{3.1}$$

where

$$\tilde{x}(t) \triangleq \begin{bmatrix} x(t) \\ x_c(t) \end{bmatrix}, \quad \tilde{A} \triangleq \begin{bmatrix} A & BC_c \\ B_cC & A_c \end{bmatrix}, \quad \tilde{A}_{\mathrm{d}} \triangleq \begin{bmatrix} A_{\mathrm{d}} & 0_{n\times n_c} \\ B_cC_{\mathrm{d}} & 0_{n_c\times n_c} \end{bmatrix}.$$

For the statement of the next result define

$$\hat{I} \triangleq \begin{bmatrix} I_n & 0_{n\times n_c} \\ 0_{n_c\times n} & 0_{n_c} \end{bmatrix}.$$

Theorem 1. *Let (A_c, B_c, C_c) be given. Suppose there exists an $\tilde{n} \times \tilde{n}$ positive-definite matrix $\tilde{P}$ and scalars $\alpha, \epsilon > 0$ such that*

$$0 = \tilde{A}^T \tilde{P} + \tilde{P}\tilde{A} + \epsilon\tilde{P} + \alpha^2 \hat{I} + \alpha^{-2}\tilde{P}\tilde{A}_{\rm d}\tilde{A}_{\rm d}^T\tilde{P} + \tilde{R}, \tag{3.2}$$

where $\tilde{R}$ is an $\tilde{n} \times \tilde{n}$ nonnegative-definite matrix. Then the function

$$V(\tilde{x}) = \tilde{x}^T\tilde{P}\tilde{x} + \alpha^2 \int_{t-\tau_{\rm d}}^{t} \tilde{x}^T(s)\hat{I}\tilde{x}(s)ds, \tag{3.3}$$

is a Lyapunov function that guarantees that the closed-loop system (3.1) is globally asymptotically stable.

Proof. First note that since $\tilde{P}$ is positive definite it follows that the Lyapunov function candidate $V(\tilde{x})$ given by (3.3) is positive definite. The corresponding Lyapunov derivative along the trajectories $\tilde{x}(t)$, $t \geq 0$, of the closed-loop system (3.1) is given by

$$\begin{aligned}\dot{V}(\tilde{x}(t)) &= \tilde{x}^T(t)(\tilde{A}^T\tilde{P} + \tilde{P}\tilde{A})\tilde{x}(t) + 2\tilde{x}^T(t-\tau_{\rm d})\tilde{A}_{\rm d}^T\tilde{P}\tilde{x}(t)\\ &\quad +\alpha^2\frac{d}{dt}\left[\int_{t-\tau_{\rm d}}^{t}\tilde{x}^T(s)\hat{I}\tilde{x}(s)ds\right], \quad t \geq 0,\end{aligned} \tag{3.4}$$

or, using

$$\frac{d}{dt}\left[\int_{t-\tau_{\rm d}}^{t}\tilde{x}^T(s)\hat{I}\tilde{x}(s)ds\right] = \tilde{x}^T(t)\hat{I}\tilde{x}(t) - \tilde{x}^T(t-\tau_{\rm d})\hat{I}\tilde{x}(t-\tau_{\rm d}),$$

(3.4) becomes

$$\begin{aligned}\dot{V}(\tilde{x}(t)) &= \tilde{x}^T(t)(\tilde{A}^T\tilde{P} + \tilde{P}\tilde{A} + \alpha^2\hat{I})\tilde{x}(t) + 2\tilde{x}^T(t-\tau_{\rm d})\tilde{A}_{\rm d}^T\tilde{P}\tilde{x}(t)\\ &\quad -\alpha^2\tilde{x}^T(t-\tau_{\rm d})\hat{I}\tilde{x}(t-\tau_{\rm d}), \quad t \geq 0.\end{aligned} \tag{3.5}$$

Furthermore, using (3.2) and grouping terms yields

$$\begin{aligned}\dot{V}(\tilde{x}(t)) &= -\epsilon\tilde{x}^T(t)\tilde{P}\tilde{x}(t) - \tilde{x}^T(t)\tilde{R}\tilde{x}(t) - [\alpha^{-1}\tilde{A}_{\rm d}^T\tilde{P}\tilde{x}(t) - \alpha\hat{I}\tilde{x}(t-\tau_{\rm d})]^T\\ &\quad \cdot[\alpha^{-1}\tilde{A}_{\rm d}^T\tilde{P}\tilde{x}(t) - \alpha\hat{I}\tilde{x}(t-\tau_{\rm d})], \quad t \geq 0.\end{aligned} \tag{3.6}$$

Since $\tilde{P}$ is positive definite it follows that $\dot{V}(\tilde{x}(t)) < 0$, $\tilde{x}(t) \neq 0$, $t \geq 0$, and hence $V(\cdot)$ is a Lyapunov function for the closed-loop system (3.1). □

Next, we consider an explicit characterization of $L(x(t), x_c(t), x(t-\tau_{\rm d}))$ in (2.5). Specifically, let $\tilde{R} = \begin{bmatrix} R_1 & 0 \\ 0 & C_c^T R_2 C_c \end{bmatrix}$, where $R_1 \geq 0$ and $R_2 > 0$, and define

$$\begin{aligned}L(x(t), x_c(t), x(t-\tau_{\rm d})) &\triangleq \tilde{x}^T(t)[\epsilon\tilde{P} + \tilde{R}_1 + \alpha^{-2}\tilde{P}\tilde{A}_{\rm d}\tilde{A}_{\rm d}^T\tilde{P}]\tilde{x}(t)\\ &\quad +u^T(t)R_2u(t) + \alpha^2\tilde{x}^T(t-\tau_{\rm d})\hat{I}\tilde{x}(t-\tau_{\rm d})\\ &\quad -2\tilde{x}^T(t-\tau_{\rm d})\tilde{A}_{\rm d}^T\tilde{P}\tilde{x}(t), \quad t \geq 0,\end{aligned} \tag{3.7}$$

where $\tilde{R}_1 \triangleq \begin{bmatrix} R_1 & 0 \\ 0 & 0 \end{bmatrix}$. Now, since $\tilde{x}(t) \to 0$ as $t \to \infty$, where $\tilde{x}(t)$, $t \geq 0$, satisfies (3.1), the performance functional (2.5) reduces to

$$\begin{aligned} J(x(t), x_c(t), x(t-\tau_d)) &= \int_0^\infty \Big[\tilde{x}^T(t)[\epsilon\tilde{P} + \tilde{R}_1 + \alpha^{-2}\tilde{P}\tilde{A}_d\tilde{A}_d^T\tilde{P}]\tilde{x}(t) \\ &\quad + u^T(t)R_2u(t) + \alpha^2\tilde{x}^T(t-\tau_d)\hat{I}\tilde{x}(t-\tau_d) \\ &\quad - 2\tilde{x}^T(t-\tau_d)\tilde{A}_d^T\tilde{P}\tilde{x}(t)\Big]dt \\ &= -\int_0^\infty \dot{V}(\tilde{x})dt \\ &= V(\tilde{x}(0)) - \lim_{t\to\infty} V(\tilde{x}(t)) \\ &= \tilde{x}^T(0)\tilde{P}\tilde{x}(0) + \Phi, \end{aligned} \tag{3.8}$$

where $\Phi \triangleq \int_{-\tau_d}^0 \phi^T(s)\phi(s)ds$ is a positive constant. With $L(x(t), x_c(t), x(t-\tau_d))$ given by (3.7) the performance functional (2.5) has the same form as the H_2 cost in standard LQG theory. Specifically, $J(\tilde{x}(0)) = \tilde{x}^T(0)\tilde{P}\tilde{x}(0) + \Phi =$ tr $\tilde{P}\tilde{x}(0)\tilde{x}^T(0) + \Phi$. Hence, we replace $\tilde{x}(0)\tilde{x}^T(0)$ by $\tilde{V} \triangleq \begin{bmatrix} V_1 & 0 \\ 0 & B_cV_2B_c^T \end{bmatrix}$, where $V_1 \geq 0$ and $V_2 > 0$, and proceed by determining controller gains that minimize tr $\tilde{P}\tilde{V} + \Phi$. This leads to the following optimization problem.

Auxiliary Minimization Problem. Determine (A_c, B_c, C_c) that minimizes $\mathcal{J}(\tilde{P}, A_c, B_c, C_c) \triangleq$ tr $\tilde{P}\tilde{V}$ where $\tilde{P} > 0$ satisfies (3.2) and such that (A_c, B_c, C_c) is minimal.

It follows from Theorem 1 that by deriving necessary conditions for the Auxiliary Minimization Problem we obtain sufficient conditions for characterizing dynamic output feedback controllers ensuring stabilization of closed-loop systems with time delay.

4 Fixed-Order Dynamic Compensation for Systems with Time Delay

In this section we present the main theorem characterizing fixed-order dynamic controllers for (2.1), (2.2). Note that for design flexibility the compensator order n_c may be less than the plant order n. We shall require for technical reasons that $C_dC_d^T = \sigma^2V_2$, where the nonnegative scalar σ is a design variable. The following lemma is required for the statement of main theorem.

Lemma 2 [24]. *Let $\hat{Q}, \hat{P}$ be $n\times n$ nonnegative-definite matrices and suppose that* rank $\hat{Q}\hat{P} = n_c$. *Then there exist $n_c \times n$ matrices G, Γ and an $n_c \times n_c$ invertible matrix M, unique except for a change of basis in $\mathbb{R}^{n_c}$, such that*

$$\hat{Q}\hat{P} = G^TM\Gamma, \qquad \Gamma G^T = I_{n_c}. \tag{4.1}$$

Furthermore, the $n \times n$ matrices $\tau \triangleq G^T\Gamma$ and $\tau_\perp \triangleq I_n - \tau$ are idempotent and have rank n_c and $n - n_c$, respectively.

For convenience in stating the main result of this section we define the notation $S \triangleq (I + \alpha^{-2}\sigma^2 Q\hat{P})^{-1}$, for arbitrary $n \times n$ nonnegative-definite matrices $Q, \hat{P}$ and

$$\begin{aligned}
Q_a &\triangleq Q[C + \alpha^{-2}C_\mathrm{d}A_\mathrm{d}^T(P + \hat{P})]^T,\\
A_\epsilon &\triangleq A + \frac{1}{2}\epsilon I_n,\\
A_{\hat{P}} &\triangleq A_\epsilon - SQ_aV_2^{-1}(C + \alpha^{-2}C_\mathrm{d}A_\mathrm{d}^T P) + \alpha^{-2}A_\mathrm{d}A_\mathrm{d}^T P,\\
A_Q &\triangleq A_\epsilon + \alpha^{-2}A_\mathrm{d}A_\mathrm{d}^T(P + \hat{P}) - \alpha^{-2}A_\mathrm{d}C_\mathrm{d}^T V_2^{-1}Q_a^T S^T\hat{P},\\
A_{\hat{Q}} &\triangleq A_\epsilon - BR_2^{-1}B^T P + \alpha^{-2}A_\mathrm{d}A_\mathrm{d}^T P,
\end{aligned}$$

for arbitrary $P, Q, \hat{P} \in \mathbb{R}^{n\times n}$ and $\alpha, \epsilon, \sigma > 0$. Note that since $Q, \hat{P}$ are nonnegative definite and the eigenvalues of $Q\hat{P}$ coincide with the eigenvalues of the nonnegative-definite matrix $Q^{1/2}\hat{P}Q^{1/2}$, it follows that $Q\hat{P}$ has nonnegative eigenvalues. Thus, the eigenvalues of $I + \alpha^{-2}\sigma^2 Q\hat{P}$ are all greater than one so that S exists.

Theorem 3. *Assume $\alpha, \epsilon, \sigma > 0$ and suppose there exist $n \times n$ nonnegative-definite matrices $P, Q, \hat{P}$, and $\hat{Q}$ satisfying*

$$\begin{aligned}
0 &= A_\epsilon^T P + PA_\epsilon + R_1 + \alpha^2 I_n + \alpha^{-2}PA_\mathrm{d}A_\mathrm{d}^T P - PBR_2^{-1}B^T P\\
&\quad + \tau_\perp^T PBR_2^{-1}B^T P\tau_\perp, &(4.2)\\
0 &= A_Q Q + QA_Q^T + V_1 - SQ_aV_2^{-1}Q_a^T S^T + \tau_\perp SQ_aV_2^{-1}Q_a^T S^T\tau_\perp^T, &(4.3)\\
0 &= A_{\hat{P}}^T\hat{P} + \hat{P}A_{\hat{P}} + \alpha^{-2}\hat{P}[\sigma^2 SQ_aV_2^{-1}Q_a^T S^T - A_\mathrm{d}C_\mathrm{d}^T V_2^{-1}Q_a^T S^T\\
&\quad - SQ_aV_2^{-1}C_\mathrm{d}A_\mathrm{d}^T]\hat{P} + \alpha^{-2}\hat{P}A_\mathrm{d}A_\mathrm{d}^T\hat{P} + PBR_2^{-1}B^T P\\
&\quad - \tau_\perp^T PBR_2^{-1}B^T P\tau_\perp, &(4.4)\\
0 &= A_{\hat{Q}}\hat{Q} + \hat{Q}A_{\hat{Q}}^T + SQ_aV_2^{-1}Q_a^T S^T - \tau_\perp SQ_aV_2^{-1}Q_a^T S^T\tau_\perp^T, &(4.5)\\
&\operatorname{rank}\hat{Q} = \operatorname{rank}\hat{P} = \operatorname{rank}\hat{Q}\hat{P} = n_c, &(4.6)
\end{aligned}$$

and let A_c, B_c, and C_c be given by

$$\begin{aligned}
A_c &= \Gamma[A - SQ_aV_2^{-1}(C + \alpha^{-2}C_\mathrm{d}A_\mathrm{d}^T P)\\
&\quad + (\alpha^{-2}A_\mathrm{d}A_\mathrm{d}^T - BR_2^{-1}B^T)P]G^T, &(4.7)\\
B_c &= \Gamma SQ_aV_2^{-1}, &(4.8)\\
C_c &= -R_2^{-1}B^T PG^T. &(4.9)
\end{aligned}$$

Then

$$\tilde{P} = \begin{bmatrix} P + \hat{P} & -\hat{P}G^T \\ -G\hat{P} & G\hat{P}G^T \end{bmatrix},$$

satisfies (3.2), (A_c, B_c, C_c) is an extremal of $\mathcal{J}(\tilde{P}, A_c, B_c, C_c)$, and $(\tilde{A}, \tilde{R})$ is detectable if and only if $\tilde{A}$ is asymptotically stable. Furthermore, the feedback interconnection of the delay system (2.1), (2.2) and the fixed-order controller (2.3), (2.4) is asymptotically stable for all $\tau_d > 0$. Finally, the cost $\mathcal{J}(\tilde{P}, A_c, B_c, C_c)$ is given by

$$\mathcal{J}(\tilde{P}, A_c, B_c, C_c) = \mathrm{tr}[(P + \hat{P})V_1 + \hat{P}SQ_a V_2^{-1} Q_a^T S^T]. \tag{4.10}$$

Proof. The proof is constructive in nature. Specifically, first we obtain necessary conditions for the Auxiliary Minimization Problem and show by construction that these conditions serve as sufficient conditions for closed-loop stability. For details of a similar proof see [26]. □

Next, we specialize Theorem 3 to the full-order case. Specifically, setting $n_c = n$ so that $\tau = G = \Gamma = I_n$ and $\tau_\perp = 0$ the last term in each of (4.2)–(4.5) is zero and (4.5) is superfluous. Hence, the following corollary is immediate.

Corollary 4. *Let $n_c = n$, assume $\alpha, \epsilon, \sigma > 0$, and suppose there exist $n \times n$ nonnegative-definite matrices P, Q, and $\hat{P}$ satisfying*

$$\begin{aligned} 0 &= A_\epsilon^T P + P A_\epsilon + R_1 + \alpha^2 I_n + \alpha^{-2} P A_d A_d^T P - P B R_2^{-1} B^T P, & (4.11)\\ 0 &= A_Q Q + Q A_Q^T + V_1 - S Q_a V_2^{-1} Q_a^T S^T, & (4.12)\\ 0 &= A_{\hat{P}}^T \hat{P} + \hat{P} A_{\hat{P}} + \alpha^{-2} \hat{P}[\sigma^2 S Q_a V_2^{-1} Q_a^T S^T - A_d C_d^T V_2^{-1} Q_a^T S^T \\ &\quad - S Q_a V_2^{-1} C_d A_d^T]\hat{P} + \alpha^{-2} \hat{P} A_d A_d^T \hat{P} + P B R_2^{-1} B^T P, & (4.13) \end{aligned}$$

and let A_c, B_c, and C_c be given by

$$\begin{aligned} A_c &= A - S Q_a V_2^{-1}(C + \alpha^{-2} C_d A_d^T P) + (\alpha^{-2} A_d A_d^T - B R_2^{-1} B^T)P, & (4.14)\\ B_c &= S Q_a V_2^{-1}, & (4.15)\\ C_c &= -R_2^{-1} B^T P. & (4.16) \end{aligned}$$

Then

$$\tilde{P} = \begin{bmatrix} P + \hat{P} & -\hat{P} \\ -\hat{P} & \hat{P} \end{bmatrix},$$

satisfies (3.2), (A_c, B_c, C_c) is an extremal of $\mathcal{J}(\tilde{P}, A_c, B_c, C_c)$, and $(\tilde{A}, \tilde{R})$ is detectable if and only if $\tilde{A}$ is asymptotically stable. Furthermore, the feedback interconnection of the delay system (2.1), (2.2) and the fixed-order controller (2.3), (2.4) is asymptotically stable for all $\tau_d > 0$. Finally, the cost $\mathcal{J}(\tilde{P}, A_c, B_c, C_c)$ is given by

$$\mathcal{J}(\tilde{P}, A_c, B_c, C_c) = \mathrm{tr}[(P + \hat{P})V_1 + \hat{P}SQ_a V_2^{-1} Q_a^T S^T]. \tag{4.17}$$

5 Full-State Feedback Control for Nonlinear Systems with Time Delay

In this section we introduce the full-state feedback control problem for nonlinear systems with delay. Specifically, given the dynamical system with nonlinear state delay

$$\begin{aligned} \dot{x}(t) &= Ax(t) + f_{\mathrm{d}}(x(t-\tau_{\mathrm{d}})) + Bu(t), \quad t \in [0,\infty), \quad \tau_{\mathrm{d}} > 0, \\ & x(t) = \phi(t), \quad t \in [-\tau_{\mathrm{d}}, 0], x(0) = \phi(0) = x_0, \quad f_{\mathrm{d}}(0) = 0, \end{aligned} \tag{5.1}$$

where $x \in \mathbb{R}^n$, $u \in \mathbb{R}^m$, $f_{\mathrm{d}} : \mathbb{R}^+ \to \mathbb{R}^n$, and $\phi : \mathbb{R}^+ \to \mathbb{R}^n$ is a continuous vector valued function specifying the initial state of the system, determine a full-state feedback control law

$$u(t) = Kx(t), \tag{5.2}$$

such that the closed-loop system (5.1), (5.2) is asymptotically stable.

Next, we show that if $f_{\mathrm{d}}(\cdot)$ in (5.1) satisfies $||f_{\mathrm{d}}(x)||_2 \le \gamma ||x||_2$, where $x \in \mathbb{R}^n$ and $\gamma > 0$, we can construct a full-state feedback control law (5.2) to stabilize the nonlinear time-delay system (5.1) independent of the delay amount τ_{d}. This result is an extension of the result in [27] where a stabilizing state feedback controller was obtained for purely linear time-delay systems.

Theorem 5. *Let $||f_{\mathrm{d}}(x)||_2 \le \gamma ||x||_2$, where $x \in \mathbb{R}^n$ and $\gamma > 0$, and suppose there exists an $n \times n$ positive-definite matrix P such that*

$$0 = A^T P + PA + \alpha^{-2} P^2 - 2PBR_2^{-1} B^T P + R_1, \tag{5.3}$$

where $\alpha > 0$, $\lambda_{\min}(R_1) > \alpha^2 \gamma^2$, and $R_2 > 0$. Furthermore, let the feedback control gain K in (5.2) be given by

$$K = -R_2^{-1} B^T P. \tag{5.4}$$

Then, for all $\tau_{\mathrm{d}} > 0$, the closed-loop system (5.1), (5.2) is globally asymptotically stable with Lyapunov function

$$V(x) = x^T P x + \alpha^2 \int_{t-\tau_{\mathrm{d}}}^{t} f_{\mathrm{d}}^T(x(s)) f_{\mathrm{d}}(x(s)) ds. \tag{5.5}$$

Proof. First note that since P is positive definite and $f_{\mathrm{d}}(x) = 0$ for $x = 0$, it follows that the Lyapunov function candidate $V(x)$ given by (5.5) is positive definite. The corresponding Lyapunov derivative along the trajectories $x(t)$, $t \ge 0$, of the closed-loop system (5.1), (5.2) is given by

$$\begin{aligned} \dot{V}(x(t)) &= \dot{x}^T(t) P x(t) + x^T(t) P \dot{x}(t) \\ & \quad + \alpha^2 \frac{d}{dt} \left[\int_{t-\tau_{\mathrm{d}}}^{t} f_{\mathrm{d}}^T(x(s)) f_{\mathrm{d}}(x(s)) ds \right], \quad t \ge 0, \end{aligned} \tag{5.6}$$

or, using (5.1) and

$$\frac{d}{dt}\left[\int_{t-\tau_d}^{t} f_d^T(x(s))f_d(x(s))ds\right] = f_d^T(x(t))f_d(x(t)) \\ -f_d^T(x(t-\tau_d))f_d(x(t-\tau_d)),$$

(5.6) becomes

$$\begin{aligned}\dot{V}(x(t)) &= x^T(t)[A^TP+PA]x(t)+u^T(t)B^TPx(t)+x^T(t)PBu(t)\\ &\quad +f_d^T(x(t-\tau_d))Px(t)+x^T(t)Pf_d(x(t-\tau_d))\\ &\quad +\alpha^2 f_d^T(x(t))f_d(x(t))\\ &\quad -\alpha^2 f_d^T(x(t-\tau_d))f_d(x(t-\tau_d)), \quad t\geq 0. \end{aligned} \tag{5.7}$$

Next, adding and subtracting $\alpha^{-2}x^T(t)P^2x(t)$, $t \geq 0$, to and from (5.7) and grouping terms yields

$$\begin{aligned}\dot{V}(x(t)) &= x^T(t)[A^TP+PA+\alpha^{-2}P^2]x(t)+u^T(t)B^TPx(t)\\ &\quad +x^T(t)PBu(t)-[\alpha f_d(x(t-\tau_d))-\alpha^{-1}Px(t)]^T\\ &\quad \cdot[\alpha f_d(x(t-\tau_d))-\alpha^{-1}Px(t)]\\ &\quad +\alpha^2 f_d^T(x(t))f_d(x(t)), \quad t\geq 0. \end{aligned} \tag{5.8}$$

Now using the control law $u(t) = -R_2^{-1}B^TPx(t)$, $t \geq 0$, in (5.8) yields

$$\begin{aligned}\dot{V}(x(t)) &= x^T(t)[A^TP+PA-P(2BR_2^{-1}B^T-\alpha^{-2}I)P]x(t)\\ &\quad -[\alpha f_d(x(t-\tau_d))-\alpha^{-1}Px(t)]^T[\alpha f_d(x(t-\tau_d))-\alpha^{-1}Px(t)],\\ &\quad +\alpha^2 f_d^T(x(t))f_d(x(t)) \quad t\geq 0. \end{aligned} \tag{5.9}$$

Finally, using (5.3), it follows that

$$\begin{aligned}\dot{V}(x(t)) &= -x^T(t)R_1x(t)+\alpha^2 f_d^T(x(t))f_d(x(t))\\ &\quad -[\alpha f_d(x(t-\tau_d))-\alpha^{-1}Px(t)]^T[\alpha f_d(x(t-\tau_d))-\alpha^{-1}Px(t)]\\ &\leq -[\lambda_{\min}(R_1)-\alpha^2\gamma^2]\|x(t)\|_2^2-[\alpha f_d(x(t-\tau_d))-\alpha^{-1}Px(t)]^T\\ &\quad \cdot[\alpha f_d(x(t-\tau_d))-\alpha^{-1}Px(t)], \quad t\geq 0. \end{aligned} \tag{5.10}$$

Since $\lambda_{\min}(R_1) > \alpha^2\gamma^2$ and $x(t) \neq 0$, $t \geq 0$, it follows that $\dot{V}(x(t)) < 0$, $x(t) \neq 0$, $t \geq 0$, and hence $V(\cdot)$ is a Lyapunov function for the closed-loop system (5.1), (5.2). □

6 Illustrative Numerical Examples

In this section we provide two numerical examples to demonstrate the proposed Riccati equation approach for delay systems. For simplicity we consider the design of full-order dynamic compensators and full-state feedback controllers. The design equations (4.11)–(4.13) were solved using a homotopy continuation algorithm. For details of a similar algorithm see [26].

Example 1. Consider the second-order system

$$\begin{aligned}\begin{bmatrix}\dot{x}_1(t)\\ \dot{x}_2(t)\end{bmatrix} &= \begin{bmatrix}0 & 1\\ -0.3 & -1\end{bmatrix}\begin{bmatrix}x_1(t)\\ x_2(t)\end{bmatrix} + \begin{bmatrix}0.145 & 0.75\\ 0.275 & -0.2\end{bmatrix}\begin{bmatrix}x_1(t-\tau_{\mathrm{d}})\\ x_2(t-\tau_{\mathrm{d}})\end{bmatrix}\\ &\quad + \begin{bmatrix}0.5\\ 0.2\end{bmatrix}u(t),\\ y(t) &= \begin{bmatrix}2.1 & 4\end{bmatrix}\begin{bmatrix}x_1(t)\\ x_2(t)\end{bmatrix} + \begin{bmatrix}6.0 & 5.0\end{bmatrix}\begin{bmatrix}x_1(t-\tau_{\mathrm{d}})\\ x_2(t-\tau_{\mathrm{d}})\end{bmatrix},\end{aligned}$$

with design data $V_1 = 0.01I_2$, $V_2 = 1$, $R_1 = 0.5I_2$, $R_2 = 1$, $\alpha = 25$, and $\sigma = 1$. Using Corollary 4 a full-order dynamic compensator was designed. To illustrate the closed-loop behavior of the system let $x(0) = \begin{bmatrix}0.4 & -6\end{bmatrix}^T$ and let $\phi(t) = \begin{bmatrix}-384t + 0.4 & -480t - 6\end{bmatrix}^T$ for $t \in [-0.025, 0]$. Figure 1 provides a comparison of the free response of the controlled system states with an LQG controller and the controller designed using Corollary 4.

Example 2. To illustrate the design of full state-feedback control for dynamic systems with nonlinear state delay consider

$$\begin{bmatrix}\dot{x}_1(t)\\ \dot{x}_2(t)\end{bmatrix} = \begin{bmatrix}0 & 1\\ 0 & 0\end{bmatrix}\begin{bmatrix}x_1(t)\\ x_2(t)\end{bmatrix} + 0.7\begin{bmatrix}\frac{x_1(t-\tau_{\mathrm{d}})}{\sqrt{1+x_1^2(t-\tau_{\mathrm{d}})}}\\ x_2(t-\tau_{\mathrm{d}})\end{bmatrix} + \begin{bmatrix}0\\ 1\end{bmatrix}u(t).$$

Furthermore, note that $||f_{\mathrm{d}}||_2 = 0.7\sqrt{\frac{x_1^2}{1+x_1^2} + x_2^2} \le \gamma\sqrt{x_1^2 + x_2^2}$, for $\gamma > 0.7$. Let $\gamma = 0.75$ and choose the design parameters $R_1 = I_2$, $R_2 = 1$, and $\alpha = 1.3$. Using Theorem 5, we obtain,

$$P = \begin{bmatrix}9.1707 & 6.0039\\ 6.0039 & 4.9379\end{bmatrix}, \qquad K = \begin{bmatrix}-6.0039 & -4.9379\end{bmatrix}.$$

To illustrate the closed-loop behavior of the system let $x(0) = \begin{bmatrix}3 & 1\end{bmatrix}^T$ and let $\phi(t) = \begin{bmatrix}100t + 3 & -200t + 1\end{bmatrix}^T$ for $t \in [-0.01, 0]$. Figure 2 provides a comparison of the free response of the controlled system states with an LQR controller and the controller designed using Theorem 5.

7 Conclusion

In this chapter we developed fixed-order dynamic output feedback controllers and full-state feedback controllers for linear and nonlinear continuous-time systems with time delays, respectively. Specifically, for linear continuous-time systems with state and measurement delay we presented sufficient conditions via fixed-order dynamic compensation. For nonlinear continuous-time systems with nonlinear state delay a constructive procedure was used to obtain full-state feedback stabilizing controllers. For both cases the controllers obtained were delay-independent. Two numerical examples were presented to illustrate the effectiveness of the proposed design approach.

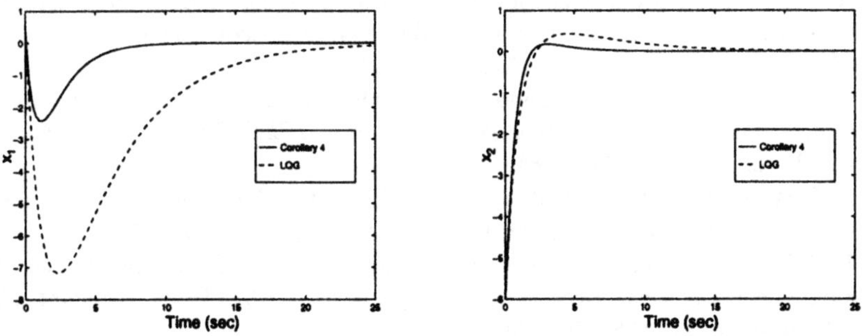

Fig. 1. Comparison of LQG and Corollary 4 Designs: Example 1

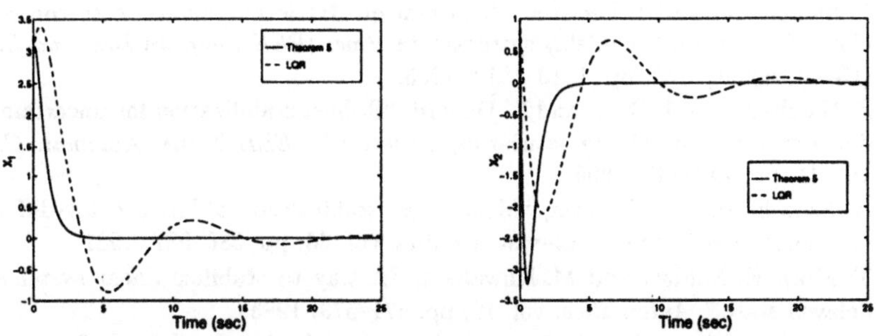

Fig. 2. Comparison of LQR and Theorem 5 Designs: Example 2

References

1. J. Cushing, "Integrodifferential equations and delay models in population dynamics," in *Lecture Notes in Bio-Mathematics*, vol. 20, Springer-Verlag, Berlin, 1977.
2. P. Schwyn and T. Bickart, "Stability of networks with distributed and nonlinear elements," *Networks*, vol. 2, pp. 45–96, 1972.
3. G. Franklin, J. Powell, and A. Emami-Naeini, *Feedback Control of Dynamic Systems*. New York: Addison Wesley, 1994.
4. A. Thowsen, "Uniform ultimate boundedness of the solutions of uncertain dynamic delay systems with state-dependent and memoryless feedback control," *Int. Journal of Control*, vol. 37, no. 5, pp. 1135–1143, 1983.
5. E. Kamen, P. Khargonekar, and A. Tannenbaum, "Stabilization of time-delay systems using finite-dimensional compensators," *IEEE Trans. Automat. Control*, vol. 30, no. 1, pp. 75–78, 1985.
6. S.-S. Wang, B.-S. Chen, and T.-P. Lin, "Robust stability of uncertain time-delay systems," *Int. Journal of Control*, vol. 46, no. 3, pp. 963–976, 1987.
7. J.-C. Shen, B.-S. Chen, and F.-C. Kung, "Memoryless stabilization of uncertain dynamic delay systems: Riccati equation approach," *IEEE Trans. Automat. Control*, vol. 36, no. 5, pp. 638–640, 1991.
8. E. Verriest, M. Fan, and J. Kullstam, "Frequency domain robust stability criteria for linear delay systems," in *Proc. IEEE Conf. on Dec. and Control*, (San Antonio, TX), 1993.
9. E. Verriest and A. Ivanov, "Robust stability of systems with delayed feedback," *Circuits, Systems, and Signal Processing*, vol. 13, pp. 213–222, 1994.
10. E. Verriest and A. Ivanov, "Robust stability of delay-difference equations," in *Proc. IEEE Conf. on Dec. and Control*, (New Orleans, LA), pp. 386–391, 1995.
11. S. Moheimani and I. Petersen, "Optimal quadratic guaranteed cost control of a class of uncertain time-delay systems," in *Proc. IEEE Conf. on Dec. and Control*, (New Orleans, LA), pp. 1513–1518, 1995.
12. S. Niculescu, J.-M. Dion, and L. Dugard, "Robust stabilization for uncertain time-delay systems containing saturating actuators," *IEEE Trans. Automat. Control*, vol. 41, pp. 742–747, 1996.
13. A. Feliachi and A. Thowsen, "Memoryless stabilization of linear delay-differential systems," *IEEE Trans. Automat. Control*, vol. 26, pp. 586–587, 1981.
14. T. Mori, E. Noldus, and M. Kuwahara, "A way to stabilize linear systems with delayed state," *Automatica*, vol. 19, pp. 571–573, 1983.
15. T. Su, P.-L. Liu, and J.-T. Tsay, "Stabilization of delay-dependence for saturating actuator systems," in *Proc. IEEE Conf. on Dec. and Control*, (Brighton, U.K.), pp. 2891–2892, 1991.
16. S. Niculescu, C. de Souza, J.-M. Dion, and L. Dugard, "Robust stability and stabilization of uncertain linear systems with state delay: Single delay case (I)," in *Proc. IFAC Workshop Robust Contr. Design*, (Rio de Janeiro, Brazil), pp. 469–474, 1994.
17. S. Niculescu, C. de Souza, J.-M. Dion, and L. Dugard, "Robust stability and stabilization of uncertain linear systems with state delay: Multiple delays case (II)," in *Proc. IFAC Workshop Robust Contr. Design*, (Rio de Janeiro, Brazil), pp. 475–480, 1994.
18. J. Hale, *Theory of Functional Differential Equations*. New York: Springer-Verlag, 1977.

19. B. Petterson, R. Robinett, and J. Werner, "Lag-stabilized force feedback damping," Tech. Rep. SAND91-0194, UC-406, Sandia National Laboratories, Albuquerque, NM, 87185, 1991.
20. C. Abdallah, P. Dorato, J. Benitez-Read, and R. Byrne, "Delayed positive feedback can stabilize oscillatory systems," in *Proc. IEEE American Control Conf.*, (San Fransisco, CA), pp. 3106–3107, 1993.
21. B. Anderson, R. Bitmead, C. Johnson, Jr., P. Kokotovic, R. Koust, I. Mareels, L. Praly, and B. Riedle, *Stability of Adaptive Systems: Passivity and Averaging Analysis*. Cambridge, MA: MIT Press, 1986.
22. S. Phoojaruenchanachai and K. Furuta, "Memoryless stabilization of uncertain linear systems including time-varying state delays," *IEEE Trans. Automat. Control*, vol. 37, no. 7, pp. 1022–1026, 1992.
23. M. Mahmoud and N.F.Al-Muthairi, "Quadratic stabilization of continuous time systems with state-delay and norm-bounded time-varying uncertainties," *IEEE Trans. Automat. Control*, vol. 39, no. 10, pp. 2135–2139, 1994.
24. D. Bernstein and W. Haddad, "Robust stability and performance via fixed-order dynamic compensation with guaranteed cost bounds," *Math. Contr. Sig. Sys.*, vol. 3, pp. 139–163, 1990.
25. K. Bhat and H. Koivo, "Modal characterizations of controllability and observability in time delay systems," *IEEE Trans. Automat. Control*, vol. 21, pp. 292–293, 1976.
26. W. Haddad and V. Kapila, "Fixed-architecture controller synthesis for systems with input-output time-varying nonlinearities," *Int. J. Robust and Nonlinear Contr.*, to appear.
27. W. Kwon and A. Pearson, "A note on feedback stabilization of a differential-difference system," *IEEE Trans. Automat. Control*, vol. 22, pp. 468–470, 1977.

Acknowledgement

This research was supported in part by the National Science Foundation under Grant ECS-9496249, the Air Force Office of Scientific Research under Grant F49620-96-1-0125, the Boeing Computer Services under contract W-300445, and ISTEC. The authors would also like to thank A. Grivas for assisting in the numerical calculations of Section 6

Nonlinear Delay Systems: Tools for a Quantitative Approach to Stabilization

J.P. Richard, A. Goubet-Bartholomeüs, P.A. Tchangani, M. Dambrine

LAIL, URA CNRS 1440, Ecole Centrale de Lille,
BP 48, 59651 Villeneuve d'Ascq Cedex - France
e-mail contact: jprichard,dambrine@ec-lille.fr

Abstract. This chapter concerns the stability of delay systems with nonlinear, time-varying, disturbed parameters, of retarded or neutral types. Two aspects of the stability problems are developed: the basic one is related to the *qualitative* stability, with conditions dependent or independent on the delay. The other aspect concerns the *quantitative* stability, with several results concerning the estimation of the decreasing rates, the robustness of the convergence with regard to the parameters bounds, the positive invariance of constraints sets and the estimation of the stability domains with regard to initial conditions. As for the first aspect, both delay-dependent and independent results are given. All the results are obtained by means of a same comparison approach linked with vector Lyapunov functions; this tool appears rather simple and powerful for the study of nonlinear models.

1 Introduction

Engineering processes often involve both nonlinear and time-delay models : on one hand, the time-delay phenomenon appears as soon as material, energetic or information transport lags cannot be neglected, in particular when the speed of closed loop controlled systems is expected to increase. On the other hand, nonlinear phenomena have to be taken into account if the process operates in a wide range of conditions. Realistic models with discrete delays can be divided in two types : retarded delay systems, and neutral ones (this last class involves a delayed derivative of the state, which can be obtained, for example, from local modelling of hyperbolic distributed parameter systems). Both cases correspond to functional differential equations (FDE) of differential-difference type: this means that, compared to ordinary differential equations (ODE), they generally appear to be very complex, since they are of infinite dimension.

Many studies have been devoted to the control of time-delay systems, and in particular to the basic question of their closed-loop stability properties. Even if the stability study of linear systems with a single constant delay now appears to be well achieved, it remains quite a difficult task for more complex (but of course more realistic) models:

- linear models with varying or non-commensurate delays, or with several tuning parameters, or with unstable memoryless feedback;
- linear models with uncertain coefficients or input perturbations (Wang et al. [32], Su and Huang [26], Niculescu et al. [23], Li and de Souza [20]);
- nonlinear models (Dambrine and Richard [2]-[3], Goubet et al. [9], Kolmanovskii [14]);
- neutral models (Kolmanovskii and Nosov[16], Tchangani et al. [27], Kolmanovskii [15], Tchangani et al. [28]).

The general stability criteria that gather these kinds of models are neither numerous nor easy-to-check. Moreover, to establish *quantitative* criteria is of practical importance: this means to complete the –yet not so simple– *qualitative* question to know whether a given operating point (or a nominal trajectory) is stable or not, attractive or not, for a given model.

This chapter aims to provide (and in an easy way if possible) workable quantitative information that is needed for the validation of a closed-loop controller:

a) *robustness with regard to the parameters*: what are the admissible bound-values of the time delays, or of the parameters of the model, that guarantee the stability ?
b) *stability domains with regard to the variables*: what are the initial conditions (which, for delay systems, are functions) that will make the state converge towards the equilibrium? (this is necessary for providing the admissible changes of operating points, or for determining whether bounded additive perturbations on the state may destabilize the closed loop system);
c) *decreasing rate*: what is the exponential rate of convergence? (this means, the velocity of the final controlled process);
d) *invariance*: how to be sure that a trajectory will not go out of a predetermined domain of constraints? (on the state, corresponding to physical security, or on the control variables, corresponding to energy considerations).

Only a few studies have considered several of these questions for time-delay systems:

- in some of them (Tokumaru et al. [29], Dambrine and Richard [2]-[3], Verriest [31], Verriest and Ivanov [30], Lehman and Shujaee [19]), the stability conditions are independent on the delay value (for such "i.o.d." criteria, the point a is answered with an infinite bound for the delay): of course, proving this kind of robustness property is very interesting because the delay margin is infinite; but, in practice, it may turn out to be rather conservative for processes involving small delays with known bounds.
- in other ones, the nominal model is linear, with possible uncertainties. Niculescu et al. [22]-[24], Su and Huang [26] corrected by Xu [33], and Wang et al. [32] developed delay-dependent criteria for points a and possibly c. In [4], the authors considered point d within the framework of the constrained stabilization of linear time-delay systems. Of course, the drawback of such

linearity hypothesis is its lack of realism, since the size of additive perturbations -point b- is supposed to have no importance.
- others approaches concerned nonlinear or time-varying systems ([Kolmanovskii [13], [14] for point a, with an approach of point b. Goubet-Bartholoméüs et al. [10] gave a first answer to points a, b, c for nonlinear time-varying delay systems, and Dambrine et al. [5] for point d.

The results that are to be presented here are in prolongation of this last item, but of course also apply to linear models. They are based on a comparison theorem and special vector Lyapunov functions: the stability (respectively asymptotic stability) of a system is proved if a linear system, the construction of which is explained, is stable (respectively asymptotically stable).

However, the results are directly workable, even without refering to the comparison concept. In our opinion, this last point (simplicity), compared with the wideness of the possible applications (points a to d), constitutes the main contribution of the present work.

If we omit section 2 which contains the notations, the remainder of the chapter is decomposed into three main sections: stability criteria for retarded systems (with conditions independent and dependent on the delay), and then for neutral systems.

Sections 3 and 4 deal with retarded systems of the following form :

$$\dot{x}(t) = A(t, x_t, d)\, x(t) + B(t, x_t, d)\, x(t - \tau(t)) \ . \tag{1.1}$$

Section 3 considers stability conditions independent of the delay, and section 4 provides delay-dependent results.

In section 5, the results are enlarged to systems of the neutral type:

$$\dot{x}(t) = A(t, x_t, d)\, x(t) + B(t, x_t, d)\, x(t - \tau(t)) + C(t, x_t, d)\, \dot{x}(t - \tau(t)) \ . \tag{1.2}$$

Throughout the presentation, the following assumptions hold:

- the delay $\tau(t)$, which depends on time only, has an upper limit τ_m, which can be finite or infinite : $0 \leq \tau(t) \leq \tau_m$; the law $\tau(t)$ may be known or not;
- the matrices $A(t, x_t, d)$ and $B(t, x_t, d)$ are bounded for $d \in S_d$ as soon as x_t is bounded;
- $\tau(t)$, $A(.)$, and $B(.)$ are such that the systems (1.1) or (1.2), for any continuous initial condition φ, admit an unique continuous solution for $t \geq t_0$.

2 Notations

- $d \in S_d$, where S_d is the set of admissible disturbances;
- $x(t) \in \mathbb{R}^n$ is the value of the solution at time t,
  x_t is the state function at time t : $x_t(s) = x(t+s)$ for $s \in [-\tau_m, 0]$;

- Initial conditions are $x_0(s) = \varphi(s)$.
- $\lambda_m(P)$: eigenvalue of the matrix P which has the minimum real part.
- $\sigma(P) = \max_i |\lambda_i(P)|$, where $\lambda_i(P)$ denotes an eigenvalue of the matrix P.
- The supremum $\sup_{\alpha \leq a \leq \beta} x(a)$ of a vector x is the vector constituted of the suprema of the different components $\sup_{\alpha \leq a \leq \beta} x_i(a)$. $|x|$ is the vector whose components are the absolute values of the components of x. The same remarks hold for matrices.
- A vector (resp. a matrix) is said to be positive if all its entries are positive.
- $\|\cdot\|$ denotes a norm of $\mathbb{R}^n$ or its induced matrix norm. $\mu(\cdot)$ is the associated matrix measure, defined by $\mu(A) = \lim_{h \to 0^+}(\| I + hA \| - 1)/h$.
- $\mathcal{C}(\mathcal{D})$ denotes the set of the continuous functions defined on $[-\tau_m, 0]$ with values in the set $\mathcal{D}$.
- $\mathcal{C}^1([-\tau, 0], \mathbb{R}^n)$ denotes the set of the differentiable and bounded functions mapping from $[\tau, 0]$ into $\mathbb{R}^n$.
- Vector and functional sets are defined by:

$$\begin{aligned} I(v,\alpha) &= \{x \in \mathbb{R}^n : |x| \leq \alpha v\}, \\ \mathcal{I}(v,\alpha) &= \mathcal{C}(I(v,\alpha)) = \{\varphi \in \mathcal{C}(\mathbb{R}^n) : |\varphi(s)| \leq \alpha v, \forall s \in [-\tau_m, 0]\}, \\ I_N(\alpha) &= \{x \in \mathbb{R}^n : \| x \| \leq \alpha\}, \\ \mathcal{I}_N(\alpha) &= \mathcal{C}(I_N(\alpha)) = \{\varphi \in \mathcal{C}(\mathbb{R}^n) : \| \varphi(s) \| \leq \alpha, \forall s \in [-\tau_m, 0]\}. \end{aligned}$$

  where v is a vector with positive components, and α is a positive constant.
- the abbreviation (.) stands for (t, x_t, d).

3 Retarded-Type Systems: Stability Criteria Independent of Delay

The stability study of a nonlinear FDE is not a trivial task. Lyapunov's second method and its extensions ([18] and [25]) are powerful theoretical tools to solve this problem. This is evidenced by the fact that the existence of a Lyapunov functional is not only sufficient but also necessary in order to prove the property of uniform asymptotic stability. But, in practice, confronted to a complex but realistic system, the efficiency of Lyapunov's theory is weakened by the difficulty of finding a suitable Lyapunov function or functional since there is no general algorithm of construction, even for linear systems: in this last case, it would need to solve coupled algebraic, ordinary and partial differential equations (see [17]).

For many systems such a difficulty may be bypassed by analyzing the system via another one, called the comparison system, simpler to analyze and for which the properties of stability imply the same characteristics for the original one: this is the comparison approach. However, the application of this method usually leads to some conservative results due to the use of linear, time-invariant comparison systems (for instance [29]) obtained at the price of strong majorations. This drawback may be removed by using more general types of comparison systems, but then the problem of their stability analysis remains.

In this section, a method proposed in [2]-[3] by some of the authors is reviewed. This method combines the advantages of being both simple and usefull: firstly, formulae are given in order to obtain the finest comparison system with respect to the structure of the initial system and this in an easy and systematic way; secondly some qualitative stability criteria that are well-suited to the structure of the comparison system are provided.

The conditions stated in the first part of this section are independent of the delay: that is, the value of the time delay has no influence on the validity of the condition. This may be a very interesting property for a controlled system, but it turns out to be also too restrictive for some applications and so, in order to remedy this problem, some enhancements will be presented in a following section.

3.1 The Comparison Approach

Let us first define what we mean by a comparison system:

Definition 1 (comparison system). A dynamic system (A) is said to be *a Comparison System* of a dynamic system (B) with regard to the property $\mathcal{P}$ (for example, stability of its zero solution), if the verification of property $\mathcal{P}$ for system (A) implies the same property for system (B).

For instance, the first-order approximation of a nonlinear ordinary differential equation may be viewed as a comparison system with regard to the uniform asymptotic stability.

Another type of comparison systems may be defined as follows:
let $V : \mathbb{R}^n \to \mathbb{R}^k_+$ with $k \leq n$ be a continuous, positive function such that $V(x) = 0 \Leftrightarrow x = 0$. Assume that, along the solutions of (1.1), the right-hand time derivative of $y(t) = V(x(t))$ satisfies the functional differential inequality

$$D^+y(t) \leq \mathcal{F}(t, y_t) \ . \tag{3.1}$$

Definition 2 (overvaluing system). The system

$$\dot{z}(t) = \mathcal{F}(t, z_t) \tag{3.2}$$

is an overvaluing system of (1.1) with respect to the function V if when the inequality

$$V(x(t)) \leq z(t)$$

holds for $t \in [t_0 - \tau_m, t_0]$, then it holds also for any $t \geq t_0$.

Using the assumptions done on V, it is very simple to prove that an overvaluing system is also a comparison system with regard to stability or asymptotic stability. Conditions on functional $\mathcal{F}$ for (3.2) to be an overvaluing system are called comparison principles. Note that traditionaly, what we call an overvaluing system is referred to as a comparison system.

3.2 Comparison Principles

We consider below that in (3.1), the functional $\mathcal{F}$ belongs to one of the two following classes:

$$\mathcal{F}(t, z_t) = g(z(t), z(t - \tau(t)), t), \tag{3.3}$$

or

$$\mathcal{F}(t, z_t) = g(z(t), \sup_{0 \le \lambda \le \tau_m} z(t - \lambda), t), \tag{3.4}$$

where g is a function defined on $\mathbb{R}^k \times \mathbb{R}^k \times [t_0, \infty[$.
The following comparison principle summarizes two results given by Tokumaru et al. in [29] and Dambrine in [1].

Theorem 3. *Assume that the function g satisfies the following conditions:*

1. $g(x, y', t) \le g(x, y'', t)$, *for any vectors x, y', $y'' \in \mathbb{R}^k$ such that $y' \le y''$ (monotonicity in the second argument),*
2. *if x^1 and x^2 are two vectors of $\mathbb{R}^k$ such that $x^1 \le x^2$ and $x^1_i = x^2_i$, then the i-th component of g satisfies the relation: $g_i(x^1, y, t) \le g_i(x^2, y, t)$ for all $y \in \mathbb{R}^k$ and $t \ge t_0$ (quasi-monotonicity in the first argument),*
3. *the solution of the differential equation*

$$\dot{z}(t) = g(z(t), \sup_{0 \le \lambda \le \tau_m} z(t - \lambda), t) + \varepsilon$$

uniquely exists for any continuous initial function $z(s)$, $(-\tau_m \le s \le 0)$ and for any sufficiently small $\varepsilon \ge 0$.

Then, (3.2) is an overvaluing system of (1.1).

The application of this theorem to the case of

$$\mathcal{F}(t, z_t) = -C\, z(t) + D \sup_{0 \le \lambda \le \tau_m} z(t - \lambda)$$

yields to the following lemma.

Lemma 4 [29]. *Let C and D be $k \times k$ matrices with real elements and let $y(t)$ be a solution of the differential inequality*

$$\dot{y}(t) \le -Cy(t) + D \sup_{0 \le \lambda \le \tau_m} y(t - \lambda), \qquad \text{for} \quad t \ge t_0$$

If $D \ge 0$, if the off-diagonal elements of C are non positive, and if $(C - D)$ is an M-matrix, then a solution $y(t)$ of this inequality is upper-bounded by the solution $z(t)$ of the differential equation

$$\dot{z}(t) = -Cz(t) + D \sup_{0 \le \lambda \le \tau_m} z(t - \lambda), \qquad \text{for} \quad t \ge t_0$$

as soon as $0 \le y_{t_0}(\theta) \le z_{t_0}(\theta)$ for $-\tau_m \le \theta \le 0$.
Moreover, if $(C - D)$ is an irreducible M-matrix, then there exist a constant $\gamma > 0$ and a constant vector $k > 0$ such that $y(t) \le ke^{-\gamma t}$, for $t \ge t_0$.
k and γ are obtained in the following way :

- *γ is the positive real solution of the equation $\lambda_m(P_\gamma) = \gamma$, where $P_\gamma = C - De^{\gamma\tau_m}$.*
- *k is a positive eigenvector of P_γ associated with $\lambda_m(P_\gamma)$, and such that $y_{t_0}(s) \leq ke^{-\gamma s}$, $\forall s \in [-\tau_m, 0]$.*

Properties of M-matrices are recalled in the appendix to this chapter. For the sake of simplicity, we stated the previous results for systems with a single delay, but they may be extended to the case of several delays. For instance, the following lemma is a generalization of Lemma 4 to systems with two delays:

Lemma 5 (Generalized Tokumaru's lemma) [11]. *Let C, D_1 and D_2 be $n \times n$ matrices with real elements and let $x(t)$ be a solution of the differential inequality*

$$\dot{x}(t) \leq -Cx(t) + D_1 \sup_{0\leq\lambda\leq\tau_1} x(t-\lambda) + D_2 \sup_{0\leq\lambda\leq\tau_2} x(t-\lambda) \text{ for } t \geq 0 .$$

If $D_1 \geq 0$, $D_2 \geq 0$, if the off-diagonal elements of C are non positive, and if $(C - D_1 - D_2)$ is an M-matrix, then a solution $x(t)$ of this inequality is overvalued by the asymptotically stable solution $z(t)$ of the differential equation

$$\dot{z}(t) = -Cz(t) + D_1 \sup_{0\leq\lambda\leq\tau_1} z(t-\lambda) + D_2 \sup_{0\leq\lambda\leq\tau_2} z(t-\lambda) \text{ for } t \geq 0,$$

as soon as $0 \leq x(\theta) \leq z(\theta)$ for $-\max(\tau_1, \tau_2) \leq \theta \leq 0$.
Moreover, if $(C - D_1 - D_2)$ is an irreducible M-matrix, then there exist a constant $\gamma > 0$ and a constant vector $k > 0$ such that $x(t) \leq ke^{-\gamma t}$, for $t \geq 0$.
k and γ are obtained in the following way :

- *γ is the positive real solution of the equation $\lambda_m(A_\gamma) = \gamma$, where $A_\gamma = C - D_1e^{\gamma\tau_1} - D_2e^{\gamma\tau_2}$.*
- *k is a positive eigenvector of A_γ associated with the eigenvalue γ.*

3.3 A Systematic Construction of Comparison Systems

The initial system (1.1) may be viewed as the interconnection of several subsystems and so (1.1) can be rewritten in the form:

$$\begin{aligned} \dot{x}^i(t) = \; & A_{ii}(t, x_t, d)\, x^i(t) + B_{ii}(t, x_t, d)\, x^i(t - \tau(t)) \\ & + \sum_{\substack{1\leq j\leq r \\ j\neq i}} (A_{ij}(t, x_t, d)\, x^j(t) + B_{ij}(t, x_t, d)\, x^j(t - \tau(t))), \qquad i = 1, \ldots, r, \end{aligned} \tag{3.5}$$

where $x^i(t) \in \mathbb{R}^{n_i}$, and $x^T(t) = [(x^1)^T(t), \ldots, (x^r)^T(t)]$.

Note that this decomposition may be natural or introduced artificially for simplifying the analysis of the system.

Consider now the following vector function $V(x) = [V_1(x^1), \ldots, V_r(x^r)]^T$, where V_i is an usual norm of $\mathbb{R}^{n_i}$, for instance, it may be one of the following Hölder's norms:

$$\begin{aligned} V_i(x^i) &= \| x^i \|_1 = \sum_{j=1}^{n_i} | x_j^i |, \\ V_i(x^i) &= \| x^i \|_2 = (\sum_{j=1}^{n_i} | x_j^i |^2)^{1/2}, \\ V_i(x^i) &= \| x^i \|_\infty = \max_{1 \le j \le n_i} | x_j^i | \ . \end{aligned} \tag{3.6}$$

Then, according to [3], the right-hand derivative of $V(x(t))$ with respect to time taken along the trajectories of (1.1) satisfies the following inequality

$$D^+V(x(t)) \le M(t, x_t, d)V(x(t)) + N(t, x_t, d)V(x(t - \tau(t))), \tag{3.7}$$

where the matrices $M(.) = \{m_{ij}(.)\}$ and $N(.) = \{n_{ij}(.)\}$ are obtained from $A(.)$ and $B(.)$ by:

$$\begin{aligned} m_{ii}(.) &= \mu_i(A_{ii}(.)), && \text{for } i = 1, \ldots, r, \\ m_{ij}(.) &= \| A_{ij}(.) \|_{ij}, && \text{for } j \ne i,\ i, j = 1, \ldots, r, \\ n_{ij}(.) &= \| B_{ij}(.) \|_{ij}, && \text{for } i, j = 1, \ldots, r \ . \end{aligned} \tag{3.8}$$

Recall that $\mu_i(X)$ denotes the measure of the $n_i \times n_i$ matrix X associated with the norm V_i (see [6]), and $\| Y \|_{ij}$ represents the matrix norm of the $n_i \times n_j$ matrix Y associated with the norms V_i and V_j, and defined by $\| Y \|_{ij} = \max_{x^j \in \mathbb{R}^{n_j} : V_j(x^j) = 1} V_i(Y x^j)$.

It is important to note that all the off-diagonal terms of $M(.)$ and all the entries of $N(.)$ are non-negative, which implies that the function $V(x(t))$ satisfies the assumptions of Theorem 3. So, the system

$$\dot{z}(t) = M(.)\, z(t) + N(.)\, z(t - \tau(t)) \tag{3.9}$$

is a comparison system of (1.1) for the properties of stability, asymptotic stability, uniform asymptotic stability, etc.

We want to emphasize that with a vector Lyapunov function composed of classical norms (3.6), it is very simple to derive a comparison system from the entries of matrices $A(.)$ and $B(.)$. For instance, if $V_i = V_j = \| \cdot \|_1$, then the formulae (3.8) reduce to:

$$\begin{aligned} m_{ii}(.) &= \max_{p \in J_i} \left[a_{pp} + \sum_{q \in J_i, q \ne p} |a_{qp}(\cdot)| \right], \\ m_{ij}(.) &= \max_{p \in J_i} \left[\sum_{q \in J_j} |a_{qp}(\cdot)| \right], \\ n_{ij}(.) &= \max_{p \in J_i} \left[\sum_{q \in J_j} |b_{qp}(\cdot)| \right], \end{aligned}$$

where J_i (resp. J_j) are the set of the row indices (resp. column indices) of the block A_{ij}, that is, $A_{ij}(.) = \{a_{pq}(.)\}_{p \in J_i,\, q \in J_j}$.

3.4 Qualitative Criteria of Stability

We assume in this part that it is possible to find for (1.1) a local comparison system of the form (3.9) associated with a vector function V. The term local means here that (3.9) is a comparison system as long as $x(t)$ belongs to a given domain $\mathcal{D}$ of $\mathbb{R}^n$.

The definition of stability in the sense of Lyapunov is just a mathematical notion: it expresses the fact that the solution does not move too far away from the equilibrium if the initial conditions belong to a given neighborhood. But, there is no condition on the size of this neighborhood. This is a weak point of the notion of stability because the diameter of this neighborhood gives an idea of the order of the size of the admissible perturbation or provides an estimate of the admissible changes of operating point. So, in order to complete this qualitative notions of stability, we define the stability domains, introduced for nonretarded systems by Grujić [7], and that extend the well-known notion of domain of attraction.

Definition 6. The set $\mathcal{D}_s$ is the *stability domain* of the zero solution of (1.1) if:

i) For any $\varepsilon > 0$, the set $\mathcal{D}_s(\varepsilon) = \{\varphi \in \mathcal{C}(\mathbb{R}^n) : \forall t \geq t_0, \quad \|x(t; t_0, \varphi)\| < \varepsilon\}$ is a neighborhood of 0 in $\mathcal{C}(\mathbb{R}^n)$ (with the uniform convergence norm)
ii) $\mathcal{D}_s = \bigcup_{\varepsilon > 0} \mathcal{D}_s(\varepsilon)$

The *asymptotic stability domain* of the zero solution of (1.1) is $\mathcal{D}_{as} = \mathcal{D}_s \cap \mathcal{D}_a$, where $\mathcal{D}_a$ is the domain of attraction of the zero solution.

A way of obtaining an estimate of the stability domain is given in the following theorem.

Theorem 7. *If there is an $\varepsilon > 0$ and a r-vector u with positive components such that*

$$[M(t,x,y) + N(t,x,y)]\, u < -\varepsilon\, u, \tag{3.10}$$

for all $t \geq t_0$, and all x, y in $\mathcal{D}$, then the zero solution of (1.1) is stable and the biggest set $\mathcal{I}_V(\alpha, u) = \{\varphi \in \mathcal{C}(\mathcal{D}) : V(\varphi(s)) \leq \alpha\, u, \forall s \in [-\tau_m, 0]\}$ is a positively invariant set with respect to (1.1), and thus is an estimate of the stability domain of its zero solution. Moreover, if the matrix $N(t,x,y)$ is bounded on $[t_0, \infty) \times \mathcal{D} \times \mathcal{D}$ then the zero solution of (1.1) is asymptotically stable, and the biggest set $\mathcal{I}_V(\alpha, u)$ contained in $\mathcal{C}(\mathcal{D})$ is an estimate of the asymptotic stability domain of the zero solution of (1.1).
If $\mathcal{D} = \mathbb{R}^n$ then stability or asymptotic stability is global.

The existence of such a vector u may be easily tested in many cases, some of them are given in the following corollaries.

Corollary 8. *If the matrix $M(.) + N(.)$ is the opposite of a constant M-matrix, then there is a vector u satisfying (3.10). If in addition the matrix $M + N$ is irreducible, then its importance vector is a possible choice for u.*

Corollary 9. *If the matrix $M(.) + N(.)$ is such that all non constant entries are located in the same row, and if there is an $\varepsilon > 0$ such that the matrix $M(t, x, y) + N(t, x, y) + \varepsilon I$ is the opposite of an M-matrix for all $t \geq t_0$, and all x, y in $\mathcal{D}$, then there is a vector u such that inequality (3.10) holds.*

Corollary 10. *If there are two matrices $D(.)$ and Z such that*

a) $D(.)$ is a diagonal matrix which elements are larger than a positive number δ, and
b) Z is the opposite of a constant M-matrix,
c) $M(.) + N(.) = D(.)\, Z$,

then there is a vector u such that inequality (3.10) holds.

Note that all the criteria proposed in this section have stated independent-of-delay conditions implying the stability of the matrix $A(.)$ in an intuitive sense.

4 Retarded-Type Systems: Stability Criteria Dependent on the Delay

In the previous section, different criteria have been given in order to test the stability of nonlinear time-delay systems and to determine stability domains. The results which have been presented are independent of the maximum value τ_m of the delay. So they can be applied to a wide class of delay systems, and give very robust conditions.

However, it may be interesting to determine stability criteria that are more precise and that take into account the value of the delay. Indeed, it is often possible to determine upper bounds on the delays in a system, and if they turn out to be small, independent-of-delay conditions may appear to be conservative. Then, stability conditions that take into account the size of the delays lead to less restrictive conditions on the values of other parameters in the system.

As in the previous section, the different results are obtained using a comparison lemma (Theorem 3), applied on a comparison principle whose coefficients depend on τ_m, and whose maximum delay is $2\tau_m$. Qualitative results will be determined (stability conditions, Theorems 11 and 13), as well as quantitative ones (estimation of stability domain, Theorem 14). These results depend on τ_m. Moreover, they allow to deal with systems whose memoryless feedback is not stable, in contrast to the i.o.d. criteria. They therefore constitute an interesting enhancement to the previous section, and allow to take into account the stabilizing effects of some of the delayed feedbacks.

4.1 Stability Criteria

As before, the system (1.1) is considered :

$$\dot{x}(t) = A(t, x_t, d)\, x(t) + B(t, x_t, d)\, x(t - \tau(t)) \ .$$

The initial time t_0 is set to 0, but the results hold for any initial times.

Let us decompose the matrix $B(.)$ into : $B(.) = B'(.) + B''(.)$.

Let $\mathcal{D}$ be a domain of $\mathbb{R}^n$ containing a neighbourhood of the origin, and let us associate with the system (1.1) the following matrix :

$$\begin{aligned} P_\sigma = & \sup_{t \geq \tau_m, \mathcal{D}, \mathcal{S}_d} \{(A(t, x_t, d) + B'(t, x_t, d))^*\} \\ & + e^{2\sigma\tau_m} \sup_{t \geq \tau_m, \mathcal{D}, \mathcal{S}_d} \left\{ \tau_m \sup_{0 \leq \lambda \leq \tau_m} \left\{ |B'(t, x_t, d)\, A(t-\lambda, x_{t-\lambda}, d)| \right.\right. \\ & \left.\left. + |B'(t, x_t, d)\, B(t-\lambda, x_{t-\lambda}, d)| \right\} + |B''(t, x_t, d)| \right\}, \end{aligned}$$

and so

$$\begin{aligned} P_0 = & \sup_{t \geq \tau_m, \mathcal{D}, \mathcal{S}_d} \{(A(t, x_t, d) + B'(t, x_t, d))^*\} \\ & + \sup_{t \geq \tau_m, \mathcal{D}, \mathcal{S}_d} \left\{ \tau_m . \sup_{0 \leq \lambda \leq \tau_m} \{ |B'(t, x_t, d) A(t-\lambda, x_{t-\lambda}, d)| \right. \\ & \left. + |B'(t, x_t, d)\, B(t-\lambda, x_{t-\lambda}, d)| \} + |B''(t, x_t, d)| \right\}, \end{aligned}$$

where the suprema $\sup_{t \geq \tau_m, \mathcal{D}, \mathcal{S}_d}$ are calculated for $t \geq \tau_m$, for functions x with values in $\mathcal{D}$, and for d in $\mathcal{S}_d$.

Theorem 11 [10]. *If P_0 is the opposite of an M-matrix on the domain $\mathcal{D} \subseteq \mathbb{R}^n$, then the equilibrium* 0 *of the system (1.1) is asymptotically stable.*

Remarks:

1. The entries of the matrix P_0 depend on the maximum value τ_m of the delay. The stability criterion is thus a delay-dependent one.
2. If $B'(.) = 0$, the criterion is delay-independent and corresponds to the delay-independent Corollary 8 given in the previous section.
3. The efficiency of this theorem depends on the decomposition of the matrix $B(.)$: $B'(.)$ must be chosen such that $\sup_{t \geq \tau_m, \mathcal{D}, \mathcal{S}_d} (A(.) + B'(.))^*$ is the opposite of an M-matrix. Moreover, the criterion is all more likely to hold that $\sup_{t \geq \tau_m, \mathcal{D}, \mathcal{S}_d} (A(.) + B'(.))^*$ has small off-diagonal elements compared to the absolute values of the diagonal ones, and that

$$\begin{aligned} \sup_{t \geq \tau_m, \mathcal{D}, \mathcal{S}_d} \{ \tau_m . & \sup_{0 \leq \lambda \leq \tau_m} \{ |B'(t, x_t, d) A(t-\lambda, x_{t-\lambda}, d)| \\ & + |B'(t, x_t, d) B(t-\lambda, x_{t-\lambda}, d)| \} + |B''(t, x_t, d)| \} \end{aligned}$$

is a small matrix. A compromise has thus to be found, which is not so difficult in many cases.
As mentionned above, Theorem 11 takes into account the fact that some delayed feedbacks can stabilize a system. Indeed, if $B(.)$ has stabilizing elements (negative diagonal elements with large absolute values), $B'(.)$ may be chosen such that $\sup\limits_{t\geq\tau_m,\mathcal{D},\mathcal{S}_d}(A(.)+B'(.))^*$ is the opposite of an M-matrix (even if $A(.)$ is not a stable matrix).

Using the same tools, an exponential rate of convergence may be determined, as well as scalar stability criteria :

Theorem 12 [10]. *If the conditions of the previous Theorem 11 are satisfied and if moreover P_0 is irreducible, then $|x(t)| \leq k.e^{-\gamma t}$ for $t \geq \tau_m$, as soon as $|x(\theta)| \leq k.e^{-\gamma\theta}$, $\theta \in [-\tau_m, \tau_m]$, where γ is the real positive solution of the equation $\lambda_m(-P_\gamma) = \gamma$ and k is an eigenvector (with positive components) of $(-P_\gamma)$ associated with γ, such that $\{x \in \mathbb{R}^n, |x| \leq k.e^{\gamma\tau_m}\} \subseteq \mathcal{D}$.*

Theorem 13 [8]. *If for every $t \geq \tau_m$,*

$$\sup_{t\geq\tau_m,\mathcal{D},\mathcal{S}_d}\Big\{\|B''(t,x_t,d)\| + \tau_m \sup_{0\leq\lambda\leq\tau_m}(\|B'(t,x_t,d)\,A(t-\lambda,x_{t-\lambda},d)\|$$

$$+\|B'(t,x_t,d)\,B(t-\lambda,x_{t-\lambda},d)\|)\Big\} + \sup_{t\geq\tau_m,\mathcal{D},\mathcal{S}_d}\{\mu(A(t,x_t,d)+B'(t,x_t,d))\} < 0,$$

then the equilibrium 0 of (1.1) is asymptotically stable.

More information about these results and other theorems can be found in the papers referenced in this chapter, as well as in [8].

The following example shows that these results, because they take into account the stabilizing effects of the delayed terms, enable the determination of a delayed static state feedback stabilizing an unstable nonlinear process. An other example will be given later, together with the determination of delay-dependent stability domains.

Example: Stabilization of an unstable open-loop process.
Let us study the nonlinear unstable system described by

$$\ddot{y}(t) - a_1(.)\,\dot{y}(t) = u(t),$$

that is to be stabilized using a feedback regulator with a delay.
The parameter $a_1(.)$ varies between 1.5 and 2. It may depend on t, on the state or eventually on a disturbance parameter. The output $y(t)$ is measured at the instant t, but its derivative is computed only after an unknown (or varying) time lag $\tau(t)$, whose upper bound is τ_m.

The problem is to obtain a relation between k_1 and τ_m that assures the stability of the closed-loop system. If the system is asymptotically stable, then

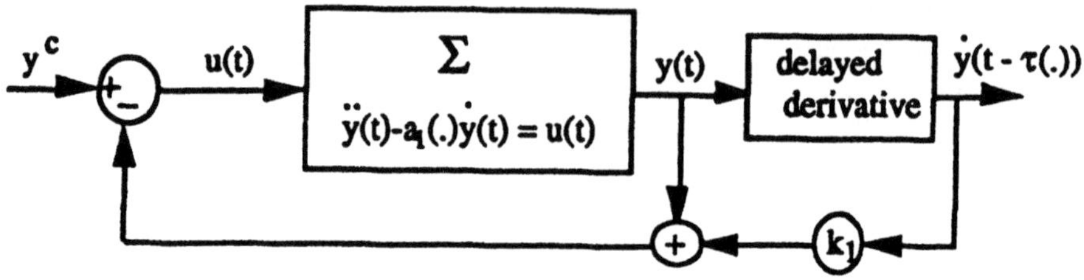

Fig. 1. Description of the process

$\lim_{t\to+\infty} y(t) = y^c$ and y^c is set to 0 during the stability study.
The state equation for the closed-loop system is:

$$\dot{z}(t) = M_z\, z(t) + N_z\, z(t-\tau),$$

with

$$M_z = \begin{bmatrix} 0 & 1 \\ -1 & a_1(.) \end{bmatrix}, \quad N_z = \begin{bmatrix} 0 & 0 \\ 0 & -k_1 \end{bmatrix}, \quad y(t) = [1 \quad 0]\, z(t)\ .$$

All delay-independent stability criteria are unable to lead to any conclusion. Let us define the change of state variables: $x = Pz$, with $P = \begin{bmatrix} 1 & \lambda \\ 0 & 1 \end{bmatrix}$, and apply the first delay-dependent stability criterion to the system that is obtained (for details, see [8] or [10]). The results providing stabilizing couples of parameters (k_1, τ_m) are shown on Fig. 2. The stabilization of the process is proved possible if the delay is not too large.

4.2 Stability Domains

Let us now lead a "quantitative" study of the stability with delay dependence. We denote $M = \sup_{[0,\tau_m],\mathcal{D},S_d} \{(A(t,x_t,.))^*\}$, $N = \sup_{[0,\tau_m],\mathcal{D},S_d} \{|B(t,x_t,.)|\}$, and $m = \sup_{[0,\tau_m],\mathcal{D},S_d} \{\mu(A(t,x_t,.)\}$, $b = \sup_{[0,\tau_m],\mathcal{D},S_d} \{\| B(t,x_t,.) \|\}$.
These suprema are this time calculated for $0 \le t \le \tau_m$, $x_t \in C(\mathcal{D})$, and when the perturbations take all their admissible values.

Theorem 14 (Stability domains) [10]. *Suppose P_0 is the opposite of an M-matrix, and let v be a positive vector such that $P_0\, v \le 0$, and $I(v,1) \subseteq \mathcal{D}$.*
An estimate of the stability domain can be found as follows:
<u>Case 1</u>: $\tau(t)$ is constant and known ($\tau(t) = \tau_m$).
Let k_1 be a positive real vector of $\mathbb{R}^n$ satisfying the following condition:

$$e^{Mt}\left(k_1 + \int_0^t e^{-Ms} N\, k_1, ds\right) \le v, t \in [0;\tau_m].$$

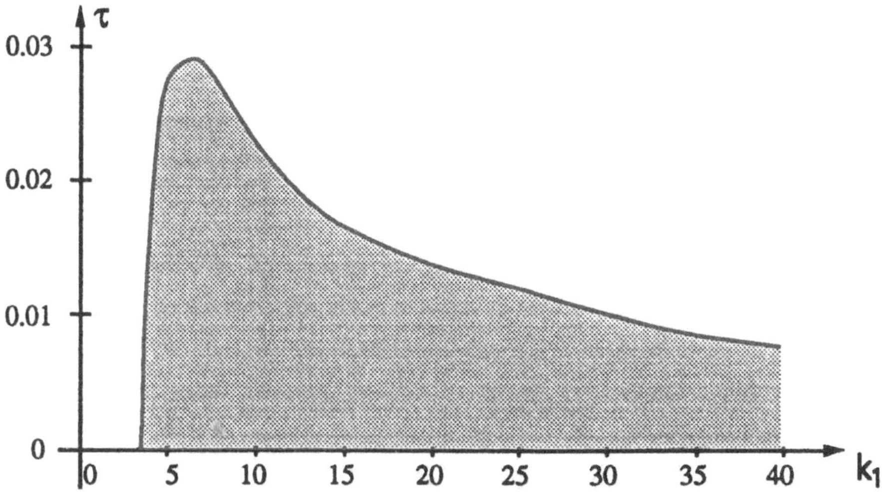

Fig. 2. Sufficient conditions of stabilization

Then $\mathcal{I}(k_1, 1)$ is an estimate of the stability domain of the equilibrium.
Case 2: $\tau(t) \leq \tau_m$ and a vector $k_2 > 0$ such that $(M + N)\, k_2 \leq 0$ can be found. Let α be a positive real such that $\alpha\, k_2 \leq v$.
Then $\mathcal{I}(\alpha\, k_2, 1)$ is an estimate of the stability domain of the equilibrium.
Case 3: cases 1 and 2 do not hold, and $-m > b$.
Let β be the unique positive solution of the equation :

$$\beta + m + \beta e^{\beta \tau_m} = 0$$

and ω_1 be the largest real positive number such that $I_N(\omega_1 e^{\beta\tau_m}) \subset I(v, 1)$.
Then $\mathcal{I}_N(\omega_1 e^{-\beta\tau_m})$ is an estimate of the stability domain of the equilibrium.
Case 4: cases 1 and 2 do not hold, and $-m \leq b$.
Let δ be the unique positive solution of the equation :

$$\delta - m - be^{-\delta\tau_0} = 0,$$

where τ_0 is the minimum possible value of the delay.
Let ω_2 be the largest real positive number such that $I_N(\omega_2 e^{\delta\tau_m}) \subset I(v, 1)$.
Then $\mathcal{I}_N(\omega_2 e^{-\delta\tau_m})$ is an estimate of the stability domain of the equilibrium.

Remarks:

1. The estimates of the stability domains depend on the supremum τ_m of the delay, and, in case 4, on the minimum τ_0.
2. If the inequality $P_0\, v \leq 0$ is strict, i.e. $P_0\, v < 0$, then the sets which are found with the above method are estimates of the asymptotic stability domain.

Example:
Let us consider the following system:

$$\dot{x}(t) = A(t,d)\,x(t) + B(t,x(t),d)\,x(t-\tau),$$
$$x(t_0+\theta) = \varphi(\theta), \quad \theta \in [-\tau, 0],$$

where

$$A(t,d) = \begin{bmatrix} -2+\alpha_1(t,d) & 0 \\ 0 & -1+\alpha_2(t,d) \end{bmatrix},$$
$$B(t,x(t),d) = \begin{bmatrix} -1+\beta_1\, g(t,x_2(t)) & 0 \\ \varepsilon & -1+\beta_2(t,d) \end{bmatrix},$$

$$|\alpha_1(t,d)| \le 1.6;\ |\beta_1| \le 0.1;\ |\alpha_2(t,d)| \le 0.05;\ |\beta_2(t,d)| \le 0.3;\ |\varepsilon| \le 1.$$

Two cases are considered for the expression of $g(t,x_2(t))$:

- $g(t,x_2(t)) = \cos(t)$. This example is the same as the one given in [20] and [23] if $\alpha_1(t,d) = \alpha_1\cos(t)$, $\alpha_2(t,d) = \alpha_2\sin(t)$, $\beta_2(t,d) = \beta_2\cos(t)$, $\varepsilon = -1$. In these last papers, the zero solution was proved to be asymptotically stable for any constant time-delay $\tau < 0.1036$ [23] and $\tau < 0.2013$ [20]. Using Theorem 11, it is proved to be stable for any time-varying delay $\tau(t) < 0.276$ (see [10] for calculations).
- In the second case, the system is nonlinear with $|g(t,x_2(t))| \le |x_2(t)|$. The asymptotic stability of this system will be studied and estimates of the stability domains will be given. They depend on the value of the delay, which is considered constant and known for simplicity.

$\mathcal{D} = \mathcal{D}_\psi = \{x \in \mathbb{R}^2 : |x_2| \le \psi\}$.

$P_0 = \begin{bmatrix} -1.4+0.1\psi+\tau(4.6+0.1\psi) & 0 \\ 1+\tau & -1.65+2.35\tau \end{bmatrix}$ with the following decomposition of the delayed matrix:

$B'(t,x(t),d) = \begin{bmatrix} -1 & 0 \\ 0 & -1 \end{bmatrix}$, $B''(t,x(t),d) = B(t,x(t),d) - B'(t,x(t),d)$.

Considering any value of τ less than 0.304, there exists a domain $\mathcal{D}_\psi$, where $\psi = \frac{14-46\tau}{1+\tau}$, such that $Pv \le 0$, $v = [\frac{1.65-2.35\tau}{1+\tau}\psi \;\; \psi]^T$. Then Theorem 14 (case 1) allows for the estimation of the stability domain of the equilibrium point with respect to the delay. Fig. 3 gives these estimates. Of course, the real estimates are the sets of functions defined on $[-\tau_m, 0]$ with values in the domains represented on Fig. 3. Two simulations for a delay $\tau = 0.01$ and for constant initial functions show that the estimation is not too conservative.

The delay was considered constant, but the same kind of results can be obtained for a time-varying delay, using the case 3 of Theorem 14 (for an example, see [10]).

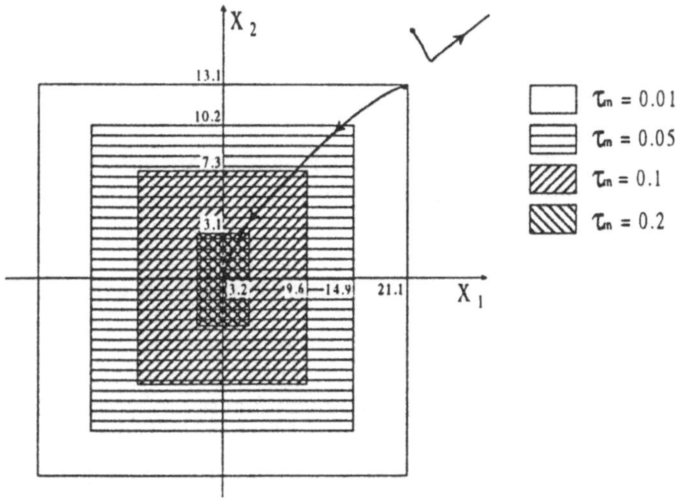

Fig. 3. Estimations of the stability domain

5 Generalization to Neutral Systems

In this section the problem of stability analysis of nonlinear FDE systems of neutral type is considered. A systematic method is given in order to compute a comparison system which is of retarded type; then a simple criterion is proposed in the linear time-invariant case.
The considered systems are now of the form:

$$\dot{x}(t) = A(t, x_t, d)\, x(t) + B(t, x_t, d)\, x(t-\tau) + C\, \dot{x}(t-\tau), \qquad t \geq t_0, \qquad (5.1)$$
$$x_{t_0}(\theta) = \varphi(\theta),\ \dot{x}_{t_0}(\theta) = \dot{\varphi}(\theta), \quad \forall\theta \in [-\tau, 0],$$

where $A(t, x_t, d)$, $B(t, x_t, d)$ are $n \times n$ matrices defined and continuous on $[t_0, \infty) \times C^1([-\tau, 0]; \mathbb{R}^n)$ for any admissible disturbances d, C is a constant $n \times n$ matrix and τ is a scalar positive constant.

This section consists of two main parts: the first part provides scalar and vector conditions for existence of a comparison system for (5.1) together with conditions that ensure stability of the zero solution of (5.1); in the second part, several examples illustrate the obtained results.

5.1 Additional Notations and Assumptions

In order to avoid the case where (5.1) reduces to a simpler retarded delay system, we assume C is non nilpotent, that is, $C^k \neq 0$ for any integer k.
In the following, we use the additional notations:

$M(.)_k = M(t - k\tau, x_{t-k\tau}, d)$,
$V(x_t)$ is the continuous function mapping from $[-\tau, 0]$ into $\mathbb{R}^r$ defined by:
$V(x_t)(\theta) = V(x(t+\theta))$, $\forall\theta \in [-\tau, 0]$.

5.2 Main Results

Scalar results

This part provides a way of constructing a scalar overvaluing system and additional conditions for this overvaluing system to be a comparison system of (5.1) with regard to stability.

Lemma 15 (construction of scalar overvaluing system) [28]. *Suppose the following conditions hold*

a) $\| C \| < 1$
b) there exists a scalar positive bounded function $\eta(.)$ such that

$$\sum_{k=1}^{\infty} \| C \|^{k-1} \| (B(.) + CA(.))_k \| \leq \eta(.) < \infty, \quad t \geq t_0,\ x_t \in \Omega,\ d \in S_d\ .$$

Then the system defined by

$$\dot{z}(t) = \mu(A(.))\, z(t) + \sum_{k=1}^{\infty} \| C \|^{k-1} \| (B(.) + CA(.))_k \| z(t - k\tau), \tag{5.2}$$

is a local overvaluing system of (5.1) with respect to the scalar norm $\| \cdot \|$ and the set Ω. Moreover, if a) holds and A and B are constant matrices, then (5.2) is a global overvaluing system of (5.1) and $\eta(.)$ can be computed as $\eta(.) = \frac{\| B + CA \|}{1 - \| C \|}$.

The following corollary gives a way for symplifying the expression of overvaluing systems.

Corollary 16 [28]. *Under conditions of lemma 15, any scalar system*

$$\dot{z}(t) = \alpha(.)\, z(t) + \sum_{k=1}^{\infty} \| C \|^{k-1} a_k(.) z(t - k\tau), \tag{5.3}$$

such that $\alpha(.) \geq \mu(A(.))$, and $a_k(.) \geq \| (B(.) + CA(.))_k \|$, for $x_t \in \Omega$; $d \in S_d$ is also a local overvaluing system of (5.1) with respect to scalar norm $\| \cdot \|$ and set Ω.

Theorem 17 (scalar stability criterion) [28]. *Let us suppose that the hypotheses a) and b) of Lemma 15 hold, and in addition*

c) $\sup\limits_{t \geq t_0, x_t \in \Omega, d \in S_d} \mu(A(.)) + \eta(.) \leq 0$

Then the zero solution of (5.1) is:

1. *locally stable;*
2. *locally asymptotically stable if the inequality in b) is strict;*
3. *if, in addition $\Omega = C$, then stability (asymptotic stability) is global.*

In the linear time-invariant case the conditions of Theorem 17 reduce to a very simple criterion. This is given in the following corollary which is important because it can be a guide for investigating a suitable overvaluing system of (5.1).

Corollary 18 [28]. *Suppose A and B are constant matrices then if $\| C \| < 1$ and if $\mu(A) + \frac{\| B+CA \|}{1-\| C \|} \leq 0$ (respectively < 0) then the zero solution of (5.1) is stable (resp. asymptotically stable).*

The following theorem summarizes the results presented in this part.

Theorem 19 (comparison principle) [28]. *Suppose the conditions a) and b) of Lemma 15 hold, then any scalar overvaluing system verifying conditions of Corollary 16 is a local comparison system of (5.1) with regard to stability (resp. asymptotic stability).*

Vector Results

This part enlarges the scalar results to the vector case by considering a Vector Lyapunov Function (VLF). Having a vector overvaluing system may be useful for the study of some properties such as estimation of asymptotic stability domains and attractors. In addition, it appears to be a suitable tool for the analysis of polyhedral constrained control problems, as is shown in [4] and [5].

As in section 3, we consider vector function V which components are of the form (3.6). With respect to this function V, we associate with a $n \times n$ matrix A, the matrices $\Gamma(A)$ and $V(A)$ defined by:

$$\begin{aligned} \Gamma(A) &= \{\Gamma(A)_{ij}\} \text{ with } \Gamma(A)_{ii} = \mu(A_{ii}) \text{ and } \Gamma(A)_{ij} = \| A_{ij} \|_{ij}, \text{ for } i \neq j, \\ V(A) &= \{V(A)_{ij}\} \text{ with } V(A)_{ij} = \| A_{ij} \|_{ij} . \end{aligned}$$

Lemma 20 (construction of vector overvaluing system) [28]. *Suppose the following conditions hold*

a) $\sigma(V(C)) < 1$,
b) there exists a matrix $\Pi(.)$ with positive bounded coefficients such that

$$\sum_{k=1}^{\infty} V(C)^{k-1} V((B(.) + CA(.))_k) \leq \Pi(.),\ t \geq t_0,\ x_t \in \Omega,\ d \in S_d . \tag{5.4}$$

then the system

$$\dot{z}(t) = \Gamma(A(.))\, z(t) + \sum_{k=1}^{\infty} V(C)^{k-1} V((B(.) + CA(.))_k)\, z(t - k\tau), \tag{5.5}$$
$$t \geq t_0, \quad z_{t_0}(\theta) = \psi(\theta), \quad \theta \in (-\infty, 0]$$

is a local overvaluing system of (5.1) with respect to VLF V and set Ω. Moreover if a) holds and A and B are constant matrices, then (5.5) is a global overvaluing system of (5.1) and $\Pi(.)$ can be expressed as:

$$\Pi(.) = (I_r - V(C))^{-1} V(B + CA) \ .$$

Corollary 21 [28]. *If the conditions of Lemma 20 hold, then any system*

$$\dot{z}(t) = \Gamma^* z(t) + \sum_{k=1}^{\infty} V(C)^{k-1} A_k z(t - k\tau)), \ z(t) \in \mathbb{R}^r, \ t \geq t_0, \tag{5.6}$$

$$z_{t_0}(\theta) = \psi(\theta), \quad \theta \in (-\infty, 0],$$

such that $\Gamma^* \geq \Gamma(A(.))$ and $A_k \geq V((B(.) + CA(.))_k)$, for $t \geq t_0$, $x_t \in \Omega$, $d \in S_d$,
is also a local overvaluing system of (5.1) with respect to VLF V and set Ω.

Theorem 22 (vector stability criterion) [28]. *Let us consider that the hypotheses a) and b) of Lemma 20 hold, and consider the two following properties:*

c) there exists a positive constant vector u such that

$$[\Gamma(A(.)) + \Pi(.)]u \leq 0, \quad \forall t \geq t_0, \ x_t \in \Omega, \ d \in S_d.$$

d) the matrix $\Gamma(A(.)) + \Pi(.)$ is less or equal to the opposite of a constant M-matrix for all $t \geq t_0$, $x_t \in \Omega$ and $d \in S_d$.

Then, the zero solution of (5.1) is:

1. *locally stable if c) holds.*
2. *locally asymptotically stable if d) holds.*

Remark: Property d) implies property c).

Corollary 23 [28]. *Suppose A and B are constant, then if $\sigma(V(C)) < 1$ and if there exists a positive vector u such that $[\Gamma(A) + (I_r - V(C))^{-1} V(B + CA)]\, u \leq 0$ (resp. $\Gamma(A) + (I_r - V(C))^{-1} V(B + CA)$ is the opposite of an M-matrix) then the zero solution of (5.1) is stable (respectively asymptotically stable).*

The following theorem is a summary of results given in this part.

Theorem 24 [28]. *Under conditions a) and b) of Lemma 20, any vector overvaluing system verifying conditions of corollary 21 is a local comparison system of (5.1) with regard to stability (resp. asymptotic stability)*

Remark: Corollary 18 and 23 are important because very easy to apply ; Corollary 18 is a generalization of [21].

5.3 Examples

Example 1: Let us consider

$$\dot{x}(t) = \begin{pmatrix} -2 & 0 \\ 0.5 & -2 \end{pmatrix} x(t) + \begin{pmatrix} 0 & 0.5 \\ 0 & 0 \end{pmatrix} x(t-\tau) + \begin{pmatrix} 0 & 0 \\ 0 & c \end{pmatrix} \dot{x}(t-\tau) \ . \quad (5.7)$$

This example given in [34] was firstly considered in [12] ; the problem is to determine the values of parameter c that guarantee the stability of zero solution of (5.7). In [34], applying complex Lyapunov function, it is shown that if $-0.9848 < c < 0.9837$ then the zero solution of (5.7) is asymptotically stable. Applying our methods yields to the following results:

1. scalar case
   with $\|\cdot\| = \|\cdot\|_1$: stability condition: $|c| < 0.5$,
   with $\|\cdot\| = \|\cdot\|_\infty$: stability condition: $|c| < 0.6$,
2. vector case with $V(x) = [|x_1|, |x_2|]^T$: stability condition $|c| < 0.5$

These results are weaker than those in [34] but immediate to obtain; they are stronger than those given by [12] ($0.25 < |c| < 0.52$).

Example 2: Consider the scalar system

$$\dot{x}(t) = -a\,x(t) - b\,x(t-\tau) + c\,\dot{x}(t-\tau), \quad t \geq 0, \qquad (5.8)$$

where $a > 0$, b, c, $\tau > 0$ are given constants, $|c| < 1$
Applying Corollary 18 proves that, if $-a + \frac{|b+ac|}{1-|c|} < 0$, then the zero solution of (5.8) is asymptotically stable.
This asymptotic stability condition, in the particular case $b = 0$, is the same as one of those given in [15] by applying the so-called "two stages method".

6 Conclusion

As we mentionned in the paper, the comparison method appears to give a simple and efficient way to answer many questions linked to stability: conditions on the parameters, on the variables (initial conditions), and on the disturbances.

Among the original contributions of this work on nonlinear delay systems, we can remark that:

- Section 3 provides very simple conditions for i.o.d. stability (including stability domains), together with a method that constitutes the keystone of the further results.
- Section 4 allows to deal with systems with unstable or stable memoryless feedbacks, which means that stabilization by means of delayed feedback can be studied by this way. The provided criteria are delay-dependent, as well as the estimated stability domains or exponentional rates of convergence.
- Section 5 allows to deal with the difficult neutral systems, using the same background. The obtained criteria turn out to be very simple when a linear comparison system can be defined (as at the end of Lemma 15).

The results of this last section can now be generalized to delay-dependent conditions for neutral systems by using the decomposition procedure of section 4: this has not be done in this presentation, but is a straightforward development of the work.

7 Appendix

Definition of an M-matrix and properties:
A matrix M is the opposite of an M-matrix if it is Hurwitz with non-negative off-diagonal elements.
If M is the opposite of an M-matrix, then the following statements hold:

- The real parts of the eigenvalues of M are negative.
- M admits a real negative eigenvalue $-\lambda_m(M)$, called the importance eigenvalue, such that for any eigenvalue λ_i of M, $\mathrm{Re}(\lambda_i) \leq -\lambda_m(M)$ holds. There is a non-negative eigenvector $k(M)$ associated with $-\lambda_m(M)$, so called importance vector. Moreover if M is irreducible then the components of $k(M)$ are positive.
- For any vector $x \geq 0, x \neq 0$, there exists an index i such that $x_i(Mx)_i < 0$.
- M verifies the Kotelyanski conditions, i.e. its successive principal minors are sign-alternate.

References

1. Dambrine, M.: Contribution à l'étude des systèmes à retards. Ph.D. Thesis, University of Sciences and Techn. of Lille, N. 1386 (Oct. 1994).
2. Dambrine, M., Richard, J. P.: Stability Analysis of Time-Delay Systems. Dynamic Systems and Applications **2** (1993) 405–414
3. Dambrine, M., Richard, J. P.: Stability and Stability Domains Analysis for Nonlinear Differential-Difference Equations. Dyn. Syst. and Applications **3** (1994) 369–378
4. Dambrine, M., Goubet, A., Richard, J. P.: New results on constrained stabilizing control of time-delay systems. *Proc. 34th IEEE Conf. on Decision and Control*, New Orleans, USA. (1995) 2052–2057
5. Dambrine, M., Richard, J. P., Borne, P.: Feedback control of time-delay systems with bounded control and state. Mathematical Problems in Engineering **1(1)** (1995) 77–87
6. Desoer, C. A. and Vidyasagar, M. *Feedback Systems: Input-output Properties*. Academic Press, New York, 1975
7. Grujić, L.J.: Novel development of Lyapunov stability of motion. Int. J. Control, **22(4)** (1975) 525–549
8. Goubet-Bartholoméüs, A.: Sur la stabilité et la stabilisation des systèmes retardés : critères dépendant des retards. Ph.D. Thesis Univ. of Lille, N. 1910 (dec 1996).
9. Goubet, A., Dambrine, M., Richard, J.P.: An extension of stability criteria for linear and nonlinear time delay systems. *Proc.IFAC Conf. System Structure and Control*, Nantes, France. (1995) 278–283

10. Goubet-Bartholomeüs, A., Dambrine, M., Richard, J.P.: Delay-dependent stability domains of nonlinear time-varying delay systems. *IEEE SMC CESA'96*, Lille, France, (1996) 801–806
11. Goubet-Bartholomeüs, A., Dambrine, M., Richard, J.P.: Stability of pertubed time-varying-delay systems. To appear in Systems and Control Letters.
12. Hmamed, A.: Stability conditions of delay-differential systems. Int. Control, 43(2) (1986) 455–463
13. Kolmanovskii, V.B.: Stability of some systems with arbitrary after effect. Reports Russ. Acad. Sci. **331** (1993) 421–425
14. Kolmanovskii, V.B.: Applications of Differential Inequalities for Stability of some Functional Differential Equations. Nonlinear Analysis, Theory, Methods and Applications **25(9–10)** (1995) 1017–1028
15. Kolmanovskii, V.B.: The stability of hereditary systems of neutral type. J. Appl. Math. Mech. **60(2)** (1996) 205–216
16. Kolmanovskii, V.B., Nosov, V.R.: *Stability of functional differential Equations*, Academic Press, New-York, 1986
17. Kolmanovskii, V.B., Shaikhet, L.E.: *Control of systems with aftereffect*, American Mathematical Society, R.I., vol. 157, 1996
18. Krasovskii, N.: *Stability of Motion*, Stanford Univ. Press, Stanford, Calif., 1963.
19. Lehman, B., Shujaee, K.: Delay independent stability conditions and decay estimates for time-varying functional differential equations. IEEE Trans. Aut. Cont. **39(8)** (1994) 1673–1676
20. Li, X., De Souza, C. E.: LMI Approach to Delay-Dependent Robust Stability and Stabilization of Uncertain Linear Delay Systems. *Proc. 34rd IEEE Conf. on Decision and Control*, New Orleans, USA (1995) 3614–3619
21. Mori, T.,Fukuma, N., Kuwahara, M. : Simple stability criteria for single and composite linear systems with time delays: Int. J. Control, 30, n 6, (1981) 1175–1184
22. Niculescu, S.I.; Dion, J.M.; Dugard, L. : Delay-dependent stability criteria for uncertain systems with delayed state : A Razumikhin based approach. *Proc. IEEE VSTL'94*, Benevento (Italy) (1994) 34–41.
23. Niculescu, S.I., de Souza, C.E., Dion, J. M., Dugard, L.: Robust Stability and Stabilization for Uncertain Linear Systems with State Delay: Single Delay Case (I). *Proc. Workshop on Robust Control Design*, Rio de Janeiro, Brazil (1994) 469–474
24. Niculescu, S.I., de Souza, C.E., Dion, J. M., Dugard, L.: Robust exponential stability of uncertain linear systems with time-varying delays. *Proc. 3rd European Contr. Conf.*, Rome, Italy (1995) 1802–1808
25. Razumikhin, B. S.: The application of Lyapunov's method to problems in the stability of systems with delay., Autom. i Telemekhanica, **21(6)** (1960), 740–748.
26. Su, T. J., Huang, C. G.: Robust stability of delay dependence for linear uncertain systems. IEEE Trans. Aut. Cont. **37(10)** (1992) 1656–1659
27. Tchangani, A.P., Dambrine, M., Richard, J. P.: Stability and stabilization of nonlinear neutral systems. *IEEE SMC CESA'96*, Lille, France, (1996) 812–815.
28. Tchangani, A. P., Dambrine, M., Richard J. P.: New stability criteria for nonlinear neutral systems. To appear in *Proc. ECC'97*, Brussels, July 1997
29. Tokumaru, H., Adachi, N., Amemiya, T.: Macroscopic stability of interconnected systems. *6-th IFAC Congress* (1975) ID44.4
30. Verriest, E. I., Ivanov, A.F.: Robust stability of systems with delayed feedback. Circuits, Systems and Signal Processing **13(2)/13(3)** (1994) 213–222

31. Verriest, E. I.: Robust stability of time-varying systems with unknown bounded delays. *Proc.33th IEEE Conf. on Decision and Control*, Lake Buena Vista, USA. (1994) 417-422
32. Wang, S.S., Chen, B. S., Lin, T. P.: Robust stability of uncertain time-delay systems. Int.J.Control **46(5)** (1987) 963–976
33. Xu, B.: Comments on: Robust stability of delay dependence for linear uncertain systems. IEEE Trans. Aut. Cont. **39(11)** (1994) 2365
34. Xu, D.-Y., Xu Z.-F.: Stability analysis of linear delay-differential systems. Control -Theory and Advanced Technology, **7(4)**, (1991) 629–642

Output Feedback Stabilization of Linear Time-delay Systems

Xi Li[1] and Carlos E. de Souza[12]

[1] Department of Electrical and Computer Engineering
The University of Newcastle, NSW 2308, Australia
[2] National Laboratory for Scientific Computing – LNCC/CNPq
Rua Lauro Müller 455, 22290-160, Rio de Janeiro, RJ, Brazil

Abstract. This chapter considers the problem of output feedback stabilization of continuous time linear systems with a constant time-delay in the state. We develop a *delay-dependent* method for designing linear dynamic output feedback controllers which ensure global uniform asymptotic stability for any time-delay not larger than a given bound. The proposed stabilization method, which is based on linear matrix inequalities, is then extended to the case of uncertain polytopic systems. We also consider the problem of *delay-dependent* robust stabilization via output feedback for state delayed systems with norm-bounded parameter uncertainty. In this situation, the solution is given in terms of a generalized eigenvalue problem. The developed stabilization methods can be implemented numerically very efficiently using existing convex and quasi-convex optimisation techniques.

1 Introduction

Time-delays are frequently encountered in many dynamic systems and very often are the source of instability and poor performance; see, e.g. [12]. The problems of stability analysis and stabilization of dynamic systems with delayed state are, therefore, of theoretical and practical importance and have attracted considerable attention for several decades. Various techniques of stability and robust stability analysis have been proposed over the past few years, including delay-independent as well delay-dependent stability criteria; see, e.g. [2], [4], [6], [9], [10], [13], [15], [17]-[19] and the references therein.

Recently, increasing attention has been devoted to the problems of stabilization and robust stabilization of linear state-delayed systems. For example, stabilization techniques which are independent of the size of the time-delay have been proposed in [3], [7], [8], [11], [16] and [19], whereas delay-dependent stabilization methods have been recently developed in [9], [10] and [15]. With exception of [3], [7] and [19], all these stabilization techniques are based on state feedback, and thus do not apply to situations where some of the state variables are not available for feedback.

This chapter is concerned with the problem of *delay-dependent* output feedback stabilization of linear systems with a constant time-delay in the state. Both the cases of systems with, or without parametric uncertainty are treated and attention is focused on the design of stabilizing linear dynamic output feedback controllers which depend on the size of the time-delay. We first consider the stabilization problem for linear time-delay systems without parameter uncertainty. We obtain conditions which ensure the system isstabilizable via a linear dynamic output feedback controller for any time-delay not larger than a given bound. A procedure for constructing stabilizing controllers is also derived. The proposed method is based on the solution of linear matrix inequalities (LMIs). This approach is then extended to the problem of robust stabilization of uncertain polytopic systems with a delayed state, where the controller is required to guarantee global uniform asymptotic stability for any time-delay not larger than a given bound and for the whole set of admissible systems. We also consider the problem of robust stabilization of linear state delayed systems with norm-bounded parameter uncertainty in all the matrices of the system state equation. We develop a controller design technique which is given in terms of a generalized eigenvalue problem. The proposed stabilization methods have the advantage that can be implemented numerically very efficiently using recently developed interior-point algorithms for solving LMIs and generalized eigenvalue problems; see, e.g. [1] and [14].

Notation. The following notation will be used throughout this chapter. Re^n denotes the n dimensional Euclidean space, $\mathrm{Re}^{n\times m}$ is the set of all $n \times m$ real matrices, $\mathrm{diag}\{\cdots\}$ denotes a block-diagonal matrix and $\|\cdot\|$ refers to the induced matrix 2-norm. The notation $X > 0$ for $X \in \mathrm{Re}^{n\times n}$ means that the matrix X is symmetric and positive definite.

2 Problem Formulation and Preliminaries

Consider the following linear time-delay system

$$\dot{x}(t) = Ax(t) + A_d x(t-\tau) + Bu(t) \tag{2.1}$$
$$x(t) = \phi(t), \quad \forall t \in [-\tau, 0] \tag{2.2}$$
$$y(t) = Cx(t) \tag{2.3}$$

where $x(t) \in \mathrm{Re}^n$ is the state, $u(t) \in \mathrm{Re}^m$ is the control input, $y(t) \in \mathrm{Re}^p$ is the output, $\tau > 0$ is the time-delay of the system, $\phi(\cdot)$ is the initial condition, and A, A_d, B and C are real constant matrices of appropriate dimensions.

In this chapter we investigate the problem of designing linear dynamic output feedback stabilizing controllers for the system (2.1)-(2.3). Attention is focused on the design of controllers which depend on the size of the time-delay.

We shall adopt the following assumption for the system of (2.1)-(2.3).

Assumption 1 *$(A + A_d, B)$ is stabilizable and $(A + A_d, C)$ is detectable.*

Note that Assumption 1, which is equivalent to the stabilizability via linear dynamic output feedback of the system (2.1)-(2.3) in the absence of time-delay, is a necessary condition for the existence of a stabilizing linear dynamic output feedback control law for the system (2.1)-(2.3).

We shall consider linear dynamic output feedback controllers for the system (2.1)-(2.3) as follows:

$$\dot{\xi}(t) = A_c\xi(t) + B_c y(t) \tag{2.4}$$

$$u(t) = C_c\xi(t) + D_c y(t) \tag{2.5}$$

where $\xi(t) \in \mathrm{Re}^n$, and A_c, B_c, C_c and D_c are matrices of appropriate dimensions.

The control problem we shall address is as follows. *Find a controller of the form of (2.4)-(2.5) for the system (2.1)-(2.3) such that the resulting closed-loop system is globally uniformly asymptotically stable for any constant time-delay τ not larger than a given positive scalar $\bar{\tau}$.*

We conclude this section by recalling two lemmas which will be used in the derivation of the main result in the next section.

Lemma 1. (see [9]) *Consider the system $\dot{x}(t) = Ax(t) + A_d x(t-\tau)$. Given a scalar $\bar{\tau} > 0$, this system is globally uniformly asymptotically stable for any constant time-delay τ satisfying $0 \le \tau \le \bar{\tau}$ if there exist a matrix $X > 0$ and a scalar $\beta > 0$ solving the following LMI*

$$\begin{bmatrix} M & XA^T & XA_d^T & A_d \\ AX & -\bar{\sigma}\beta I & 0 & 0 \\ A_d X & 0 & -\bar{\sigma}(1-\beta)I & 0 \\ A_d^T & 0 & 0 & -\bar{\sigma}I \end{bmatrix} < 0$$

where $\bar{\sigma} = 1/\bar{\tau}$ and

$$M = X(A + A_d)^T + (A + A_d)X.$$

Lemma 2. (see, e.g., [5]) *Given matrices $G = G^T \in \mathrm{Re}^{m\times m}$, $Y \in \mathrm{Re}^{r\times m}$ and $Z \in \mathrm{Re}^{s\times m}$, then there exists a matrix $\Theta \in \mathrm{Re}^{r\times s}$ satisfying*

$$G + Y^T\Theta Z + Z^T\Theta^T Y < 0$$

if and only if

$$\mathcal{N}_Y^T G \mathcal{N}_Y < 0, \qquad \mathcal{N}_Z^T G \mathcal{N}_Z < 0$$

where $\mathcal{N}_Y$ and $\mathcal{N}_Z$ are any matrices whose columns form bases of the null spaces of Y and Z, respectively.

3 Output Feedback Stabilization

Motivated by the LMI approach to $\mathcal{H}_\infty$ control proposed in [5], in the sequel we develop an LMI based method for solving the delay-dependent output feedback stabilization problem for the system (2.1)-(2.3).

Theorem 3. *Consider the system (2.1)-(2.3) satisfying Assumption 1. Given a scalar $\bar{\tau} > 0$, this system is stabilizable via an output feedback controller (2.4)-(2.5) for any constant time-delay τ satisfying $0 \le \tau \le \bar{\tau}$, if there exist $n \times n$ symmetric positive definite matrices R, S and P, and a scalar $\beta > 0$ satisfying the following LMIs:*

$$\left[\begin{array}{c|c} \begin{matrix}\mathcal{N}_{B1} \\ \mathcal{N}_{B2}\end{matrix} & 0 \\ \hline 0 & I \end{array}\right]^T H_S(S,P,\beta) \left[\begin{array}{c|c} \begin{matrix}\mathcal{N}_{B1} \\ \mathcal{N}_{B2}\end{matrix} & 0 \\ \hline 0 & I \end{array}\right] < 0 \tag{3.1}$$

$$\left[\begin{array}{c|c} \mathcal{N}_C & 0 \\ \hline 0 & I \end{array}\right]^T H_R(R,\beta) \left[\begin{array}{c|c} \mathcal{N}_C & 0 \\ \hline 0 & I \end{array}\right] < 0 \tag{3.2}$$

$$\begin{bmatrix} R & I \\ I & S \end{bmatrix} > 0 \tag{3.3}$$

where $\begin{bmatrix} \mathcal{N}_{B1} \\ \mathcal{N}_{B2} \end{bmatrix}$ *and* $\mathcal{N}_C$ *are any matrices whose columns form bases of the null spaces of* $[B^T \;\; B^T]$ *and* C*, respectively, and*

$$H_S(S,P,\beta) = \left[\begin{array}{cc|ccc} Q_S & SA^T & A+A_d & SA_d^T & A_d \\ AS & -\bar{\sigma}\beta I & A & 0 & 0 \\ \hline A^T + A_d^T & A^T & -\bar{\sigma}P & A_d^T & 0 \\ A_d S & 0 & A_d & -\bar{\sigma}(1-\beta)I & 0 \\ A_d^T & 0 & 0 & 0 & -\bar{\sigma}I \end{array}\right] \tag{3.4}$$

$$H_R(R,\beta) = \left[\begin{array}{c|ccc} Q_R & A^T & A_d^T & RA_d \\ \hline A & -\bar{\sigma}\beta I & 0 & 0 \\ A_d & 0 & -\bar{\sigma}(1-\beta)I & 0 \\ A_d^T R & 0 & 0 & -\bar{\sigma}I \end{array}\right] \tag{3.5}$$

$$Q_S = S(A+A_d)^T + (A+A_d)S, \tag{3.6}$$

$$Q_R = (A+A_d)^T R + R(A+A_d), \tag{3.7}$$

$$\bar{\sigma} = 1/\bar{\tau}. \tag{3.8}$$

Proof. The proof technique is inspired by that used in [5] to prove Theorems 4.2 and 4.3. The closed-loop system of (2.1)-(2.3) with the controller (2.4)-(2.5) can be described by the following state-space model

$$\dot{x}_c(t) = \hat{A}x_c(t) + \bar{A}_d x_c(t-\tau) \tag{3.9}$$

where

$$x_c = [\, x^T \ \ \xi^T \,]^T, \qquad \hat{A} = \bar{A} + \bar{B}\Theta\bar{C}, \tag{3.10}$$

$$\Theta = \begin{bmatrix} A_c & B_c \\ C_c & D_c \end{bmatrix}, \tag{3.11}$$

$$\bar{A} = \begin{bmatrix} A & 0 \\ 0 & 0 \end{bmatrix}, \quad \bar{A}_d = \begin{bmatrix} A_d & 0 \\ 0 & 0 \end{bmatrix}, \tag{3.12}$$

$$\bar{B} = \begin{bmatrix} 0 & B \\ I & 0 \end{bmatrix}, \quad \bar{C} = \begin{bmatrix} 0 & I \\ C & 0 \end{bmatrix}. \tag{3.13}$$

Applying Lemma 1 to the system (3.9), it follows that the controller (2.4)-(2.5) solves the stabilization problem for a given $\bar{\tau} > 0$ if there exist a $2n \times 2n$ matrix $X > 0$ and a scalar $\beta > 0$ such that

$$G_X + Y^T\Theta Z_X + Z_X^T\Theta^T Y < 0 \tag{3.14}$$

where

$$G_X = \begin{bmatrix} Q_X & X\bar{A}^T & X\bar{A}_d^T & \bar{A}_d \\ \bar{A}X & -\bar{\sigma}\beta I & 0 & 0 \\ \bar{A}_d X & 0 & -\bar{\sigma}(1-\beta)I & 0 \\ \bar{A}_d^T & 0 & 0 & -\bar{\sigma}I \end{bmatrix}, \tag{3.15}$$

$$Q_X = X(\bar{A} + \bar{A}_d)^T + (\bar{A} + \bar{A}_d)X, \tag{3.16}$$

$$Y = \begin{bmatrix} \bar{B}^T & \bar{B}^T & 0 & 0 \end{bmatrix}, \tag{3.17}$$

$$Z_X = \begin{bmatrix} \bar{C}X & 0 & 0 & 0 \end{bmatrix}. \tag{3.18}$$

By Lemma 2, the inequality (3.14) is equivalent to

$$\mathcal{N}_Y^T G_X \mathcal{N}_Y < 0, \qquad \mathcal{N}_{Z_X}^T G_X \mathcal{N}_{Z_X} < 0 \tag{3.19}$$

where $\mathcal{N}_Y$ and $\mathcal{N}_{Z_X}$ are any matrices whose columns form bases of the null spaces of Y and Z_X, respectively.

Note that by defining

$$Z \triangleq [\bar{C} \quad 0 \quad 0 \quad 0], \tag{3.20}$$

we have

$$Z = Z_X J_X^{-1}$$

where

$$J_X = \begin{bmatrix} X & 0 & 0 & 0 \\ 0 & I & 0 & 0 \\ 0 & 0 & I & 0 \\ 0 & 0 & 0 & I \end{bmatrix}.$$

Hence, the columns of

$$\mathcal{N}_Z = J_X \mathcal{N}_{Z_X}$$

form a basis of the null space of Z. This implies that

$$\mathcal{N}_{Z_X}^T G_X \mathcal{N}_{Z_X} = \mathcal{N}_{Z_X}^T J_X G_{X^{-1}} J_X \mathcal{N}_{Z_X} = \mathcal{N}_Z^T G_{X^{-1}} \mathcal{N}_Z$$

where

$$G_{X^{-1}} = \begin{bmatrix} Q_{X^{-1}} & \bar{A}^T & \bar{A}_d^T & X^{-1}\bar{A}_d \\ \bar{A} & -\bar{\sigma}\beta I & 0 & 0 \\ \bar{A}_d & 0 & -\bar{\sigma}(1-\beta)I & 0 \\ \bar{A}_d^T X^{-1} & 0 & 0 & -\bar{\sigma} I \end{bmatrix} \tag{3.21}$$

$$Q_{X^{-1}} = (\bar{A} + \bar{A}_d)^T X^{-1} + X^{-1}(\bar{A} + \bar{A}_d).$$

Hence, it follows that $\mathcal{N}_{Z_X}^T G_X \mathcal{N}_{Z_X} < 0$ is equivalent to $\mathcal{N}_Z^T G_{X^{-1}} \mathcal{N}_Z < 0$.

Next, we shall express the conditions $\mathcal{N}_Y^T G_X \mathcal{N}_Y < 0$ and $\mathcal{N}_Z^T G_{X^{-1}} \mathcal{N}_Z < 0$ in terms of the plant parameters. To this end, we shall partition X and X^{-1} as

$$X = \begin{bmatrix} S & N \\ N^T & V \end{bmatrix}, \quad X^{-1} = \begin{bmatrix} R & M \\ M^T & U \end{bmatrix} \tag{3.22}$$

where R, S, M, N, U and V are $n \times n$ real matrices. Note that in view of (3.3) and considering that R, S, M and N satisfy

$$MN^T = I - RS$$

it results that M and N are non-singular matrices. Moreover, it can be easily established that given any non-singular matrices $R > 0$, $S > 0$ and N, there exist unique matrices M, U and V such that $X > 0$. Indeed, we have that

$$M = (I - RS)N^{-T}, \quad U = N^{-1}(SRS - S)N^{-T}, \quad V = N^T(S - R^{-1})^{-1}N. \tag{3.23}$$

With the partition as in (3.22) and considering (3.12), G_X and $G_{X^{-1}}$ of (3.15) and (3.21), respectively, can be rewritten as:

$$G_X = \begin{bmatrix} G_{11} & G_{12} & G_{13} & G_{14} \\ G_{12}^T & Q_1 & 0 & 0 \\ G_{13}^T & 0 & Q_2 & 0 \\ G_{14}^T & 0 & 0 & Q_3 \end{bmatrix} \tag{3.24}$$

$$G_{X^{-1}} = \begin{bmatrix} \hat{G}_{11} & \hat{G}_{12} & \hat{G}_{13} & \hat{G}_{14} \\ \hat{G}_{12}^T & Q_1 & 0 & 0 \\ \hat{G}_{13}^T & 0 & Q_2 & 0 \\ \hat{G}_{14}^T & 0 & 0 & Q_3 \end{bmatrix} \tag{3.25}$$

where

$$G_{11} = \begin{bmatrix} Q_S & (A + A_d)N \\ N^T(A + A_d)^T & 0 \end{bmatrix}, \quad G_{12} = \begin{bmatrix} SA^T & 0 \\ N^T A^T & 0 \end{bmatrix},$$

$$G_{13} = \begin{bmatrix} SA_d^T & 0 \\ N^T A_d^T & 0 \end{bmatrix}, \quad G_{14} = \begin{bmatrix} A_d & 0 \\ 0 & 0 \end{bmatrix},$$

$$\hat{G}_{11} = \begin{bmatrix} Q_R & M^T(A + A_d) \\ (A + A_d)^T M & 0 \end{bmatrix}, \quad \hat{G}_{12} = \begin{bmatrix} A^T & 0 \\ 0 & 0 \end{bmatrix},$$

$$\hat{G}_{13} = \begin{bmatrix} A_d^T & 0 \\ 0 & 0 \end{bmatrix}, \quad \hat{G}_{14} = \begin{bmatrix} RA_d & 0 \\ M^T A_d & 0 \end{bmatrix},$$

$$Q_1 = \begin{bmatrix} -\bar{\sigma}\beta I & 0 \\ 0 & -\bar{\sigma}\beta I \end{bmatrix}, \quad Q_2 = \begin{bmatrix} -\bar{\sigma}(1-\beta)I & 0 \\ 0 & -\bar{\sigma}(1-\beta)I \end{bmatrix},$$

$$Q_3 = \begin{bmatrix} -\bar{\sigma}I & 0 \\ 0 & -\bar{\sigma}I \end{bmatrix}$$

and the matrices Q_S and Q_R are as in (3.6) and (3.7), respectively.

On the other hand, considering that

$$Y = [\, \bar{B}^T \quad \bar{B}^T \quad 0 \quad 0 \,]$$

$$= \begin{bmatrix} 0 & I & \vdots & 0 & I & \vdots & 0 & 0 & \vdots & 0 & 0 \\ B^T & 0 & \vdots & B^T & 0 & \vdots & 0 & 0 & \vdots & 0 & 0 \end{bmatrix}$$

and

$$Z = [\, \bar{C} \quad 0 \quad 0 \quad 0 \,] = \begin{bmatrix} 0 & I & \vdots & 0 & 0 & \vdots & 0 & 0 & \vdots & 0 & 0 \\ C & 0 & \vdots & 0 & 0 & \vdots & 0 & 0 & \vdots & 0 & 0 \end{bmatrix},$$

it follows that $\mathcal{N}_Y$ and $\mathcal{N}_Z$ are of the form as below:

$$\mathcal{N}_Y = \begin{bmatrix} \mathcal{N}_{B1} & 0 & 0 & 0 & 0 & 0 \\ 0 & I & 0 & 0 & 0 & 0 \\ \cdots & \cdots & \cdots & \cdots & \cdots & \cdots \\ \mathcal{N}_{B2} & 0 & 0 & 0 & 0 & 0 \\ 0 & -I & 0 & 0 & 0 & 0 \\ \cdots & \cdots & \cdots & \cdots & \cdots & \\ 0 & 0 & I & 0 & 0 & 0 \\ 0 & 0 & 0 & I & 0 & 0 \\ \cdots & \cdots & \cdots & \cdots & \cdots & \cdots \\ 0 & 0 & 0 & 0 & I & 0 \\ 0 & 0 & 0 & 0 & 0 & I \end{bmatrix} \tag{3.26}$$

and

$$\mathcal{N}_Z = \begin{bmatrix} \mathcal{N}_C & 0 & 0 & 0 & 0 & 0 & 0 \\ 0 & 0 & 0 & 0 & 0 & 0 & 0 \\ \cdots & \cdots & \cdots & \cdots & \cdots & \cdots & \cdots \\ 0 & I & 0 & 0 & 0 & 0 & 0 \\ 0 & 0 & I & 0 & 0 & 0 & 0 \\ \cdots & \cdots & \cdots & \cdots & \cdots & \cdots & \cdots \\ 0 & 0 & 0 & I & 0 & 0 & 0 \\ 0 & 0 & 0 & 0 & I & 0 & 0 \\ \cdots & \cdots & \cdots & \cdots & \cdots & \cdots & \cdots \\ 0 & 0 & 0 & 0 & 0 & I & 0 \\ 0 & 0 & 0 & 0 & 0 & 0 & I \end{bmatrix} \tag{3.27}$$

where $\begin{bmatrix} \mathcal{N}_{B1} \\ \mathcal{N}_{B2} \end{bmatrix}$ and $\mathcal{N}_C$ are any matrices whose columns form bases of the null spaces of $[B^T \quad B^T]$ and C, respectively.

By considering (3.24) and (3.26), it can be easily established that the condition $\mathcal{N}_Y^T G_X \mathcal{N}_Y < 0$ is equivalent to

$$\begin{bmatrix} \mathcal{N}_{B1} & 0 & 0 & 0 \\ \mathcal{N}_{B2} & 0 & 0 & 0 \\ 0 & I & 0 & 0 \\ 0 & 0 & I & 0 \\ 0 & 0 & 0 & I \end{bmatrix}^T \bar{H}_S \begin{bmatrix} \mathcal{N}_{B1} & 0 & 0 & 0 \\ \mathcal{N}_{B2} & 0 & 0 & 0 \\ 0 & I & 0 & 0 \\ 0 & 0 & I & 0 \\ 0 & 0 & 0 & I \end{bmatrix} < 0 \tag{3.28}$$

where

$$\bar{H}_S = \begin{bmatrix} Q_S & SA^T & (A+A_d)N & SA_d^T & A_d \\ AS & -\bar{\sigma}\beta I & AN & 0 & 0 \\ N^T(A+A_d)^T & N^TA^T & -\bar{\sigma}\beta I & N^TA_d^T & 0 \\ A_dS & 0 & A_dN & -\bar{\sigma}(1-\beta)I & 0 \\ A_d^T & 0 & 0 & 0 & -\bar{\sigma}I \end{bmatrix}.$$

Now, multiplying (3.28) on the left and on the right by H^T and H, respectively, where

$$H = \begin{bmatrix} I & 0 & 0 & 0 \\ 0 & N^{-1} & 0 & 0 \\ 0 & 0 & I & 0 \\ 0 & 0 & 0 & I \end{bmatrix}$$

and introducing the new variable

$$P = \beta(NN^T)^{-1} > 0,$$

it follows that (3.28) is equivalent to (3.1).

On the other hand, by considering (3.25) and (3.27) it can be easily shown that the condition $\mathcal{N}_Z^T G_{X^{-1}} \mathcal{N}_Z < 0$ is equivalent to the inequality of (3.2) which completes the proof. ∇∇∇

In the case when the conditions of Theorem 3 are fulfilled, an output feedback controller that solves the stabilization problem can be easily obtained. Indeed, assuming that the LMIs (3.1)-(3.3) are satisfied for some (not necessarily unique) matrices R, S and P, and scalar β, a stabilizing controller can be found as follows:

1. Compute an $n \times n$ non-singular matrix N such that

$$NN^T = \beta P^{-1};$$

2. Find a matrix $X > 0$, which satisfies the inequality (3.14), by using (3.22) and (3.23), i.e.

$$X = \begin{bmatrix} S & N \\ N^T & N^T(S - R^{-1})^{-1}N \end{bmatrix};$$

3. Compute the controller matrices, A_c, B_c, C_c and D_c, by solving the LMI of (3.14) for $\Theta = \begin{bmatrix} A_c & B_c \\ C_c & D_c \end{bmatrix}$, i.e.

$$G_X + Y^T \Theta Z_X + Z_X^T \Theta^T Y < 0$$

where G_X, Y and Z_X are as in (3.15)-(3.18).

Remark 1. *Theorem 3 provides a delay-dependent sufficient condition for output feedback stabilization of linear time-delay systems in terms of the solvability of linear matrix inequalities. Observe that since this stabilizability condition includes information on the size of the time-delay, in general, it is expected to be less conservative than the delay-independent result of [19], especially when the time-delay is small. The proposed stabilization method has also the advantage that it can be implemented numerically very efficiently by using interior point algorithms, which have been recently developed for solving LMIs; see, e.g. [1] and [14].*

Remark 2. *It should be remarked that although Assumption 1 has not been explicitly used in the proof of Theorem 3, it is necessary for the inequalities (3.1) and (3.2) to hold. Indeed, it can be easily verified that if $(A + A_d, B)$ is not stabilizable, then there exists a vector v such that $\mathcal{N}_{B1}v \in \mathrm{Ker}(B^T)$, $\mathcal{N}_{B2}v = 0$ and*

$$v^*\mathcal{N}_{B1}^T Q_S \mathcal{N}_{B1} v \geq 0$$

where Ker(M) denotes the kernel of the matrix M and the superscript '$*$' stands for complex conjugate transpose. The above inequality implies that (3.1) cannot be satisfied. Similarly, if $(A + A_d, C)$ is not detectable, there exists a vector w such that $\mathcal{N}_C w \in \mathrm{Ker}(C)$ and $w^*\mathcal{N}_C^T Q_R \mathcal{N}_C w \geq 0$, which contradicts (3.2).

Remark 3. *The problem of finding the largest $\bar{\tau}$ which ensures output feedback stabilization using the method of Theorem 3 can be easily solved without the need of carrying out iterations for increasing $\bar{\tau}$. Indeed, the largest $\bar{\tau}$ can be computed by solving the following quasi-convex optimization problem in R, S, β and $\bar{\sigma}$:*

$$\begin{array}{ll} \textit{minimize} & \bar{\sigma} \\ \textit{subject to} & R > 0, \quad S > 0, \quad P > 0, \quad \beta > 0, \quad \bar{\sigma} > 0, \quad \textit{and} \quad (3.1) - (3.3). \end{array}$$

The largest value of τ, namely $\bar{\tau}^*$, is given by $\bar{\tau}^* = 1/\bar{\sigma}^*$, where $\bar{\sigma}^*$ is the optimal value of $\bar{\sigma}$. Note that the above optimization problem has the form of a generalized eigenvalue problem, which is known to be solvable numerically very efficiently; see, e.g. [1] and [14].

4 Robust Output Feedback Stabilization

In this section, we extend the output feedback stabilization method of Section 3 to the case of linear state delayed systems with parameter uncertainty in the state equation. We shall consider uncertain polytopic systems as well as systems with norm-bounded uncertainty.

4.1 Polytopic Uncertain Case

Consider uncertain linear polytopic systems described by

$$\dot{x}(t) = A(t)x(t) + A_d(t)x(t-\tau) + Bu(t) \tag{4.1}$$

$$x(t) = \phi(t), \quad \forall t \in [-\tau, 0] \tag{4.2}$$

$$y(t) = Cx(t) \tag{4.3}$$

where $x(t) \in \mathrm{Re}^n$ is the state, $u(t) \in \mathrm{Re}^m$ is the control input, $y(t) \in \mathrm{Re}^p$ is the output, $\tau > 0$ is the time-delay of the system, $\phi(\cdot)$ is the initial condition. $A(t)$, $A_d(t)$, B and C are real matrices of appropriate dimensions, with $A(t)$ and $A_d(t)$ being uncertain matrices satisfying

$$\begin{bmatrix} A(t) & A_d(t) \end{bmatrix} \in \Omega, \quad \forall t \geq 0 \tag{4.4}$$

where Ω is a polytope with L vertices described by

$$\Omega = \left\{ [A \quad A_d] : [A \quad A_d] = \sum_{i=1}^{L} \lambda_i [A_i \quad A_{di}]; \; \lambda_i \geq 0, \; \sum_{i=1}^{L} \lambda_i = 1 \right\}. \tag{4.5}$$

The system (4.1)-(4.3) is supposed to satisfy the following assumption.

Assumption 2 *$(A_i + A_{di}, B)$ is stabilizable and $(A_i + A_{di}, C)$ is detectable for $i = 1, \ldots, L$.*

Note that Assumption 2, which is equivalent to the quadratic stabilizability via linear dynamic output feedback of the system (4.1)-(4.3) in the absence of time-delay, is a necessary condition for the existence of a robust stabilizing linear dynamic output feedback control law for the system (4.1)-(4.3).

We shall address the following robust stabilization problem: *Given a scalar $\bar{\tau} > 0$, find a controller of the form of (2.4)-(2.5) for the system (4.1)-(4.3) such that the resulting closed-loop system is globally uniformly asymptotically stable for any constant time-delay τ satisfying $0 \leq \tau \leq \bar{\tau}$ and for all $A(t)$ and $A_d(t)$ satisfying (4.4)-(4.5).*

Theorem 4. *Consider the system (4.1)-(4.3) satisfying Assumption 2. Given a scalar $\bar{\tau} > 0$, this system is robustly stabilizable via an output feedback controller (2.4)-(2.5) for any constant time-delay τ satisfying $0 \leq \tau \leq \bar{\tau}$, if there exist $n \times n$ symmetric positive definite matrices R, S and P, and a scalar $\beta > 0$ satisfying the following LMIs:*

$$\left[\begin{array}{c|c} \mathcal{N}_{B1} & \\ \mathcal{N}_{B2} & 0 \\ \hline 0 & I \end{array}\right]^T H_{Si}(S, P, \beta) \left[\begin{array}{c|c} \mathcal{N}_{B1} & \\ \mathcal{N}_{B2} & 0 \\ \hline 0 & I \end{array}\right] < 0, \quad i = 1, \ldots, L \tag{4.6}$$

$$\left[\begin{array}{c|c} \mathcal{N}_C & 0 \\ \hline 0 & I \end{array}\right]^T H_{Ri}(R, \beta) \left[\begin{array}{c|c} \mathcal{N}_C & 0 \\ \hline 0 & I \end{array}\right] < 0, \quad i = 1, \ldots, L \tag{4.7}$$

$$\begin{bmatrix} R & I \\ I & S \end{bmatrix} > 0 \tag{4.8}$$

where $\begin{bmatrix} \mathcal{N}_{B1} \\ \mathcal{N}_{B2} \end{bmatrix}$ *and* $\mathcal{N}_C$ *are any matrices whose columns form bases of the null spaces of* $[B^T \;\; B^T]$ *and* C*, respectively, and*

$$H_{Si}(S,P,\beta) = \left[\begin{array}{cc|ccc} Q_{Si} & SA_i^T & A_i + A_{di} & SA_{di}^T & A_{di} \\ A_iS & -\bar{\sigma}\beta I & A_i & 0 & 0 \\ \hline A_i^T + A_{di}^T & A_i^T & -\bar{\sigma}P & A_{di}^T & 0 \\ A_{di}S & 0 & A_{di} & -\bar{\sigma}(1-\beta)I & 0 \\ A_{di}^T & 0 & 0 & 0 & -\bar{\sigma}I \end{array}\right]$$

$$H_{Ri}(R,\beta) = \left[\begin{array}{c|ccc} Q_{Ri} & A_i^T & A_{di}^T & RA_{di} \\ \hline A_i & -\bar{\sigma}\beta I & 0 & 0 \\ A_{di} & 0 & -\bar{\sigma}(1-\beta)I & 0 \\ A_{di}^T R & 0 & 0 & -\bar{\sigma}I \end{array}\right]$$

$$Q_{Si} = S(A_i + A_{di})^T + (A_i + A_{di})S,$$

$$Q_{Ri} = (A_i + A_{di})^T R + R(A_i + A_{di}),$$

$$\bar{\sigma} = 1/\bar{\tau}.$$

Proof. Multiplying (4.6) and (4.7) by the weight λ_i and summing for $i = 1, \ldots, L$, we obtain

$$\left[\begin{array}{c|c} \mathcal{N}_{B1} & 0 \\ \mathcal{N}_{B2} & \\ \hline 0 & I \end{array}\right]^T \sum_{i=1}^{L} \lambda_i H_{Si}(S,P,\beta) \left[\begin{array}{c|c} \mathcal{N}_{B1} & 0 \\ \mathcal{N}_{B2} & \\ \hline 0 & I \end{array}\right] < 0 \tag{4.9}$$

$$\left[\begin{array}{c|c} \mathcal{N}_C & 0 \\ \hline 0 & I \end{array}\right]^T \sum_{i=1}^{L} \lambda_i H_{Ri}(R,\beta) \left[\begin{array}{c|c} \mathcal{N}_C & 0 \\ \hline 0 & I \end{array}\right] < 0. \tag{4.10}$$

With (4.9) and (4.10) and in view of (4.5), the result follows immediately from Theorem 3. ∇∇∇

Remark 4. Theorem 4 establishes a delay-dependent condition for robust output feedback stabilization of uncertain polytopic systems with a delayed state. The proposed result is given in terms of the solution of linear matrix inequalities.

We observe that the computation of a robust output feedback controller can be carried out using the same procedure of Section 3 for the controller design.

4.2 Norm-Bounded Uncertain Case

In this subsection, as an extension to method of Section 3, we shall develop a delay-dependent robust output feedback stabilization method for linear time-delay systems with norm-bounded parameter uncertainty. We consider systems described by the differential delay equation

$$\dot{x}(t) = [A + \Delta A(t)]x(t) + [A_d + \Delta A_d(t)]x(t-\tau) + [B + \Delta B(t)]u(t) \tag{4.11}$$

$$x(t) = \phi(t), \quad \forall t \in [-\tau, 0] \tag{4.12}$$

$$y(t) = Cx(t) \tag{4.13}$$

where $x(t) \in \mathrm{Re}^n$ is the state, $u(t) \in \mathrm{Re}^m$ is the control input, $y(t) \in \mathrm{Re}^p$ is the output, $\tau > 0$ is the time-delay of the system, $\phi(\cdot)$ is the initial condition, A, A_d, B and C are known real constant matrices of appropriate dimensions which describe the nominal system of (4.11)-(4.13), and $\Delta A(\cdot)$, $\Delta A_d(\cdot)$ and $\Delta B(\cdot)$ are unknown real norm-bounded matrix functions which represent time-varying parameter uncertainties. The admissible uncertainties are assumed to be of the form

$$[\Delta A(t) \quad \Delta B(t)] = DF(t)[E_a \quad E_b], \qquad \Delta A_d(t) = D_d F_d(t) E_d \tag{4.14}$$

where $F(t) \in \mathrm{Re}^{i \times j}$ and $F_d(t) \in \mathrm{Re}^{i_d \times j_d}$ are unknown real time-varying matrices with Lebesgue measurable elements satisfying

$$\|F(t)\| \le 1; \quad \|F_d(t)\| \le 1, \quad \forall t \tag{4.15}$$

and D, D_d, E_a, E_b and E_d are known real constant matrices which characterize how the uncertain parameters in $F(t)$ and $F_d(t)$ enter the nominal matrices A, A_d and B.

The robust stabilization problem to be investigated in this subsection is as follows. *Given a scalar $\bar{\tau} > 0$, find a controller of the form of (2.4)-(2.5) for the system (4.11)-(4.13) such that the resulting closed-loop system is globally uniformly asymptotically stable for any constant time-delay τ satisfying $0 \le \tau \le \bar{\tau}$ and for all admissible uncertainties $\Delta A(t)$, $\Delta A_d(t)$ and $\Delta B(t)$.*

In order to derive a solution to the robust output feedback stabilization problem, the following delay-dependent robust stability result will be needed.

Lemma 5. *Consider the system (4.11)-(4.12) with $u(t) \equiv 0$. Given a calar $\bar{\tau} > 0$, this system is robustly stable for any constant time-delay τ satisfying $0 \le \tau \le \bar{\tau}$ and for all admissible uncertainties $\Delta A(t)$ and $\Delta A_d(t)$ if there exist a matrix $X > 0$ and scalars $\alpha_i > 0$, $i = 1, \ldots, 5$, solving the following LMI:*

$$\begin{bmatrix} Q(X) & X\hat{M}^T & L^T(\alpha_1, X) & \hat{N} \\ \hat{M}X & -\bar{\sigma}\hat{U}_1 & 0 & 0 \\ L(\alpha_1, X) & 0 & -\alpha_1 I & 0 \\ \hat{N}^T & 0 & 0 & -\bar{\sigma}\hat{U}_2 \end{bmatrix} < 0 \tag{4.16}$$

where $\bar{\sigma} = 1/\bar{\tau}$ *and*

$$Q(X) = X(A + A_d)^T + (A + A_d)X$$

$$L^T(\alpha_1, X) = \begin{bmatrix} \alpha_1 D & \alpha_1 D_d & X E_a^T & X E_d^T \end{bmatrix}$$

$$\hat{M}^T = \begin{bmatrix} A^T & E_a^T & A_d^T & E_d^T \end{bmatrix}$$

$$\hat{N} = \begin{bmatrix} A_d & D_d \end{bmatrix}$$

$$\hat{U}_1 = \operatorname{diag}\{\alpha_2 I - \alpha_3 D D^T, \quad \alpha_3 I, \quad (1 - \alpha_2)I - \alpha_4 D_d D_d^T, \quad \alpha_4 I\}$$

$$\hat{U}_2 = \operatorname{diag}\{I - \alpha_5 E_d^T E_d, \quad \alpha_5 I\}.$$

Proof. The result follows immediately from Theorem 3.1 of [9].

∇∇∇

Hence, we have the following robust output feedback stabilization result.

Theorem 6. *Consider the system (4.11)-(4.13) satisfying Assumption 1. Given a scalar* $\bar{\tau} > 0$, *this system is robustly stabilizable via an output feedback controller (2.4)-(2.5) for any constant time-delay* τ *satisfying* $0 \le \tau \le \bar{\tau}$, *if there exist* $n \times n$ *symmetric positive definite matrices* R, S *and* P, *and scalars* $\alpha_i > 0$, $i = 1, \ldots, 5$, *satisfying the following inequalities:*

$$\left[\begin{array}{c|c} \mathcal{N}_S & 0 \\ \hline 0 & I \end{array}\right]^T \left[\begin{array}{cc|cc} Q_S & S\hat{F}_1^T & \hat{F}_2 & \hat{F}_3^T \\ \hat{F}_1 S & -\hat{T}_1 & \hat{F}_1 & 0 \\ \hline \hat{F}_2^T & \hat{F}_1^T & -\bar{\sigma}P & \hat{F}_4^T \\ \hat{F}_3 & 0 & \hat{F}_4 & -\hat{T}_2 \end{array}\right] \left[\begin{array}{c|c} \mathcal{N}_S & 0 \\ \hline 0 & I \end{array}\right] < 0 \tag{4.17}$$

$$\left[\begin{array}{c|c} \mathcal{N}_C & 0 \\ \hline 0 & I \end{array}\right]^T \left[\begin{array}{c|ccc} Q_R & A^T & E_a^T & \hat{F}_5^T \\ \hline A & -\bar{\sigma}J_1 & 0 & 0 \\ E_a & 0 & -\alpha_3\bar{\sigma}I & 0 \\ \hat{F}_5 & 0 & 0 & -\hat{T}_3 \end{array}\right] \left[\begin{array}{c|c} \mathcal{N}_C & 0 \\ \hline 0 & I \end{array}\right] < 0 \tag{4.18}$$

$$\begin{bmatrix} R & I \\ I & S \end{bmatrix} > 0 \tag{4.19}$$

where $\mathcal{N}_S$ *and* $\mathcal{N}_C$ *are any matrices whose columns form bases of the null spaces of* $[B^T \quad B^T \quad E_b^T \quad E_b^T]$ *and* C, *respectively,* Q_S *and* Q_R *are as in (3.6) and (3.7), respectively,* $\bar{\sigma} = 1/\bar{\tau}$ *and*

$$\hat{F}_1^T = \begin{bmatrix} A^T & E_a^T & E_a^T \end{bmatrix}$$

$$\hat{F}_2 = A + A_d$$

$$\hat{F}_3^T = \begin{bmatrix} SA_d^T & SE_d^T & \alpha_1 D & \alpha_1 D_d & SE_d^T & A_d & D_d \end{bmatrix}$$

$$\hat{F}_4^T = \begin{bmatrix} A_d^T & E_d^T & 0 & 0 & E_d^T & 0 & 0 \end{bmatrix}$$

$$\hat{F}_5^T = \begin{bmatrix} A_d^T & E_d^T & \alpha_1 RD & \alpha_1 RD_d & E_a^T & E_d^T & RA_d & RD_d \end{bmatrix}$$

$$\hat{T}_1 = \operatorname{diag}\{\bar{\sigma}J_1, \ \alpha_3\bar{\sigma}I, \ \alpha_1 I\}$$

$$\hat{T}_2 = \operatorname{diag}\{\bar{\sigma}J_2, \ \alpha_4\bar{\sigma}I, \ \alpha_1 I, \ \alpha_1 I, \ \alpha_1 I, \ \bar{\sigma}J_3, \ \alpha_5\bar{\sigma}I\}$$

$$\hat{T}_3 = \operatorname{diag}\{\bar{\sigma}J_2, \ \alpha_4\bar{\sigma}I, \ \alpha_1 I, \ \alpha_1 I, \ \alpha_1 I, \ \alpha_1 I, \ \bar{\sigma}J_3, \ \alpha_5\bar{\sigma}I\}$$

$$J_1 = \alpha_2 I - \alpha_3 DD^T$$

$$J_2 = (1-\alpha_2)I - \alpha_4 D_d D_d^T$$

$$J_3 = I - \alpha_5 E_d^T E_d.$$

Proof. The proof is along the same lines as that of Theorem 3, except that Lemma 5 is used *in lieu* of Lemma 1. Similarly to the proof of Theorem 3, the closed-loop system of (4.11)-(4.13) with the controller (2.4)-(2.5) can be described by

$$\dot{x}_c(t) = [\hat{A} + \bar{D}F(t)\hat{E}_a]x_c(t) + [\bar{A}_d + \bar{D}_d F_d(t)\bar{E}_d]x_c(t-\tau) \tag{4.20}$$

where

$$x_c = \begin{bmatrix} x \\ \xi \end{bmatrix}, \quad \bar{D} = \begin{bmatrix} D \\ 0 \end{bmatrix}, \quad \bar{D}_d = \begin{bmatrix} D_d \\ 0 \end{bmatrix},$$

$$\hat{E}_a = \bar{E}_a + \bar{E}_b\Theta\bar{C}, \quad \bar{E}_a = [\,E_a \quad 0\,], \quad \bar{E}_b = [\,0 \quad E_b\,], \quad \bar{E}_d = [\,E_d \quad 0\,]$$

and $\hat{A}$, $\bar{A}_d$, $\bar{C}$ and Θ are as in (3.10)-(3.13).

By Lemma 5, the closed-loop system (4.20) is robustly stable for any constant time-delay τ satisfying $0 \le \tau \le \bar{\tau}$ if there exist a matrix $X > 0$ and scalars $\alpha_i > 0$, $i = 1, \ldots, 5$, satisfying the following inequality

$$\bar{G}_X + \bar{Y}^T\Theta\bar{Z}_X + \bar{Z}_X^T\Theta^T\bar{Y} < 0 \tag{4.21}$$

where

$$\bar{Y} = \begin{bmatrix} \bar{B}^T & \bar{B}^T & \bar{E}_b^T & 0 & 0 & 0 & 0 & \bar{E}_b^T & 0 & 0 & 0 \end{bmatrix}$$

$$\bar{Z}_X = \begin{bmatrix} \bar{C}X & 0 & 0 & 0 & 0 & 0 & 0 & 0 & 0 & 0 & 0 \end{bmatrix}$$

$$\bar{G}_X = \begin{bmatrix} \bar{Q}(X) & X\bar{M}^T & \bar{L}^T(\alpha_1, X) & \bar{N}^T \\ \bar{M}X & -\bar{\sigma}\bar{U}_1 & 0 & 0 \\ \bar{L}(\alpha_1, X) & 0 & -\alpha_1 I & 0 \\ \bar{N} & 0 & 0 & -\bar{\sigma}\bar{U}_2 \end{bmatrix}$$

$$\bar{Q}(X) = X(\bar{A}+\bar{A}_d)^T + (\bar{A}+\bar{A}_d)X$$

$$\bar{L}^T(\alpha_1, X) = \begin{bmatrix} \alpha_1\bar{D} & \alpha_1\bar{D}_d & X\bar{E}_a^T & X\bar{E}_d^T \end{bmatrix}$$

$$\bar{M}^T = \begin{bmatrix} \bar{A}^T & \bar{E}_a^T & \bar{A}_d^T & \bar{E}_d^T \end{bmatrix}$$

$$\bar{N}^T = \begin{bmatrix} \bar{A}_d & \bar{D}_d \end{bmatrix}$$

$$\bar{U}_1 = \operatorname{diag}\left\{\alpha_2 I - \alpha_3\bar{D}\bar{D}^T,\ \ \alpha_3 I,\ \ (1-\alpha_2)I - \alpha_4\bar{D}_d\bar{D}_d^T,\ \ \alpha_4 I\right\}$$

$$\bar{U}_2 = \operatorname{diag}\left\{I - \alpha_5\bar{E}_d^T\bar{E}_d,\ \ \alpha_5 I\right\}.$$

The result can then be obtained after lengthly manipulations using similar arguments as in the proof of Theorem 3 and defining $P = \alpha_2(NN^T)^{-1}$. ∇∇∇

Remark 5. Theorem 6 provides a delay-dependent condition for robust output feedback stabilization of linear uncertain time-delay systems. Note that the inequalities (4.17)-(4.19) can be solved numerically very efficiently by using interior-point algorithms for generalized eigenvalue problems; see, e.g. [1].

The computation of a robust output feedback controller can be carried out using a procedure similar to that of Section 3 for the controller design. More specifically, when the LMIs (4.17)-(4.19) are satisfied for some matrices R, S and P, and scalars α_i, $i = 1, \ldots, 5$, an output feedback controller that solves the robust stabilization problem can be obtained as follows:

1. Compute an $n \times n$ non-singular matrix N such that

$$NN^T = \alpha_2 P^{-1};$$

2. Find a matrix $X > 0$, which satisfies the inequality (4.21), by using (3.22) and (3.23), i.e.

$$X = \begin{bmatrix} S & N \\ N^T & N^T(S-R^{-1})^{-1}N \end{bmatrix};$$

3. Compute the controller matrices, A_c, B_c, C_c and D_c, by solving the LMI of (4.21) for $\Theta = \begin{bmatrix} A_c & B_c \\ C_c & D_c \end{bmatrix}$.

5 An Example

Consider the linear time-delay system

$$\dot{x}(t) = \begin{bmatrix} 0 & 0 \\ 0 & 1 \end{bmatrix} x(t) + \begin{bmatrix} -2 & -0.5 \\ 0 & -1 \end{bmatrix} x(t-\tau) + \begin{bmatrix} 0 \\ 1 \end{bmatrix} u(t) \tag{5.1}$$

$$y(t) = \begin{bmatrix} 1 & 1 \end{bmatrix} x(t).$$

We observe that the above system with $\tau = 0$ is not asymptotically stable.

It should be noted that the delay-independent output feedback stabilization methods of [3], [7] and [19] cannot be applied to system (5.1) as the pair (A, B) is not stabilizable. On the other hand, applying Theorem 3 to system (5.1), it was found using the software package MATLAB–LMI Lab that this system is stabilizable via linear dynamic output feedback for any constant time-delay τ satisfying $0 \leq \tau \leq 0.2650$. Moreover, a stabilizing output feedback controller is given by

$$\begin{aligned} \dot{\xi}(t) &= A_c\xi(t) + B_c y(t) \\ u(t) &= C_c\xi(t) + D_c y(t) \end{aligned}$$

where

$$\left[\begin{array}{c|c} A_c & B_c \\ \hline C_c & D_c \end{array}\right] = \left[\begin{array}{cc|c} -40.6723 & -110.2708 & 10.1577 \\ -20.8760 & -60.7428 & 5.4553 \\ \hline 4.3109 & -12.7778 & -0.6107 \end{array}\right].$$

We emphasize that smaller controller gain could be obtained by reducing the maximum allowed time-delay.

6 Conclusions

This chapter focused on the design of output feedback controllers for continuous time linear systems with a delayed state. Both the cases of systems without, or with parameter uncertainty have been treated. Polytopic and norm-bounded uncertainties have been considered. Delay-dependent sufficient conditions for stabilization and robust stabilization via linear dynamic output feedback have been obtained and the design of such controllers have been discussed. The proposed stabilization approach as well as the robust stabilization method for uncertain polytopic systems are based on linear matrix inequalities, whereas the robust stabilization method for norm-bounded uncertainties involves the solution of a generalized eigenvalue problem.

References

1. S. Boyd, L. El Ghaoui, E. Feron and V. Balakrishnan, *Linear Matrix Inequalities in Systems and Control Theory*, Studies in Applied Mathematics, Vol. 15, SIAM, Philadelphia, 1994.
2. S.D. Brierly, J.N. Chiasson, E.B. Lee and S.H. Zak, "On stability independent of delay for linear systems," *IEEE Trans. Automat. Control*, **AC-27**, 252–254, 1982.
3. H.H. Choi and M.J. Chung, "Observer-based $\mathcal{H}_\infty$ controller design for state delayed linear systems," *Automatica*, **32**, 1073–1075, 1996.
4. C.E. de Souza and X. Li, "Delay-dependent stability of linear time-delay systems: an LMI approach," *Proc. 3rd IEEE Mediterranean Symposium on New Directions in Control and Automation*, Limassol, Cyprus, July 1995.

5. P. Gahinet and P. Apkarian, "A linear matrix inequality approach to $\mathcal{H}_\infty$ control," *Int. J. of Robust and Nonlinear Control*, **4**, 421–448, 1994.

6. J. Hale, *Theory of Functional Differential Equations*, Springer-Verlag, New York, 1977.

7. E.T. Jeung, D.C. Oh, J.H. Kim and H.B. Park, "Robust controller design for uncertain systems with time delays: LMI approach," *Automatica*, **32**, 1229–1231, 1996.

8. J.H. Lee, S.W. Kim and W.H. Kwon, "Memoryless $\mathcal{H}_\infty$ controllers for state delayed systems," *IEEE Trans. Automat. Control*, **AC-39**, 159–162, 1994.

9. X. Li and C.E. de Souza, "LMI approach to delay-dependent robust stability and stabilization of uncertain linear delay systems," *Proc. 34th IEEE Conf. on Decision and Control*, New Orleans, LA, Dec. 1995.

10. X. Li and C.E. de Souza, "Robust stabilization and H_∞ control for uncertain linear time-delay systems," *Proc. 13th IFAC World Congress*, San Francisco, CA, June 1996.

11. M.S. Mahmoud and N.F. Al-Muthairi, "Quadratic stabilization of continuous-time systems with state-delay and norm-bounded time-varying uncertainties," *IEEE Trans. Automat. Control*, **AC-39**, 2135–2139, 1994.

12. M. Malek-Zavarei and M. Jamshidi, *Time Delay Systems: Analysis, Optimization and Applications*, North-Holland, 1987.

13. T. Mori, and H. Kokame, "Stability of $\dot{x}(t) = Ax(t) + Bx(t-\tau)$," *IEEE Trans. Automat. Control*, **AC-34**, 460–462, 1989.

14. Yu. Nesterov and A. Nemirovsky, *Interior Point Polynomial Methods in Convex Programming*, Studies in Applied Mathematics, Vol. 13, SIAM, Philadelphia, 1994.

15. S.I. Niculescu, C.E. de Souza, J.M. Dion and L. Dugard, "Robust stability and stabilization of uncertain linear systems with state delay: Single delay case," *Proc. IFAC Symp. Robust Control Design*, Rio de Janeiro, Brazil, Sept. 1994.

16. J.C. Shen, B.-S. Chen and F.-C. Kung, "Memoryless stabilization of uncertain dynamic delay systems: Riccati equation approach," *IEEE Trans. Automat. Control*, **AC-36**, 638–640, 1991.

17. J.-H. Su, "Further results on the robust stability of linear systems with a single time delay," *Systems & Control Letts.*, **23**, 375–379, 1994.

18. T.J. Su and C.G. Huang, "Robust stability of delay dependence for linear uncertain systems," *IEEE Trans. Automat. Control*, **AC-37**, 1656–1659, 1992.

19. L. Xie and C.E. de Souza, "Output feedback control of uncertain time-delay systems," *Proc. 1993 European Control Conf.*, Gröningen, The Netherlands, July 1993.

Robust Control of Systems with A Single Input Lag

Gilead Tadmor

ECE Department, Northeastern University
Boston, MA 02115, USA
e-mail: tadmor@cdsp.neu.edu

Abstract. A state space design methodology is developed for various H_∞ problems and gap optimization in systems with a single input lag. The main contribution is in converting associated operator Riccati equation and abstract model compensator realizations to algebraic and differential matrix Riccati equations of a fixed order and finite dimensional, integro-differential realizations.

1 Introduction.

This chapter presents a state space solution method for certain H_∞ and gap robustness optimization problems, in systems with a single, pure input lag at the control port. Systems with a single input lag form what is probably the simplest and yet one of most frequently encountered class of distributed parameter models. Examples of the use of such models include those systems where the presence of delay is justified by a physical phenomenon (such as in process control systems), systems where an input delay is used as a simplified representation of more complex phenomena (such as point to point wave propagation), or systems where a delay provides a conceptually simple, approximation–over–a–band of a phase–lag due to high order components. The vast use of this class of systems motivates the search for effective, tailored–to–measure design methods that make full advantage of its relative simplicity.

H_∞ optimization in the general context of distributed parameter systems, as well as in the framework of delay systems, has been investigated by several authors. A few examples are results based on state space analysis [34, 25], the skew Toeplitz approach [17] and direct reduction to a commutant lifting / operator interpolation type results [8, 7, 13, 21, 22, 35]. The problem of robustness optimization in the gap metric has been long established to be equivalent to an H_∞ problem [9] and its variant in a system with a single input lag has already been treated in [6, 16, 14].

State space solutions of H_∞ and gap optimization problems in ordinary systems are well established since the late 1980's [1, 5, 9], and have been extended early on to distributed parameter systems (see e.g. in [26, 34] and references therein). Typical to the distributed parameter case is the difficulty to solve associated infinite dimensional operator Riccati equations. [15] addressed this challenge – in the context of systems with a single input or output lag – by viewing the problem as constrained by an essentially periodic system (with the delay as its period); using a "lifting" technique the problem is transferred to an equivalent setting in terms of a distributed input and output, LTI discrete time system over a finite dimensional state space. This technique, and the resulting generic solution, are akin to what has been previously done in sampled data H_∞ and H_2 control (compare, e.g. with [24]). The solution is based on algebraic matrix Riccati equations that arise in an allied LTI problem and a differential matrix Riccati equation that stems from I/O norm evaluation of a certain (continuous time) system over the delay interval. A marked disadvantage of this solution is that the periodic structure is reflected in the generic compensator, which turns out to be periodic, time varying, even when the original system is LTI.

In this chapter we suggest a way to overcome this difficulty by a combination of a direct appeal to continuous time, abstract evolution models over the state space $M_2 = \mathbb{R}^n \times L_2[-1,0]$, and a two steps, finite dimensional analysis of an associated differential game. Our method is based, jointly, on the observation that the solution of the operator Riccati equation (that arises from the abstract model formulation) is the kernel of the M_2 quadratic form for the game's optimal value, in terms of initial data, and on our capacity to reduce the game and solve it in a finite dimensional setting. The resulting compensators are all of the usual "two loop" form, where the "central compensator" is based on an integro-differential equation of a neutral type.

This chapter presents the statements of the main results in several generic H_∞ and gap optimization problems, and reviews the proof for the case of the one block problem; the main ideas, jointly used in all variant problems, will be presented in that proof. The current presentation builds on our work in [28, 29, 30, 31, 27], where complete arguments can be found, for all the results mentioned. We shall also limit our reference list to a minimum, and exclude important references to significant, both recent and earlier work on solutions to optimal and robust control problems in distributed parameter systems. More references and leads are provided in the author's cited papers.

2 A Basic Abstract Model

This section presents some basic features of M_2 models for systems with a pure control delay, as captured in a representative example. The principles of such models and their general forms are well known and the purpose of this section is mainly to serve as a brief review. For a background on semigroups and abstract model representations of distributed parameter systems, in general, one may consult [3, 4, 18]. Examples of Hilbert state space representations and LQ optimization in general linear delay systems are [10, 11, 19]. Directly relevant details and more related references are provided in [30, 31, 34].

As an example of the type of systems that this work addresses, consider now the standard system form

$$
\begin{aligned}
\dot{x}(t) &= Ax(t) + B_1 w(t) + B_2 u(t-1) \\
z(t) &= C_1 x(t) + D_{11} w(t) + D_{12} u(t-1) \\
y(t) &= C_2 x(t) + D_{21} w(t)
\end{aligned} \tag{2.1}
$$

with $x \in \mathbb{R}^n$, the exogenous input $w \in \mathbb{R}^{m_1}$, control $u \in \mathbb{R}^{m_2}$, controlled output $z \in \mathbb{R}^{l_1}$ and observation $y \in \mathbb{R}^{l_2}$. By standard nomenclature, the relevant control history is denoted $u_t(\cdot)$ (where $u_t(\theta) = u(t+\theta)$, $\theta \in [-1,0]$) and is embedded in $L_2[-1,0]$. A complete state of the system must account for both the Euclidean $x(t) \in \mathbb{R}^n$ and for $u_t \in L_2[-1,0]$.

The following abstract model will be shown to provide a realization of the I/O mapping in (2.1).

$$
\begin{aligned}
\dot{f} &= \mathcal{A} + \mathcal{B}_1 w + \mathcal{B}_2 u \\
z &= \mathcal{C}_1 f + \mathcal{D}_{11} w \\
y &= \mathcal{C}_2 f + \mathcal{D}_{21} w
\end{aligned} \tag{2.2}
$$

with the state $f = (f^0, f^1) \in M_2$ and with the following coefficients –

$$
\begin{aligned}
\mathcal{B}_1 w &= (B_1 w, 0), & \mathcal{B}_2 u &= (0, \delta_0(\cdot)u) \\
\mathcal{C}_1 f &= C_1 f^0 + D_{12} f^1(-1), & \mathcal{C}_2 f &= C_2 f^0
\end{aligned} \tag{2.3}
$$

$$
\mathcal{A} f = (Af^0 + B_2 f^1(-1),\ \tfrac{d}{ds} f^1)
$$

Here δ_0 is Dirac's function, centered at zero) and $\mathcal{A}$ is defined over the dense domain

$$
\mathcal{D}(\mathcal{A}) = \left\{ f \in M_2 \ : \ f^1(s) = \int_s^0 \phi(r)dr,\ \phi \in L_2 \right\} \tag{2.4}
$$

The following analysis is a summary of some basic facts, relating (2.1) and (2.2). The main results of this chapter use this and several other associations of abstract models with delay systems or systems with neutral FDE realizations.

Lemma 1. *Let $\mathcal{S}(t)$ be the family of linear operators over M_2, as defined by the homogeneous dynamics in (2.1) and the relation $(x(t), u_t) = \mathcal{S}(t)(x(0), u_0)$. Then $\mathcal{S}(t)$ is a c_0 – semigroup over M_2 with $\mathcal{A}$ as its infinitesimal generator.*

Results similar to Lemma 1 can be found in the literature cited above concerning the use of M_2 models in the treatment of LQ optimization in delay systems, and we shall be content with a review of some main points in the proof. For more details and justifications of other such associations, the reader is referred to [30, 31, 27, 32] and references therein.

Outline of the proof. The fact that $\mathcal{S}(t)$ is a c_0 – semigroup (that is, that $\mathcal{S}(t)\mathcal{S}(r) = \mathcal{S}(t+r)$ and that $\mathcal{S}(t)$ is strongly continuous in t [4]) stems from the basic properties of solution operators in ODEs.

To motivate the stated form of the generator of $\mathcal{S}$, one may evaluate the time derivative of $(x(t), u_t(s))$ where it exists. Complete proofs can be obtained, e.g. by adaptations of either the proofs of [4] Theorem 2.4.6, [2] Theorem 2.3, or of [32] Theorems A and B. Guidelines for the adaptation of proofs from [2] and [32] will now follow.

It is first noted that, both here and in later instances, one can write the underlying homogeneous, integro–differential delay equation in the standard form of a neutral FDE –

$$\frac{d}{dt}\mathcal{E}z_t = \mathcal{F}z_t, \qquad t > 0 \tag{2.5}$$

where $z = (x, u)$ and in terms of bounded linear operators $\mathcal{E}$ and $\mathcal{F}$ $: W_2^1([-1,0], \mathbb{R}^{n+m_2}) \mapsto \mathbb{R}^{n+m_2}$. Precisely, in the case of $\mathcal{S}(t)$ we have $\mathcal{E}z_t = z_t(0)$ and $\mathcal{F}(x_t, u_t) = (Ax_t(0) + B_2u_t(-1)\ ,\ 0)$.

In the framework of the cited papers this would have called for the use of the higher dimensional "M_2" state space $\mathbb{R}^{n+m_2} \times L_2([-1,0], \mathbb{R}^{n+m_2})$, with the complete state $(\mathcal{E}z_t, z_t)$. The simplification in the current lower dimensional setting is due to the following specific features: (a) The dependence of both $\mathcal{E}z_t$ and $\mathcal{F}z_t$ on the component x_t of z_t is restricted to $x(t) = x_t(0)$. Particularly, the vector formed by the first n entries of $\mathcal{E}(x_t, u_t)$ is $x(t)$. This allows to replace the component x_t in the complete state by $x(t)$, without losing necessary information. (b) The last m_2 entries of $\mathcal{F}z_t$ vanish, making the subspace of $(\mathcal{E}z_t, z_t)$ where the last m entries of $\mathcal{E}(x_t, u_t)$ are zero, an invariant subspace under (2.5). Focusing on that subspace, the last m entries of $\mathcal{E}(x_t, u_t)$ can be removed from the state, ending with the current choice of $(x(t), u_t)$. □

The analysis leading to the results in this chapter utilizes several other M_2 semigroups. When relating to such semigroups we shall be content with providing, without proof, the forms of their generators and their respective domains. In each of these cases one will be able to draw on arguments from [2, 32] to verify the association of the semigroup and the generator in a manner similar to the proof outline, above.

For later reference we write down the details form of the relationship $f(t) = \mathcal{S}(t)f(0)$, as defined by an explicit solution of (2.1) –

$$\begin{aligned} f^0(t) &= e^{At}f^0(0) + \int_0^{\min(t,1)} e^{A(t-s)}B_2 f^1(0)(s-1)ds \\ f^1(t,\theta) &= \begin{cases} f^1(0, t+\theta) & -1 \le \theta \le -t, \quad 0 \le t < 1 \\ 0 & \text{else} \end{cases} \end{aligned} \tag{2.6}$$

It is a standard observation that a restriction of $\mathcal{S}(t)$ to the dense subspace $\mathcal{D}(\mathcal{A}) \subset M_2$ defines a c_0 – semigroup over $\mathcal{D}(\mathcal{A})$, relative to the stronger graph$(\mathcal{A})$

topology. Also, the definition of $\mathcal{S}(t)$ extends, by dense injection, to a c_0 – semigroup over the larger space $\mathcal{D}(\mathcal{A}')' \supset M_2$. Such extensions and restrictions are used extensively (cf. the more general discussions in [20, 19]). The definition of the restriction to $\mathcal{D}(\mathcal{A})$ is obvious. The following details concern the adjoint semigroup and the of $\mathcal{S}(t)$ to $\mathcal{D}(\mathcal{A}')'$. The adjoint Hilbert space M_2' will be identified with M_2, throughout.

Lemma 2. *The adjoint semigroup $g(t) = \mathcal{S}(t)'g(0)$ is define via —*

$$g^0(t) = e^{A't}g^0(0),$$

$$g^1(t,s) = \begin{cases} g^1(0, s-t), & s \in (t-1, 0],\ t \in [0,1] \\ B_2' e^{A'(t-s-1)} g^0(0), & \text{else} \end{cases} \tag{2.7}$$

The infinitesimal generator of $\mathcal{S}(t)'$ is —

$$\mathcal{A}'g = (A'g^0, -\frac{d}{ds}g^1) \tag{2.8}$$

over the domain

$$\mathcal{D}(\mathcal{A}') = \{g \in M_2 \ :\ g^1 \in W_2^1[-1,0] \text{ and } g^1(-1) = B_2'g^0\} \tag{2.9}$$

Proof: The expressions (2.7) readily follow from the expression (2.6), for $\mathcal{S}(t)$, and the definition of the adjoint operator via $\langle \mathcal{S}(t)f, g\rangle_{M_2} = \langle f, \mathcal{S}(t)'g\rangle_{M_2}$.

The domain $\mathcal{D}(\mathcal{A}')$ is characterized by the fact that $g \in \mathcal{D}(\mathcal{A}')$ and $h = \mathcal{A}'g$ means that $\forall f \in \mathcal{D}(\mathcal{A})$, $\langle \mathcal{A}f, g\rangle_{M_2} = \langle f, h\rangle_{M_2}$. Indeed, for $f \in \mathcal{D}(\mathcal{A})$ we can write $f^1(s) = -\int_s^0 \frac{d}{dr} f^1(r)dr$. On the one hand, for such selections and any $g \in M_2$, one has –

$$\begin{aligned} \langle \mathcal{A}f, g\rangle_{M_2} &= \langle (Af^0 + B_2 f^1(-1)\ ,\ \tfrac{d}{ds} f^1(s))\ ,\ g\rangle_{M_2} \\ &= \langle (f^0, \tfrac{d}{ds} f^1(s))\ ,\ (A'g^0, -B_2 g^0 + g^1(s))\rangle_{M_2} \end{aligned} \tag{2.10}$$

On the other hand, if also $g \in \mathcal{D}(\mathcal{A}')$ and $h = \mathcal{A}'g$, the expression (2.10) must be equal to –

$$\begin{aligned} \langle \mathcal{A}f, g\rangle_{M_2} &= \langle f, \mathcal{A}'g\rangle_{M_2} \\ &= \langle (f^0, -\textstyle\int_s^0 \frac{d}{dr} f^1(r)dr)\ ,\ h\rangle_{M_2} \\ &= \langle (f^0, \tfrac{d}{ds} f^1(s))\ ,\ (h^0, -\textstyle\int_0^s h^1(r)dr)\rangle_{M_2} \end{aligned} \tag{2.11}$$

Comparing the right hand sides of (2.10) and (2.11), both the stated forms of $\mathcal{A}'$ and of $\mathcal{D}(\mathcal{A}')$ follow. □

We have just seen that an element $g \in \mathcal{D}(\mathcal{A}')$ can be identified with the pair $(g^0, \frac{d}{ds}g^1) \in M_2$. A norm on $\mathcal{D}(\mathcal{A}')$ that is consistent with the graph topology of $\mathcal{A}'$, is $\|g\|_{\mathcal{D}(\mathcal{A}')} = \|(g^0, \frac{d}{ds}g^1)\|_{M_2}$. The adjoint space $\mathcal{D}(\mathcal{A}')'$ can therefore be identify with pairs $(h^0, h^1) \in M_2$, via —

$$\langle h, g\rangle_{\mathcal{D}(\mathcal{A}')',\mathcal{D}(\mathcal{A}')} = \langle (h^0, h^1)\ ,\ (g^0, \frac{d}{ds}g^1)\rangle_{M_2} \tag{2.12}$$

Associated with that representation is the norm $\|h\|_{\mathcal{D}(\mathcal{A}')'} = \|(h^0, h^1)\|_{M_2}$.

We shall maintain these representations and introduce the continuous injection $\imath : M_2 \hookrightarrow \mathcal{D}(\mathcal{A}')'$ and its (unbounded, left) inverse $\pi : \mathcal{D}(\mathcal{A}')' \mapsto M_2$, as follows –

$$\begin{aligned} \imath(f) &= (f^0 + B_2 \textstyle\int_{-1}^{0} f^1(r)dr\ ,\ \int_s^0 f^1(r)dr) \\ \pi(h) &= (h^0 - B_2 h^1(-1)\ ,\ -\tfrac{d}{ds}h^1) \end{aligned} \tag{2.13}$$

Integration by parts provides the following equality, holding for all $f \in M_2$ and $g \in \mathcal{D}(\mathcal{A}')$

$$\langle f, g\rangle_{M_2} = \langle \imath(f)\ ,\ (g^0, \frac{d}{ds}g^1)\rangle_{M_2}, \tag{2.14}$$

which justifies (2.13). The (unbounded) adjoint mapping π' defines the embedding of $\mathcal{D}(\mathcal{A}')$ with the M_2 structure as explained earlier; its bounded left inverse is $\imath'$. Explicitly –

$$\begin{aligned} \pi'(\phi) &= (\phi^0\ ,\ \tfrac{d}{ds}\phi^1) \\ \imath'(\psi) &= (\psi^0\ ,\ B_2'\psi^0 + \textstyle\int_{-1}^{s} \psi^1(r)dr) \end{aligned} \tag{2.15}$$

The definition of the semigroup $\mathcal{S}(t)$ over the entire $\mathcal{D}(\mathcal{A}')'$ is made by continuous extension of $\imath \circ \mathcal{S}(t) \circ \pi$ from the dense submanifold $\imath(M_2)$. Its infinitesimal generator will be denoted $\mathcal{A}^e$. The following lemma provides their precise forms.

Lemma 3. *Let $\mathcal{D}(\mathcal{A}')'$ be embedded with the M_2 structure, as explained above. For $h(0) \in \mathcal{D}(\mathcal{A}')'$ let $h(t) = \mathcal{S}(t)h(0)$ be the trajectory of the extended semigroup. Then –*

$$\begin{aligned} h^0(t) &= e^{At}h^0(0) - \int_0^{\min(t,1)} e^{A(t-r)}AB_2h^1(0, r-1)dr, \\ h^1(t,s) &= \begin{cases} h^1(0, t+s), & s \in [-1, -t),\ \ t \in [0,1], \\ 0 & \text{else} \end{cases}, \end{aligned} \tag{2.16}$$

The generator of the extended semigroup, $\mathcal{A}^e$, is defined over the domain –

$$\mathcal{D}(\mathcal{A}^e) = \{h\ :\ h^1 \in W_2^1[-1,0],\ h^1(0) = 0\} = \imath(M_2) \tag{2.17}$$

via –

$$\mathcal{A}^e h = (A(h^0 - B_2h^1(-1))\ ,\ \frac{d}{ds}h^1) \tag{2.18}$$

In particular, $\mathcal{A}^e$ is continuous over $\imath(M_2)$ relative to the M_2 topology. (Also noted is the equality $\mathcal{A}^e h = \imath \circ \mathcal{A} \circ \pi h$ for $h \in \imath(\mathcal{D}(\mathcal{A}))$.)

Proof: We compute $h(t) = \mathcal{S}(t)h(0)$ for $h(0) = \imath(f(0))$ in the dense submanifold $\imath(M_2) \subset \mathcal{D}(\mathcal{A}')'$. By definition of the extended semigroup, this means that $h(t) = \imath(f(t))$ for $f(t) = \mathcal{S}(t)f(0)$. Using the explicit expressions for the injection $\imath$ in (2.13), this first implies –

$$\begin{aligned} h^1(t,s) &= \int_s^0 f^1(t,r)dr = \int_s^{\max(s,-t)} f^1(0,t+r)dr \\ &= \int_{\min(t+s,0)}^0 f^1(0,r)dr = \begin{cases} h^1(0,t+s), & s \in [-1,-t),\ t \in [0,1], \\ 0 & \text{else} \end{cases} \end{aligned} \tag{2.19}$$

The expression (2.6) provides the explicit value of $f^1(t)(s)$ in $f(t) = \mathcal{S}(t)f(0)$. We use that expression, the definition (2.13), the definition of the extended semigroup and integration by parts, to obtain –

$$\begin{aligned} h^0(t) &= f^0(t) + B_2 h^1(t,-1) \\ &= e^{At} f^{(}0) + \int_0^{\min(t,1)} e^{A(t-r)} B_2 f^1(0,r-1)dr \\ &\qquad + B_2 \int_{\min(t-1,0)}^0 f^1(0,r)dr \\ &= e^{At} h^0 - \int_0^{\min(t,1)} e^{A(t-r)} A B_2 h^1(0,r-1)dr \end{aligned} \tag{2.20}$$

which completes the proof of (2.16).

The expression just derived for the extension of $\mathcal{S}(t)$ to $\mathcal{D}(\mathcal{A}')'$ are very similar to the expressions (2.6) for $\mathcal{S}(t)$ over its original domain. That analogy must therefore carry in the form of the infinitesimal generator $\mathcal{A}^e$ of the extended semigroup, and of its domain. □

The definitions of the input coefficients operators in (2.1) and their adjoints adapt to the state space extension from M_2 to $\mathcal{D}(\mathcal{A}')'$, as follows. The bounded operator $\mathcal{B}_1 : \mathbb{R}^{m_1} \mapsto M_2$ extends to a bounded operator : $\mathbb{R}^{m_1} \mapsto \mathcal{D}(\mathcal{A}')'$ via $\imath \circ \mathcal{B}_1 w = (B_1 w, 0)$. The adjoint operator is $\mathcal{B}_1' f = B_1' f^0$, $f \in M_2$, and its restriction (via $\imath'$) to $\mathcal{D}(\mathcal{A}')$ is $\mathcal{B}_1' g = B_1' g^0$, $g \in \mathcal{D}(\mathcal{A}')$.

The operator $\mathcal{B}_2$ takes values in $\mathcal{D}(\mathcal{A}')'$. We shall now derive an expression for $\mathcal{B}_2$, based on the representation of $\mathcal{D}(\mathcal{A}')$ by members of M_2, as explained earlier: fix $u \in \mathbb{R}^{m_2}$ and $g \in \mathcal{D}(\mathcal{A}')$; then –

$$\begin{aligned} &\langle \mathcal{B}_2 u, g\rangle_{\mathcal{D}(\mathcal{A}')',\mathcal{D}(\mathcal{A}')} = \langle u, g^1(0)\rangle_e = \langle u, g^1(-1) + \int_{-1}^0 \tfrac{d}{ds} g^1(s)\rangle_e \\ &= \langle u\ ,\ B_2' g^0 + \int_{-1}^0 \tfrac{d}{ds} g^1(s)\rangle_e = \langle (B_2 u, 1(s)u)\ ,\ (g^0, \tfrac{d}{ds} g^1(s))\rangle_{M_2} \end{aligned} \tag{2.21}$$

where $1(s)$ is the unit–valued constant function. The adjoint operator is $\mathcal{B}_2' g = g^1(0)$, $g \in \mathcal{D}(\mathcal{A}')$.

The state equation in (2.2) should be understood in the context of the extended state space. The validity of (2.2) as an abstract model realization of the inhomogeneous dynamics of (2.1) will now be established.

Lemma 4. *Fix initial data $f(0) = (x(0), u_0) \in M_2$ and inputs $u, w \in L_{2\,loc}[0,\infty)$. Let $x(t)$ be the state trajectory in (2.1) that corresponds to these initial data and inputs and denote $f(t) = (x(t), u_t)$. Finally, let $h(0) = \imath(f(0))$. Then the following mild evolution of (2.2) –*

$$h(t) = \mathcal{S}(t)h(0) + \int_0^t \mathcal{S}(t-r)\left(\mathcal{B}_1 w(r) + \mathcal{B}_2 u(r)\right)\, dr \tag{2.22}$$

is such that $h(t) = \imath(f(t))$ and $\mathcal{C}_i h(t) = \mathcal{C}_i f(t) = C_i x(t)$, $t \geq 0$. If, moreover, $u \in W^1_{2\,loc}[-1,\infty)$, then for any $g \in \mathcal{D}(\mathcal{A}')$ there holds –

$$\frac{d}{dt}\langle h(t), g\rangle_{\mathcal{D}(\mathcal{A}')',\mathcal{D}(\mathcal{A})} = \langle \mathcal{A}^e h(t) + \mathcal{B}_1 w(t) + \mathcal{B}_2 u(t)\ ,\ g\rangle_{\mathcal{D}(\mathcal{A}')',\mathcal{D}(\mathcal{A})} \tag{2.23}$$

for a.e. $t \geq 0$.

Proof: The variations of parameters formula represents effects of initial data, the exogenous input w and the control input u. Each will now be analyzed separately.

From the original definition of $\mathcal{S}(t)$ and the relation $\mathcal{S}(t) \circ \imath = \imath \circ \mathcal{S}(t)$ in the extended semigroup, the validity of the component of (2.22) that involves initial data, is verified.

By (2.16) –

$$\mathcal{S}(t-r)\mathcal{B}_1 w(r) = (e^{A(t-r)} B_1 w(r)\ ,\ 0) \in M_2$$

Thus the claim concerning effects of w is also true.

To analyze control effects set $h(0) = 0$ and $w = 0$ and denote $\chi(t,r) = \mathcal{S}(t-r)\mathcal{B}_2 u(r)$, so that

$$h(t) = \int_0^t \chi(t,r)dr$$

In what follows we use the equalities $\mathcal{B}_2 u = (B_2 u, 1(s)u)$ (for the $\mathcal{D}(\mathcal{A}')'$ value of $\mathcal{B}_2 u$) and (2.16). The $L_2[-1,0]$ component of $\chi(t,r)$ is –

$$\chi(t,r)^1(s) = \begin{cases} u(r), & t-r+s,\ s \in [-1,0] \\ 0 & \text{else} \end{cases}$$

Thus –

$$h^1(t,s) = \int_{\max(t+s,0)}^{t} u(r)dr = \int_{\max(s,-t)}^{0} u_t(r)dr, \tag{2.24}$$

in agreement with the $L_2[-1,0]$ component of $\imath(f(t))$.

The $\mathbb{R}^n$ component of $\chi(t,r)$ is –

$$\begin{aligned} \chi(t,r)^0 &= \left(e^{A(t-r)} - \int_0^{\min(t-r,1)} e^{A(t-r-s)}Ads\right) B_2 u(r) \\ &= \begin{cases} B_2 u(r), & r \in [\max(0,t-1)\ ,\ t] \\ e^{A(t-r-1)} B_2 u(r), & t > 1,\ r \in [0, t-1) \end{cases} \end{aligned}$$

Therefore –

$$h^0(t) = \int_0^{\max(0,t-1)} e^{A(t-r-1)}B_2u(r)dr + B_2\int_{\max(0,t-1)}^t u(r)dr$$

$$= \int_{\min(1,t)}^t e^{A(t-r)}B_2u(r-1)dr \tag{2.25}$$

$$+B_2\int_{\max(-1,-t)}^0 u_t(r)dr = x(t) + B_2h^1(t,-1)$$

Here too we observe the asserted agreement with the definition of $\imath(f(t))$. We have thus established that $h(t) = \imath(x(t), u_t)$ in (2.22).

It remains to establish (2.23). Indeed, fix $g \in \mathcal{D}(\mathcal{A}')$, an initial value $x(0)$ and inputs $w \in L_{2\,loc}[0,\infty)$ and $u \in W^1_{2\,loc}[-1,\infty)$ in (2.1) and denote $f(t) = (x(t), u_t)$ and $h(t) = \imath(f(t))$. Then –

$$\tfrac{d}{dt}\langle h(t), g\rangle_{\mathcal{D}(\mathcal{A}')',\mathcal{D}(\mathcal{A}')} = \tfrac{d}{dt}\langle f(t), g\rangle_{M_2}$$

$$= \tfrac{d}{dt}\left(\langle x(t), g^0\rangle_e + \int_{-1}^0 \langle u(t+s), g^1(s)\rangle_e ds\right)$$

$$= \langle Ax(t) + B_1w(t) + B_2u(t-1), g^0\rangle_e + \int_{-1}^0 \langle \tfrac{d}{ds}u(t+s), g^1(s)\rangle_e ds$$

$$= \langle Ax(t) + B_1w(t), g^0\rangle_e + \langle u(t), g^1(0)\rangle_e - \int_{-1}^0 \langle u(t+s), \tfrac{d}{ds}g^1(s)\rangle_e ds$$

$$= \langle \mathcal{A}^e\imath(x(t), u_t) + \mathcal{B}_1w(t) + \mathcal{B}_2u(t)\ ,\ g\rangle_{\mathcal{D}(\mathcal{A}')',\mathcal{D}(\mathcal{A}')} \tag{2.26}$$

where the first equality is due to (2.14) and the equality $h(t) = \imath(f(t))$; the second equality merely writes the previous term explicitly; the third equality is obtained by invoking the state equation in (2.1) and the assumed differentiability of u; the fourth equality is the result of integration by parts, using the fact that $g^1 \in W_2^1[-1,0]$ and that $g^1(-1) = B_2'g^0$ for $g \in \mathcal{D}(\mathcal{A}')$; the fifth equality builds on the previously computed expression for $\mathcal{A}^e$ and the definitions of $\mathcal{B}_i$. This completes the proof. □

The output operator $\mathcal{C}_2$ is bounded over M_2, but its extension to $\mathcal{D}(\mathcal{A}')'$ is not. The output operator $\mathcal{C}_2$ is already unbounded over M_2, and only its restriction to $\mathbb{R}^n \times W_2^1[-1,0]$ (and certainly, to $\mathcal{D}(\mathcal{A})$) is bounded. Using the previous lemma, however, it is noticed that when the initial state and the inputs to (2.1) and (2.2) coincide then the outputs of the two systems coincide as well. Thus when the input and output trajectories are embedded with the $L_{2\,loc}[-1,\infty)$ topology, the mapping $(f(0), w, u) \mapsto (z, y)$ is continuous.

In closing it is notice that, while other variants may fall into the framework of the Pritchard-Salamon class ([19]), the model (2.2) does not.

3 A One Block Problem

Let $\mathbf{P} = [A, B, C, D]$ be a minimal realization of a rational $\mathbf{P} \in L_\infty(\jmath\mathbb{R})$. The following is standard. Let

$$\mathbb{R}^n = V_{st} \times V_{as}$$

be a direct sum partition of $\mathbb{R}^n$ into a stable and an anti-stable eigenspaces. Let X_{as} and Y_{as} be the positive definite controllability and observability Grammians of the respective restrictions of $[A, B]$ and $[A, C]$ to V_{as}, and define the $n \times n$, positive semidefinite matrices

$$P = \begin{bmatrix} 0_{st} & 0 \\ 0 & X_{as}^{-1} \end{bmatrix} \quad \text{and} \quad Q = \begin{bmatrix} 0_{st} & 0 \\ 0 & Y_{as}^{-1} \end{bmatrix}$$

A left co-prime factorization $\mathbf{P} = \mathbf{M}_l^{-1}\mathbf{N}_l$ is then provided by

$$\mathbf{N}_l = [A - QCC', B - QC'D, C, D] \quad \text{and} \quad \mathbf{M}_l = [A - QC'C, -QC', C, I]$$

Several optimization problems, such as the weighted sensitivity minimization, involving plants with a pure input lag, can be shown to reduce to a model matching form, defined in terms of an allied system "**P**" and its factorization, as follows

$$\inf\{\|\mathbf{N}_l(s) - \mathbf{M}_l(s)e^{-s}\Theta(s)\|_\infty \ : \ \Theta \in H_\infty\} \tag{3.1}$$

The optimal value of (3.1) is denoted γ_0 and for $\gamma > \gamma_0$, the suboptimal set is denoted $\Theta_\gamma = \{\Theta \in H_\infty \ : \ \|\Theta\|_\infty < \gamma\}$. To avoid issues of well posedness, we restrict our attention to transfer functions of system with atomic neutral FDE realizations [32]. (This does not affect γ_0!)

The following definitions are used in the statement of our first result. The first is $\rho_0 = max\,\{\rho(X_{as}Y_{as}), \rho(D'D)\}$ (where "$\rho(M)$" is the spectral radius).

The rest of these definitions are made for $\gamma^2 > \rho_0$, as follows

$$
\begin{aligned}
Z_{as} &= Y_{as}^{-1} - \tfrac{1}{\gamma^2} X_{as}, \\
R &= \begin{bmatrix} 0_{st} & 0 \\ 0 & Z_{as}^{-1} \end{bmatrix}, \\
E_{22} &= -(\gamma^2 I - D'D)^{-1} D', \\
E_{12} &= -\gamma(\gamma^2 I - DD')^{-\frac{1}{2}}, \\
E_{21} &= (\gamma^2 I - D'D)^{-\frac{1}{2}}, \\
H_1 &= \gamma(\gamma^2 I - DD')^{-\frac{1}{2}} C, \\
H_2 &= (\gamma^2 I - D'D)^{-1} D'C, \\
G_1 &= (B - QC'D)(\gamma^2 - D'D)^{-\frac{1}{2}} \\
G_2 &= (\gamma^2 QC' - BD')(\gamma^2 I - DD')^{-1}
\end{aligned}
$$

In these terms we have

Theorem 5. *$\gamma > \gamma_0 \;\Leftrightarrow\;$ (a) $\gamma^2 > \rho_0$ and (b) $\exists R_0(t) \geq 0 \in L_\infty[0,1]$ such that*

$$
\begin{aligned}
&\dot{R}_0 + R_0(A - G_2C) + (A - G_2C)'R_0 \\
&\qquad + R_0 G_1 G_1' R_0 + H_1' H_1 = 0, \quad R_0(1) = R
\end{aligned}
\tag{3.2}
$$

Set $\gamma > \gamma_0$. Then the I/O mappings for $\Theta \in \Theta_\gamma \;\Leftrightarrow\;$ are defined by realization of the following forms

$$
\begin{aligned}
\dot{x}_c(t) &= A_c x_c(t) + B_{c1} w(t) + B_{c2} u(t-1) \\
u(t) &= C_{c1} x_c(t) + D_{c12} \phi(t) + D_{c13} u_t \\
\psi(t) &= C_{c2} x_c(t) + D_{c21} w(t) + D_{c23} u_t, \qquad \phi = \Theta_0 \psi
\end{aligned}
\tag{3.3}
$$

where Θ_0 is selected subject to the restriction that it is defined as the I/O mapping in a stable, neutral FDE and satisfies the norm constraint $\|\Theta_0\|_\infty < 1$, and where the coefficients are as follows: $\Phi(t,s)$ is the transition generated by $A_0 =$

$A - G_2C + G_1G_1'R_0 \cdot$; *in these terms*

$$
\begin{aligned}
A_c &= A - QC'C, \\
B_{c1} &= B - QC'D, \\
B_{c2} &= QC', \\
C_{c1} &= \left(C(I - QR) + \tfrac{1}{\gamma^2}DB'R\right)\Phi_0(1,0), \\
D_{c12} &= (2E_{12}'E_{12})^{-\frac{1}{2}}, \\
C_{c2} &= -G_1'R_0(0) - \tfrac{1}{\gamma}D'H_1, \\
D_{c21} &= E_{21}^{-1}, \\
\Xi(r) &= \textstyle\int_0^r \Phi_0(r,s)G_1G_1'\Phi_0(r,s)'ds, \\
D_{c13}u_t &= (C(I - QR) + \tfrac{1}{\gamma^2}DB'R)\textstyle\int_{-1}^0 \Phi_0(1, s+1)\cdot \\
&\quad \cdot((I + \Xi(s+1)R_0(s+1)G_2 + \Xi(s+1)H_1'E_{12})u_t(s)ds, \\
D_{c22}u_t &= -\tfrac{1}{\gamma}D'E_{12}u(t-1) \\
&\quad -G_1'\textstyle\int_{-1}^0 \Phi_0(s+1,0)'(R_0(s+1)G_2 + H_1'E_{12})u_t(s)ds
\end{aligned}
$$

It is easy to see that the "central solution" – the one with $\Theta_0 = 0$ – is defined by an integro-differential equation that adheres to the general pattern of a neutral FDE [32].

4 Gap Optimization

This section concerns robustness optimization in the gap metric of the standard negative feedback loop of Fig. 1, in systems with a single input lag. (A recent solution from a different perspective is [6] and ideas similar to ours were explored in [14].) We shall thus consider a plant $\mathbf{P}(s) = \mathbf{P_0}(s)e^{-s}$ where $\mathbf{P_0}(s)$ is rational and, for simplicity, strictly proper, with a minimal realization $\mathbf{P_0} = [A, B, C, 0]$. It has been established [9] that this problem is equivalent to a search for stabilizing compensators $\mathbf{C}$ that minimize the H_∞ norm of

$$\mathbf{F_C} = \begin{bmatrix} \mathbf{I} \\ \mathbf{P} \end{bmatrix} (\mathbf{I} + \mathbf{CP})^{-1}[\mathbf{I}\ \mathbf{C}]$$

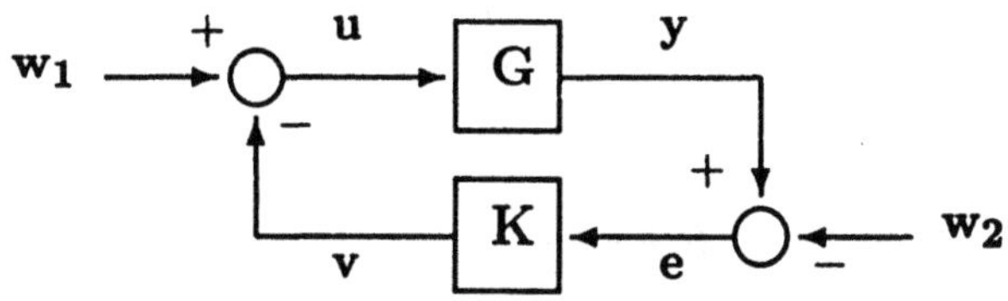

Fig. 1. Closed Loop Configuration

Following the standard convention, we denote

$$\gamma_0 = \inf\left\{\|\mathbf{F_C}\|_\infty \ : \ \mathbf{C} \text{ is a stabilizing compensator}\right\}$$

For $\gamma > \gamma_0$ denote by $\mathbf{C}_\gamma$ the set of strictly γ suboptimal, stabilizing compensators. The following is our result, pertaining to this problem.

Theorem 6. *Let $X, Y > 0$ be the stabilizing solutions of the LQG Riccati equations*

$$\begin{aligned} XA + A'X - XBB'X + C'C = 0 \\ AY + YA' - YC'CY + BB' = 0 \end{aligned} \tag{4.1}$$

Then $\gamma > \gamma_0 \ \Leftrightarrow \ \gamma > 1$ and $\exists Z(t) \geq 0, \in L_\infty[0,1]$, satisfying

$$\dot{Z} = ZA' + AZ + \frac{1}{\gamma^2 - 1} ZC'CZ + BB' \tag{4.2}$$

subject to $Z(0) = Y$ and $(\gamma^2 - 1)I > X^{\frac{1}{2}} Z(1) X^{\frac{1}{2}}$. Given $\gamma > \gamma_0$ and the associated solution of (4.2), let matrices and matrix functions U^i and V^i be defined below. Then the set $\mathbf{C}_\gamma$ comprises compensators that can be realized as follows

$$\begin{aligned} \dot{x}_c(t) &= Ax_c(t) - Bu(t-1) + YC'(e(t) - Cx_c(t)) \\ u(t) &= U^0 x_c(t) + \textstyle\int_{-1}^{0} U^1(s) u_t(s) ds + b(t) \\ c(t) &= V^0 x_c(t) + \textstyle\int_{-1}^{0} V^1(s) u_t(s) ds - e(t); \quad b = \mathbf{C}_0 c \end{aligned} \tag{4.3}$$

where the free design parameter is the stably realizable (neutral) system $\mathbf{C}_0$, selected freely, subject to the norm bound $\|\mathbf{C}_0\| < \sqrt{\gamma^2 - 1}$, and where the coefficients are defined as follows:

$$A_Z(t) = A + \frac{1}{\gamma^2 - 1} Z(t) C'C$$

generates the transition matrix $\Phi_Z(t,s)$:

$$R = \gamma^2((\gamma^2 - 1)I - XZ(1))^{-1} X,$$

The matrix function $\Xi(t)$ solves

$$\frac{d}{dt}\Xi + A_Z{}'\,\Xi + \Xi A_Z + \frac{\gamma^2}{\gamma^2-1}C'C = 0, \quad \Xi(1) = R$$

In these terms

$$\begin{aligned} U^0 &= B'R\Phi_Z(1,0), \\ U^1(s) &= B'R\Phi_Z(1,s+1)B, \\ V^0 &= \tfrac{1}{\gamma^2-1}C(Y\Xi(0)+\gamma^2 I), \\ V^1(s) &= \tfrac{1}{\gamma^2-1}CY\Phi_Z(s+1,0)'\,\Xi(s+1)B \end{aligned}$$

5 The Standard Problem

Here we consider the standard (four block) problem, as defined in terms of the system (2.1). The optimal value γ_0 is now the infimal H_∞ norm of the mapping $\mathbf{T}_{zw} : w \mapsto z$ with stabilizing compensation $u = \mathbf{C}y$. Given $\gamma > \gamma_0$, the set $\mathbf{C}_\gamma$ contains the stabilizing, strictly γ suboptimal compensators. Again, to avoid issues of well posedness, and without any effect on γ_0, we restrict our attention to compensators with atomic, neutral FDE realizations. This problem will be considered under the following, standard assumptions.

Assumption 5.1 *1. The pairs $[A, B_1]$ and $[A, B_2]$ are stabilizable .*
2. The pairs $[A, C_1]$ and $[A, C_2]$ are detectable.
3. $D_{12}'[C_1\ D_{12}] = [0\ \ I]$
4. $D_{21}[B_2'\ D_{21}'] = [0\ \ I]$

The pertinent result follows.

Theorem 7. $\gamma > \gamma_0 \;\Leftrightarrow\; \exists X, Y \geq 0$ *and* $Z(t) \geq 0 \in L_\infty[0,1]$, *satisfying the following.*

$$XA + A'X + X(\frac{1}{\gamma^2}B_1B_1' - B_2B_2')X + C_1'C_1 = 0 \tag{5.1}$$

and

$$A_1 = A + (\frac{1}{\gamma^2}B_1B_1' - B_2B_2')X$$

is stable;

$$\dot{Z} + ZA + A'Z + \frac{1}{\gamma^2}ZB_1B_1'Z + C_1'C_1 = 0, \quad Z(1) = X; \tag{5.2}$$

Set

$$A_Z(t) = A + \frac{1}{\gamma^2}B_1B_1'Z(t)$$

let $\Phi_Z(t,s)$ be the generated transition matrix and

$$G(t) = \frac{1}{\gamma^2}\int_0^t \Phi_Z(t,s)B_1B_1'\Phi_Z(s,t)'ds$$

In these terms

$$\begin{aligned}&(A+\tfrac{1}{\gamma^2}B_1B_1'Z(0))Y+Y(A+\tfrac{1}{\gamma^2}B_1B_1'Z(0))'\\&\quad +Y\left(\tfrac{1}{\gamma^2}\Phi_Z(1,0)'XB_2B_2'X\Phi_Z(1,0)-C_2'C_2\right)Y+B_1B_1'=0\end{aligned}\tag{5.3}$$

such that

$$A_2 = A+\frac{1}{\gamma^2}B_1B_1'X+Y(\frac{1}{\gamma^2}\Phi_Z(1,0)'XB_2B_2'X\Phi_Z(1,0)-C_1'C_1)$$

is stable. Assume that, indeed, $\gamma > \gamma_0$, let X, $Z(s)$ and Y be the said solutions of the Riccati equations (5.1), (5.2) and (5.3). Then the family of stabilizing, strictly γ - attenuating compensators $u = \mathbf{C}y$ is parameterized in terms of the following realization.

$$\begin{aligned}\dot{x}_c(t) &= A_{c00}x_c(t) + A_{c01}u_t + B_{c1}y(t) + B_{c2}v(t)\\ u(t) &= A_{c10}x_c(t) + A_{c11}u_t + v(t)\\ q(t)) &= -C_2x_c(t) + y(t), \quad v = \mathbf{C}_0 q\end{aligned}\tag{5.4}$$

where the free design parameter is the stably realizable (atomic neutral FDE) system $\mathbf{C}_0$, subject to the $L_2(0,\infty)$ - induced norm constraint $\|\mathbf{C}_0\| < \gamma$, and where the coefficients in (5.4) are defined as follows:

$$\begin{aligned}A_{c00} &= A+\tfrac{1}{\gamma^2}B_1B_1'Z(0)-YC_2'C_2,\\ A_{c01}u_t &= \tfrac{1}{\gamma^2}\textstyle\int -1^0 B_1B_1'\Phi_Z(s+1,0)'Z(s+1)B_2u_t(s)ds + B_2u(t-1),\\ B_{c1} &= YC_2',\\ B_{c2} &= \tfrac{1}{\gamma^2}Y\Phi_Z(1,0)'XB_2,\\ A_{c10} &= -B_2'X\Phi_Z(1,0),\\ A_{c11}u_t &= -\textstyle\int_{-1}^0 B_2'X\Phi_Z(1,s+1)(I+G(s+1)Z(s+1))B_2u_t(s)ds\end{aligned}$$

6 Proof of Theorem 5

Notwithstanding various differences in detail and even some changes in important features in the makings of the three H_∞ problems that are represented by the results stated above, the proofs of these results share the same fundamental structure. we shall thus be content with an outline of the proof for the first of the three stated theorems, Theorem 5.

The transfer function $N_l(s) - M_l(s)e^{-s}\Theta(s)$ has a stable realization of the form (2.1), with the coefficient substitutions $A - QC'C \mapsto A$, $B - QC'D \mapsto B_1$, $QC' \mapsto B_2$, $C \mapsto C_1$, $D \mapsto D_{11}$, $-I \mapsto D_{12}$, $0 \mapsto C_2$ and $I \mapsto D_{22}$, and with the mapping $u = \Theta y$ assuming the compensator's role.

A Differential Game. The analysis is an adaptation from that of the regular case in [33].

Lemma 8. *If $\gamma > \gamma_0$ in (3.1) then for any $(x(0), u_0) \in M_2$ there exists a unique solution w^*, u^* for the (open loop) game*

$$\inf_{w \in L_2[0,\infty)} \left\{ \gamma^2 \|w\|_2^2 - \inf_{u \in L_2[0,\infty)} \|z\|_2^2 \right\} \tag{6.1}$$

The game (6.1) is analyzed in two steps. First considered is its restriction to $[1, \infty)$. The problem's data is then $x(1)$ and the optimization is over the pertinent input selections $u\Big|_{[0,\infty)}$ and $w\Big|_{[1,\infty)}$. Substituting $\tilde{u}(t) = u(t-1)$, the delay is eliminated and results from the ordinary case apply.

Lemma 9. *The $[1, \infty)$ restriction of (6.1) is solvable if and only if $\gamma > \hat{\gamma} = \rho(X_{as}Y_{as})$. For $\gamma > \hat{\gamma}$ the matrix $R = \begin{bmatrix} 0_{st} & 0 \\ 0 & Z_{as}^{-1} \end{bmatrix}$ satisfies*

$$RA + A'R + R\left(\frac{1}{\gamma^2}B'B - QC'CQ\right)R = 0 \tag{6.2}$$

and the matrix

$$A_1 = A + \left(\frac{1}{\gamma^2}B'B - QC'CQ\right)R$$

is stable. The optimal value of the restricted game is $-\langle x(1), Rx(1)\rangle$ and the optimal trajectories w^*, $\tilde{u}^*$ and x^* satisfy the equations –

$$w^* = \frac{1}{\gamma^2}B'Rx^*$$

and

$$\tilde{u}^* = (C(I - QR) + \frac{1}{\gamma^2}DB'R)x^*$$

Now (6.1) reduces to the $L_2[0, 1]$ optimization problem

$$\inf_{w} \left\{ \gamma^2 \|w\|_{L_2[0,1]}^2 - \|z\|_{L_2[0,1]}^2 - \langle x(1), Rx(1)\rangle \right\} \tag{6.3}$$

considered for $\gamma > \hat{\gamma}$ with the data $(x(0), u_0)$. Extending results from [12, 23] one has

Lemma 10. *The problem (6.3) is solvable* $\Leftrightarrow \gamma > \rho_0$ *and* $\exists R_0$ *as stated in Theorem 5. Given the definitions in Section 3, then* $\exists! x, p \in L_2[0,1]$ *s.t.*

$$\begin{aligned} \dot{x} &= (A - G_2C)x + G_1G_1'p + G_2\tilde{u} \\ \dot{p} &= -H_1'H_1x - (A - G_2C)'p - H_1'E_{12}\tilde{u} \end{aligned} \tag{6.4}$$

where $\tilde{u}(t) = u_0(t-1)$ *is part of the initial data and* $p(1) = Rx(1)$. *The optimal state is then* $x^* = x$ *and the optimal input is*

$$w^* = H_2x^* + E_{21}G_1'p^* + E_{22}\tilde{u}$$

This completes the proof of necessity in Theorem 5.

In preparation for the proof of sufficiency and of the validity of the parameterization (3.3), assume the necessary conditions are met, whereby, in particular, (6.1) is solved, as explained above. We shall now compute complete state feedback expressions for the solution of (6.1) (i.e. we shall seek formulae that determine the optimal inputs at the time t in terms of $(x(t), u_t)$) and the optimal value of the game, as a quadratic form in the complete initial state.

Both tasks pend on solving (6.4). Let R_0 be the solution of (3.2), which existence is now assumed. Setting $q = R_0x - p$ and invoking (3.2), the Hamilton-Jacobi-Bellman system (6.4) assumes an equivalent, upper block triangular form

$$\begin{aligned} \dot{x} &= A_0x - G_1G_1'q + G_2\tilde{u} \\ \dot{q} &= - A_0'q + (R_0G_2 + H_1'E_{12})\tilde{u} \end{aligned} \tag{6.5}$$

with $q(1) = 0$. Eqs. (6.5)-(6.4) are easily solved, first for q (which is independent of x), then for x and finally, for p. Direct variations of parameters computation yields

$$\begin{aligned} x(1) &= \Phi_0(1,0)x(0) + \textstyle\int_0^1 \Phi_0(1,s)\cdot \\ &\qquad \cdot((I + \Xi R_0)G_2 + \Xi H_1'E_{12})(s)u_0(s-1)ds \\ p(0) &= R_0(0)x(0) + \textstyle\int_0^1 \Phi(s,0)'\cdot \\ &\qquad \cdot(R_0G_2 + H_1'E_{12})(s)u_0(s-1)ds \end{aligned} \tag{6.6}$$

Evidently, the restriction of (6.1) to any ray $[t, \infty)$, given $(x(t), u_t)$, will be solved in complete analogy to the solution of that problem in its original setting (i.e., over $[0, \infty)$, given $(x(0), u_0)$). Let the "$*$" super-script denote the solution of the latter. The uniqueness of that solution implies, furthermore, that the solution of the restriction of (6.1) to any ray $[t, \infty)$, given the initial data $(x^*(t), u_t^*)$, must coincide with the restriction to $[t, \infty)$ of the original solution, over $[0, \infty)$. (This is the standard argument in any dynamic programming solution.) In reference to optimal trajectories, one can thus substitute $x(t)$, u_t, $x(t+1)$ and $p(t)$ for $x(0)$, u_0, $x(1)$ and $p(0)$, in (6.6) –

$$\begin{aligned} x^*(t+1) &= \Phi_0(1,0)x^*(t) + \textstyle\int_0^1 \Phi_0(1,s)\cdot \\ &\quad \cdot((I+\Xi R_0)G_2 + \Xi H_1' E_{12})(s)u_t^*(s-1)ds \\ p^*(t) &= R_0(0)x^*(t) + \textstyle\int_0^1 \Phi(s,0)'\cdot \\ &\quad \cdot(R_0G_2 + H_1'E_{12})(s)u_t^*(s-1)ds \end{aligned} \tag{6.7}$$

These expressions can be then used in the formulae for the optimal solutions along the first time unit, which have been provided in Lemmas 9 (for u^*) and 10 (for w^*) –

$$\begin{aligned} w^*(t) &= H_2x^*(t) + E_{21}G_1'p^*(t) + E_{22}u^*(t-1) \\ u^*(t) &= (C(I-QR) + \tfrac{1}{\gamma^2}DB'R)x^*(t+1) \end{aligned} \tag{6.8}$$

Once (6.7) is substituted in (6.8), the desired complete state feedback formulae are obtained. For convenience we introduce the abbreviated notations

$$\bar{w}^*(t) = L^0x(t) + L^1u_t, \quad u^*(t) = K^0x(t) + K^1u_t$$

where $\bar{w} = E_{21}^{-1}w(t) - \frac{1}{\gamma}D'(H_1x(t) + E_{12}u(t-1))$.

To compute the optimal value of (6.1), make the following definitions: the mapping $\mathcal{N}(x(0),u_0) \mapsto (x(0),p(0),u_0)$ translates the boundary value problem data in (6.4) to the data in an allied initial value problem, utilizing (6.6); the mapping $\mathcal{M}(x(0),p(0),u_0) \mapsto (x(1),x(\cdot),p(\cdot),u_0(\cdot+1))$ is defined in terms of the variations of parameters solution of the said initial value problem; set a matrix

$$\mathcal{J} = \begin{bmatrix} R & 0 & 0 & 0 \\ 0 & H_1'H_1 & 0 & H_1'E_{12} \\ 0 & 0 & -G_1'G_1 & 0 \\ 0 & E_{12}'H_1 & 0 & E_{12}'E_{12} \end{bmatrix};$$

finally, define $\mathcal{R} = \mathcal{N}'\mathcal{M}'\mathcal{J}\mathcal{M}\mathcal{N}$. In these terms, the optimal value of (6.1) is $\langle (x(0),u_0), \mathcal{R}(x(0),u_0)\rangle$.

Explicit computations, based on the definitions, above, show that $\mathcal{R}$ is defined by the unique solution of (6.4), via

$$\begin{aligned} \mathcal{R}(x(0),u_0) &= (p(0),\ E_{12}'(H_1x(\cdot+1) \\ &\quad +E_{12}u_0(\cdot)) + G_2'p(\cdot+1)) \end{aligned} \tag{6.9}$$

Details of these computations can be found in the author's papers that are cited above. They are based on straightforward, albeit somewhat lengthy manipulations of the Hamilton-Jacobi-Bellman system.

The differential game: An abstract model based solution. Let an abstract model of the form (2.2) be associated with our system; it is constructed by first bringing the system to the form (2.1), as explained earlier, and then on the association of (2.1) with (2.2).

Lemma 11. *(i) For any L_2 inputs and the associated state, denote*

$$w^{\nabla}(t) = \bar{w}(t) - L^0 x(t) - L^1 u_t \quad \text{and} \quad u^{\nabla} = (2E_{12}'E_{12})^{\frac{1}{2}}(u(t) - K^0 x(t) - K^1 u_t)$$

and $f(t) = (x(t), u_t)$. Define also mappings

$$\begin{aligned} \bar{\mathcal{B}}_1 w^{\nabla} &= (G_1 w^{\nabla}, 0) \\ \bar{\mathcal{B}}_2 u^{\nabla} &= (2E_{12}'E_{12})^{-\frac{1}{2}} \mathcal{B}_2 u^{\nabla} \\ \bar{\mathcal{C}}_1 f &= H_1 f^0 + E_{12} f^1(-1) \end{aligned}$$

Then these definitions imply

$$w^{\nabla}(t) = \bar{w} - \bar{\mathcal{B}}_1' \mathcal{R} f \quad \text{and} \quad u^{\nabla} = 2\bar{\mathcal{B}}_2' \mathcal{R} f$$

(ii) $\forall t > 0$

$$\begin{aligned} &\gamma^2 \|w\|^2_{L_2[0,t]} - \|z\|^2_{L_2[0,t]} - \langle f(t), \mathcal{R} f(t) \rangle_{M_2} \\ &\qquad = \|w^{\nabla}\|^2_{L_2[0,t]} - \|u^{\nabla}\|^2_{L_2[0,t]} - \langle f(0), \mathcal{R} f(0) \rangle_{M_2} \end{aligned} \tag{6.10}$$

(iii) Let $\mathcal{S}_1(t)(x(0), u_0) = (x(t), u_t)$ be defined in terms of shifts along optimal solutions of (6.1). Then $\mathcal{S}_1$ is an exponentially stable c_0 - semigroup over M_2, generated by

$$\mathcal{A}_1 f = ((A - G_2 C + G_1 L^0) f^0 + G_1 L^1 f^1 + G_2 f^1(-1)\, ,\ \frac{d}{ds} f^1)$$

over the domain

$$\mathcal{D}(\mathcal{A}_1) = \left\{ f \in M_2 \ :\ \frac{d}{ds} f^1 \in L_2[-1, 0],\ \ f^1(0) = K^0 f^0 + K^1 f^1 \right\}$$

(iv) Trajectories of $f(t) = (x(t), u_t)$ are governed by the abstract model

$$\begin{aligned} \dot{f} &= \mathcal{A}_1 f \ + \ \bar{\mathcal{B}}_1 w^{\nabla} \ + \ \bar{\mathcal{B}}_2 u^{\nabla} \\ \bar{z} &= \bar{\mathcal{C}}_1 f \end{aligned} \tag{6.11}$$

(v) The following integral, operator Riccati equation is satisfied over M_2

$$\langle f, \mathcal{R} f \rangle_{M_2} = \int_0^{\infty} \langle f\, ,\ \mathcal{S}_1(t)' \left(\bar{\mathcal{C}}_1' \bar{\mathcal{C}}_1 - \mathcal{R} \bar{\mathcal{B}}_1 \bar{\mathcal{B}}_1' \mathcal{R} \right) \mathcal{S}_1(t) f \rangle_{M_2} dt \tag{6.12}$$

Part (i) of Lemma 11 follows directly from the definitions of w^{∇}, u^{∇} and $\bar{\mathcal{B}}_i$, and from the explicit form of $\mathcal{R}$, as computed above. Since existence and uniqueness of solutions of (6.1) are established, the algebraic semigroup property of $\mathcal{S}_1$ (in part (iii)) follows. The complete–state feedback formulae for the optimal w and u show that optimal trajectories satisfy well posed integro-differential equations,(equivalently, a well posed neutral FDE [32]). Continuous dependence on the data and strong continuity in t, follow immediately. From the analysis of the restriction of (6.1) to $[1,\infty)$ it followed that optimal trajectories of $x(t)$, $t>1$, are generated by the exponentially stable ODE

$$\dot{x} = A_1 x$$

In particular

$$\|x(t)\|_{\mathbb{R}^n} \le \alpha e^{-\beta(t-1)}\|x(1)\|_{\mathbb{R}^n}, \qquad t>1$$

for some positive α and β. The optimal inputs satisfy

$$w(t) = \frac{1}{\gamma^2}BRx(t), \quad u(t-1) = \left(C(I-QR) + \frac{1}{\gamma^2}DB'R\right)x(t), \quad t>1$$

Hence the exponential decay of $(x(t), u_t)$ relative to $x(1)$ and, eventually, relative to $(x(0), u_0)$. Consequently, $\mathcal{S}_1$ is exponentially stable. The form of the generator, $\mathcal{A}_1$, and its domain, are obtained by standard associations [32] of neutral FDEs and their semigroup representations. The association of the current case with the general setting of [32] is similar to what is explained in the outlined proof of Lemma 1. The same applies to the abstract model (6.11), for the inhomogeneous system. The equalities (6.12) and then, (6.10) are obtained by a laborious and yet straightforward manipulation of the integral variation of parameters formula in (6.11).

The proof of Theorem 5 will be complete with the following lemma.

Lemma 12. *If γ satisfies the necessary conditions in Theorem 5 then $\gamma > \gamma_0$ and $\Theta \in \Theta_\gamma \Leftrightarrow \Theta$ admits the following realization*

$$\begin{aligned} \dot{f}_c &= \mathcal{A}_c f_c + \mathcal{B}_{c1} w + \mathcal{B}_{c2}\phi \\ u &= \mathcal{C}_{c1} f_c + \mathcal{D}_{c12}\phi \\ \psi &= \mathcal{C}_{c2} f_c + \mathcal{D}_{c21} w, \qquad \phi = \Theta_0 \psi \end{aligned} \tag{6.13}$$

where

$$\mathcal{A}_c f_c = \left((A - QC'C)f_c^0 + QC'f_c^1(-1),\ \frac{d}{ds}f_c^1\right)$$

is defined over the domain

$$\mathcal{D}(\mathcal{A}_c) = \left\{ f_c \in M_2 \ :\ \frac{d}{ds}f_c^1 \in L_2[-1,0],\ f_c^1(0) = K^0 f_c^0 + K^1 f_c^1 \right\}$$

where the remaining coefficients are

$$\begin{aligned}
\mathcal{B}_{c1} &= \mathcal{B}_1 \\
\mathcal{B}_{c2} &= \bar{\mathcal{B}}_2, \\
\mathcal{C}_{c1} f_c &= K^0 f_c^0 + K^1 f_c^1, \\
\mathcal{D}_{c12} &= (2E_{12}' E_{12})^{-\frac{1}{2}}, \\
\mathcal{C}_{c2} f_c &= -(L^0 + \tfrac{1}{\gamma} D' H_1) f_c^0 - L^1 f_c^1 - \tfrac{1}{\gamma} D' E_{12} f_c^1(-1), \\
\mathcal{D}_{c21} &= E_{21}^{-1}
\end{aligned}$$

and where the free design parameter, the mapping Θ_0, is defined by the I/O mapping in a stable, neutral FDE with the $L_2[0,\infty)$ induced norm bound $\|\Theta_0\| <$ 1. Moreover, trajectories of (6.13) correspond to trajectories of (3.3) via $f_c(t) = (x_c(t), u_t)$. Thus the parameterizations (6.13) and (3.3) are identical.

The system (3.3) is built of the closed loop interconnection of two well posed, atomic integro-differential equations of neutral type: one is Θ_0 and the other is the strictly proper system which governs the dynamics of x_c and u, with the exogenous input ϕ and output ψ. This interconnection is thus a well posed neutral integro-differential equation in its own right. The last statement in the Lemma 12, namely, the association of (3.3) with (6.13) is yet another standard association of an integro-differential equation of neutral type with an abstract model [32]. This association thus implies, in particular, that the abstract model (6.13) is well posed and that $\mathcal{A}_c$ is the infinitesimal generator of a c_0 - semigroup over M_2. Furthermore, close inspection shows that the dynamics is identical to the associated dynamics of the original system (that is, wiιh the state x).

The following is an outline of the proof of stability of (6.13) and the induced norm bound in. The notation "$\phi = \Theta_0\psi$" in (6.13) represents the I/O mapping in a stable, neutral FDE. Let f_0 be the state in a stable M_2 realization of Θ_0. Including the contribution of the homogeneous part (and the initial state) to the output, in that system, we denote $\phi = \Upsilon_0 f_0(0) + \Theta_0\psi$; in these terms Υ_0 and Θ_0 are bounded operators from M_2 and $L_2[0,\infty)$ into $L_2[0,\infty)$, respectively. The condition $\|\Theta_0\| < 1$ allows us to introduce the notation of $\lambda^2 = 1 - \|\Theta_0\|^2$.

To establish stability set $w = 0$ and select a combined initial state $(f_c(0), f_0(0))$ in (6.13) and the said realization of Θ_0. Then $w^\nabla = \psi = \mathcal{C}_{c2} f_c$ and $u^\nabla = \phi$. Using these equalities, (6.10), and the induced norms $\|\Upsilon_0\|$ and $\|\Theta_0\|$, one obtains the inequality

$$\begin{aligned}
0 \leq \langle f_c(0), \mathcal{R} f_c(0)\rangle_{M_2} &+ \|\Upsilon_0\|^2 \|f_0\|_{M_2}^2 \\
&+ 2\|\Theta_0\|\|\Upsilon_0\|\|f_0\|_{M_2}\|\psi\|_{L_2[0,t]} - \lambda^2 \|\psi\|_{L_2[0,t]}^2
\end{aligned} \tag{6.14}$$

where the right hand side is a quadratic expression in $\|\psi\|_{L_2[0,t]}$. This leads to a bound of $\|\psi\|_{L_2[0,\infty)}$ in terms of $\|f_c(0)\|_{M_2}$ and $\|f_0(0)\|_{M_2}$. Since $\|\phi\|_{L_2[0,\infty)}$ is bounded in terms of $\|f_0(0)\|_{M_2}$ and $\|\psi\|_{L_2[0,\infty)}$, the continuity of the mapping $(f_c(0), f_0(0)) \mapsto (\psi, \phi, f_0)$ $M_2 \times M_2 \mapsto L_2 \times L_2 \times L_2$ is established. The stable state equation in (6.11) is valid with f_c, ψ and ϕ substituting f, w^∇ and u^∇. This verifies the continuity of $(f_c(0), \psi, \phi) \mapsto f_c$. Consequently the mapping $(f_c(0), f_0(0)) \mapsto (f_c, f_0)$ $M_2 \times M_2 \mapsto L_2 \times L_2$ is continuous. As is well known [4], that last continuity is equivalent to exponential stability.

To establish γ suboptimality of (6.13) select $w \neq 0$ with the zero initial data . Then again, $f_c(t) = (x_c(t), u_t)$ is a trajectory of the original system (2.1) (with the interpretation specified in the beginning of the proof). Using the established stability, let $t \to \infty$ in (6.10) –

$$\gamma^2\|w\|_2^2 - \|z\|_2^2 = \|w^\nabla\|_2^2 - \|\Theta_0 w^\nabla\|_2^2 \geq \lambda^2\|w^\nabla\|_2^2 \tag{6.15}$$

The mapping $w^\nabla \mapsto w$ is governed by the stable (6.11), the relation $u^\nabla = \Theta_0 w^\nabla$ and the output equation $w = \mathcal{D}_{c21}^{-1}(w^\nabla - \mathcal{C}_{c2}f)$. That mapping is therefore continuous and $\lambda^2\|w^\nabla\|_2^2$ can be bounded below by $\mu^2\|w\|_2^2$, with some fixed $\mu \neq 0$. Thus the closed loop L_2 induced I/O norm is $< \sqrt{\gamma^2 - \mu^2}$.

The argument for completeness of the Parameterization will be briefly outlined, to complete the proof. Given any stable, γ suboptimal Θ, one can realize the closed loop mapping $w^\nabla \mapsto u^\nabla$, denoted Θ_0, by an appropriate, well posed perturbation of the original closed loop system. Relying heavily on (6.10), it can be shown (in similarity to arguments used in [26, 33]) that Θ is a stabilizing closed loop compensator in the latter system, and that it renders the closed loop mapping Θ_0 a strict L_2 contraction. It is then easy to reconstruct the closed loop mapping $\Theta : w \mapsto u$ in terms of the original system and of Θ_0 is. That reconstruction is (6.13). □

References

1. J. A. Ball and N. Cohen. Sensitivity minimization in an H_∞ norm: parameterization of suboptimal solutions. *Int. J. Control*, 46:785 – 816, 1987.
2. J. A. Burns, T. L. Herdman, and H. W. Stetch. Linear functional differential equations as semigroups on product spaces. *SIAM J. Math. Anal.*, 14:98 – 116, 1983.
3. R. F. Curtain and A. J. Pritchard. *Infinite Dimensional Linear Systems Theory*. Springer, 1978.
4. R. F. Curtain and H. J. Zwart. *An Introduction to Infinite Dimensional Linear Systems Theory*. Springer, 1995.
5. J.C. Doyle, K. Glover, P.P. Khargonekar, and B.A. Francis. State space solutions to standard H_2 and H_∞ control problems. *IEEE Transactions on Automatic Control*, AC – 34:831–847, 1989.
6. H. Dym, T. T. Georgiou, and M. C. Smith. Explicit formulas for optimally robust controllers for delay systems. *IEEE Transactions on Automatic Control*, AC – 40:656 – 669, 1995.
7. D. Flamm and S. Mitter. H_∞ Sensitivity minimization for delay systems. *System and Control letters*, 9:17–24, 1987.

8. C. Foias, A. Tannenbaum, and G. Zames. Weighted sensitivity minimization for delay systems. *IEEE Transactions on Automatic Control*, AC – 31:763–766, 1986.
9. T. T. Georgiou and M. C. Smith. Optimal robustness in the gap metric. *IEEE Transactions on Automatic Control*, AC – 35:673–686, 1990.
10. A. Ichikawa. Optimal control and filtering of evolution equations with delays in conrol and observation. Technical Report 53, Control Theory centre, university of warwick, 1977.
11. A. Ichikawa. Quadratic control of evolution equations with delays in control. *SIAM J. Control and Optim*, pages 645 – 668, 1982.
12. P. P. Khargonekar, K. M. Nagpal, and K. R. Poolla. H_∞ control with transients. *SIAM J. Control and Optim*, 29:1373–1393, 1991.
13. P. P. Khargonekar and K. Zhou. on the weighted sensitivity minimization problem for delay systems. *System and Control letters*, 8:307 – 312, 1987.
14. A. Kojima and S. Ishijima. Robust controller design for delay systems in the gap metric. In *Proceedings of the American Control Conference*, pages 1939 – 1944, 1994.
15. K. M. Nagpal and R. Ravi. H_∞ Control and estimations problems with delayed measurements: state space solutions. In *Proceedings of the American Control Conference*, pages 2379–2383, 1994.
16. O. Toker and H. Özbay. Suboptimal robustness in the gap metric for MIMO delay systems. In *Proceedings of the American Control Conference*, pages 3183–3187, 1994.
17. H. Özbay and A. Tannenbaum. A skew Toeplitz approach to the H_∞ optimal control of multivariable distributed systems. *SIAM J. Control and Optim*, 28:653–670, 1990.
18. A. Pazy. *Semigroups of Linear Operators and Relations to Differential Equations*. Springer, 1983.
19. A. J. Pritchard and D. Salamon. The linear-quadratic problem for retarded systems with delays in the control and observation. *IMA Journal of Mathematical Control and Information*, pages 335–362, 1985.
20. D. Salamon. *Control And Observation of Neutral Systems*. Pitman, 1984.
21. G. Tadmor. An interpolation problem associated with H_∞ optimization in systems with distributed lags. *System and Control letters*, 8:313–319, 1987.
22. G. Tadmor. H_∞ Interpolation in systems with commensurate input lags. *SIAM J. Control and Optim*, 27:511–526, 1989.
23. G. Tadmor. Worst case design in the time domain: the maximum principle and the standard H_∞ problem. *Math. Control, Signals and Systems*, 3:301–324, 1990.
24. G. Tadmor. H_∞ Optimal sampled data control in continuous time systems. *Int. J. Control*, 56:99–141, 1992.
25. G. Tadmor. The standard H_∞ problem and the maximum principle: the general linear case. *SIAM J. Control and Optim*, 31:831–846, 1993.
26. G. Tadmor. The standard H_∞ problem and the maximum principle: the general linear case. *SIAM J. Control and Optim*, 31:831–846, 1993.
27. G. Tadmor. The standard H_∞ problem in systems with a single input delay. *Technical Report*, 1994.
28. G. Tadmor. H_∞ Control in systems with a single input lag. In *Proceedings of the American Control Conference*, pages 321 – 325, 1995.
29. G. Tadmor. The nehari problem in systems with distributed input delays is inherently finite dimensional. *System and Control letters*, 26:11 – 16, 1995.

30. G. Tadmor. Robust control in the gap: a state space solution in the presence of a single input delay. *IEEE Transactions on Automatic Control*, in press.
31. G. Tadmor. Weighted sensitivity minimization in systems with a single input delay: a state space solution. *SIAM J. Control and Optim*, in press.
32. G. Tadmor and J. Turi. Neutral equations and associated semigroups. *J. Differential Eqs.*, 116:59–87, 1995.
33. G. Tadmor and M. Verma. Factorization and the Nehari theorem in time varying systems. *Math. Control, Signals and Systems*, 5:419–452, 1992.
34. B. van Keulen. *H_∞ control for infinite dimensional systems: a state space approach.* PhD thesis, University of Groningen, 1993.
35. K. Zhou and P. P. Khargonekar. On the weighted sensitivity minimization problem for delay systems. *System and Control letters*, 8:307–312, 1987.

Robust Guaranteed Cost Control for Uncertain Linear Time-delay Systems

Huaizhong Li[1], Silviu-Iulian Niculescu[2], Luc Dugard[1] and Jean-Michel Dion[1]

[1] Laboratoire d'Automatique de Grenoble (CNRS-INPG-UJF)
ENSIEG, BP 46
38402 Saint-Martin-d'Hères, France
e-mail: {lihz,dugard,dion}@lag.ensieg.inpg.fr
[2] Laboratoire de Mathématiques Appliquées, Ecole Nationale Supérieure de Techniques Avancées, 32, Blvd. Victor, 75739 Paris, France
e-mail: silviu@ensta.fr

Abstract. This chapter is concerned with robust guaranteed cost control for uncertain linear time-delay systems with quadratically constrained uncertainty using a linear matrix inequality (LMI) approach. We only consider the case of using memoryless static state feedback in this chapter. Two specific problems are considered in this chapter, namely the robust guaranteed cost control problem for linear systems with single state delay and the one for systems with mixed state and input delays. We show that feasibility of some LMIs guarantees the solvability of the corresponding robust guaranteed cost control problem.

1 Introduction

Stability and stabilization of dynamical systems which include time-delays in their physical models are problems of recurring interest since the existence of delays often induce instability and/or undesired performance (see, e.g. [7, 6, 11, 10, 15]).

Although the last decade has witnessed significant advances on the robust control theory [20], the robust control problem for linear systems with delayed state and/or delayed control input has not been fully investigated. There are, however, some results on robust control of time-delay systems available in literature. For example, robust memoryless controllers have been considered in [17, 18] (delay-independent closed-loop stability) or in [12, 9] (delay-dependent closed-loop stability) using the Lyapunov's second method based on Lyapunov-Krasovskii functional approach [12, 13], or on the Lyapunov-Razumikhin function approach [9, 17, 13].

In this chapter we consider a class of uncertain linear systems described by differential equations with delayed state as well as delayed control input. The focal point of the chapter is to design finite dimensional memoryless static state feedback controllers that make the closed-loop systems uniformly asymptotically stable for all admissible uncertainties and guarantee an adequate level of performance. The performance index considered in the chapter is an integral quadratic cost function as in the LQ regulator problem, see e.g. [14, 16]. Reza Moheiami and Petersen [16] have already considered this problem in the delay case, but handling uncertainties only on the non-delayed state, which is a particular form of the proposed framework.

The approach adopted here is based on the Lyapunov-Krasovskii functional technique [13] combined with an LMI technique [1]. The obtained conditions are *delay-independent* (do not include any information on the size of delay) and handle only the case of a *single* and *constant* state delay. Using an appropriate choice of the Lyapunov-Krasovskii functional, these results can be easily extended to multiple state delays case or to the time-varying delay case.

The chapter is organized as follows: in Section 2, some preliminary results are given. Section 3 is devoted to robust performance analysis problem. The single state delay case is treated in Section 4, and the mixed state and input delays case is considered in Section 5. We illustrate our design procedure using examples in section 6. Some final remarks conclude the chapter.

Notations. The following notations are used throughout the whole chapter: $\mathcal{C}_\tau = \mathcal{C}([-\tau, 0], \mathbb{R}^n)$ denotes the Banach space of continuous vector functions which maps the interval $[-\tau, 0]$ into $\mathbb{R}^n$ with the topology of uniform convergence; $\|\phi\|_\mathcal{C} = \sup_{-\tau\leq t\leq 0}\|\phi(t)\|$ is the norm of a function $\phi \in \mathcal{C}_\tau$; $\mathcal{C}_\tau^\nu$ is the set defined by $\mathcal{C}_\tau^\nu = \{\phi \in \mathcal{C}_\tau : \|\phi\|_\mathcal{C} < \nu\}$, where ν is a positive real number. The rest of the notations follow the convention.

2 Preliminaries and Definitions

Next, we introduce and define the $\mathcal{S}$-procedure.

Definition 1. [19]
Denote a space $\mathbb{H}$ and let $\mathcal{F}(g), \mathcal{Y}_1(g), \ldots, \mathcal{Y}_k(g)$, $g \in \mathbb{H}$, be some functionals or functions. Further define domain $\mathbb{F}$:

$$\mathbb{F} = \{g \in \mathbb{H} : \mathcal{Y}_1(g) \geq 0, \ldots, \mathcal{Y}_k(g) \geq 0\} \tag{2.1}$$

and two conditions:

(A) $\mathcal{F}(g) > 0$, $\forall g \in \mathbb{F}$;
(B) $\exists \epsilon_1 \geq 0, \ldots, \epsilon_k \geq 0$ such that

$$\mathcal{S}(\epsilon, g) = \mathcal{F}(g) - \sum_{j=1}^{k} \epsilon_j \mathcal{Y}_j(g) > 0, \qquad \forall g \in \mathbb{H}. \tag{2.2}$$

Then (B) implies (A). The procedure of replacing (A) by (B) is called the $\mathcal{S}$-procedure.

Definition 2. [19] The $\mathcal{S}$-procedure for the condition (A) is said to be lossless if (A) is equivalent to (B) and lossy otherwise.

In fact, (B) implies (A) is rather trivial as the $\mathcal{S}$-procedure in this case is almost equivalent to Lagrange multipliers method which is frequently used in optimization. However, the $\mathcal{S}$-procedure lossless property is not trivial. One of the important $\mathcal{S}$-procedure results we will use in this chapter is the following *$\mathcal{S}$-procedure lossless lemma*:

Lemma 3. *[19] If $k = 1$, $\mathbb{H}$ is a real linear space and $\mathcal{F}$, $\mathcal{Y}_1$ are quadratic functionals, the $\mathcal{S}$-procedure is lossless.*

We also recall the following linear matrix inequality result:

Lemma 4. *[1, 4] Given a symmetric matrix $\Psi \in \mathbb{R}^{m\times m}$ and two matrices $U \in \mathbb{R}^{i_1\times m}$ $V \in \mathbb{R}^{i_2\times m}$. There exists a matrix Θ of compatible dimension such that*

$$\Psi + U^T\Theta^T V + V^T\Theta U < 0 \tag{2.3}$$

if and only if

$$U_\perp^T \Psi U_\perp < 0 \tag{2.4}$$

$$V_\perp^T \Psi V_\perp < 0 \tag{2.5}$$

where $U_\perp \in \mathbb{R}^{m\times j_1}$ and $V_\perp \in \mathbb{R}^{m\times j_2}$ are any matrices whose columns form bases of the null spaces of U and V, respectively.

3 Robust Performance Analysis

Consider the following uncertain linear time-delay system

$$\begin{aligned} \dot{x}(t) &= Ax(t) + A_d x(t-\tau_1) + B_d x(t-\tau_2) \\ &\quad + H_1\zeta_1(t) + H_2\zeta_2(t) + H_3\zeta_3(t) && (3.1)\\ z_1(t) &= E_1 x(t) + E_{21}\zeta_1(t) && (3.2)\\ z_2(t) &= E_{d1} x(t-\tau_1) + E_{22}\zeta_2(t) && (3.3)\\ z_3(t) &= E_{d2} x(t-\tau_2) + E_{23}\zeta_3(t) && (3.4) \end{aligned}$$

with the initial condition

$$x(t_0+\theta) = \phi(\theta), \quad \forall\theta \in [-\max\{\tau_1,\tau_2\}, 0]; \quad (t_0,\phi) \in \mathbb{R}^+ \times \mathcal{C}^{\nu}_{\max\{\tau_1,\tau_2\}}$$

where $x(t) \in \mathbb{R}^n$ is the state, $z_i(t) \in \mathbb{R}^{k_i}$, $i = 1,2,3$, the fictitious outputs, and $\zeta_i(t) \in \mathbb{R}^{k_i}$, $i = 1,2,3$, the uncertain variables. We call the uncertainties as admissible if the uncertain variables satisfying the following quadratic constraint

$$\sum_{i=1}^{3} \|\zeta_i(t)\|^2 \le \sum_{i=1}^{3} \|z_i(t)\|^2, \qquad \forall t \ge t_0. \tag{3.5}$$

In the above, A, A_d, B_d, H_1, H_2, H_3, E_1, E_{21}, E_{22}, E_{23}, E_{d1} and E_{d2} are known constant matrices of appropriate dimension.

Remark 1. Note that (3.5) allows dynamic, time-varying and nonlinear uncertain structures. For $H_1 \neq H_2 \neq H_3$, the uncertainties added to the state matrix, the delayed state matrices have different structures.

Remark 2. The following well-known uncertain linear systems with delayed state

$$\dot{x}(t) = (A + \Delta A)x(t) + (A_d + \Delta A_d)x(t - \tau_1) + (B_d + \Delta B_d)x(t - \tau_2) \quad (3.6)$$

with norm-bounded uncertainty

$$\left[\Delta A \quad \Delta A_d \quad \Delta B_d \right] = H_1 F(t) \left[E_1 \ E_{d1} \ E_{d2} \right], \qquad F^T(t)F(t) \leq I, \ \ \forall t \geq t_0 \quad (3.7)$$

is a special case of system (3.1)-(3.5) with $E_{2i} = 0, i = 1, 2, 3$ and $H_1 = H_2 = H_3$.

Remark 3. It seems to be nature that the uncertainties are admissible if the uncertain variables satisfying the following quadratic constraints:

$$||\zeta_i(t)||^2 \leq ||z_i(t)||^2, \quad i = 1, 2, 3, \quad \forall t \geq t_0. \quad (3.8)$$

However, we notice that all uncertainties satisfying (3.8) will also satisfy (3.5), the reverse is not necessary true. Therefore, (3.5) allows a broader class of uncertainties than (3.8). On the other hand, (3.5) permits the so-called non-generic uncertainties, i.e., the uncertain variables and the fictitious outputs could have the following relation

$$||\zeta_i(t)||^2 \geq ||z_i(t)||^2 \quad (3.9)$$

for a specific fictitious output signal $z_i(t)$ at a specific time instant t as long as the overall constraint (3.5) is satisfied for all time instant. It is generally difficult to describe and treat the non-generic uncertainties using other uncertainty descriptions like norm-bounded uncertainty.

Without loss of generality, we assume $t_0 = 0$ in the sequel.

Associated with system (3.1)-(3.5) is the following quadratic cost function:

$$\mathcal{J} = \int_0^\infty x^T(t)Qx(t)dt, \qquad Q \in \mathbb{R}^{n \times n}, \ Q > 0, \quad \forall t > 0. \quad (3.10)$$

Now, we address the robust performance analysis problem associated with the uncertain system (3.1)-(3.5) as follows:

Determine if the system (3.1)-(3.4) is uniformly asymptotically stable and find an upper bound for the cost function (3.10) for all admissible uncertainty satisfying (3.5).

We then have the following result for the robust performance analysis problem:

Theorem 5. *Consider the system (3.1)-(3.4) with uncertainty satisfying (3.5), the robust performance analysis problem associated with the uncertain time-delay system (3.1)-(3.5) is solvable if there exist matrices* $P \in \mathbb{R}^{n \times n}$, $S_1 \in \mathbb{R}^{n \times n}$,

$S_2 \in \mathbb{R}^{n\times n}$, $P > 0$, $S_1 > 0$, $S_2 > 0$ *and scaling scalar parameter* $\epsilon > 0$ *such that the following LMI is feasible:*

$$\mathcal{L}_1 = \begin{bmatrix} \mathcal{P}_{11} & \mathcal{P}_{12} & \mathcal{P}_{13} \\ \mathcal{P}_{12}^T & \mathcal{P}_{22} & \mathcal{P}_{23} \\ \mathcal{P}_{13}^T & \mathcal{P}_{23}^T & \mathcal{P}_{33} \end{bmatrix} < 0. \tag{3.11}$$

where

$$\begin{aligned}
\mathcal{P}_{11} &= A^TP + PA + \epsilon E_1^T E_1 + Q + S_1 + S_2 \\
\mathcal{P}_{12} &= [PA_d \;\; PB_d] \\
\mathcal{P}_{13} &= [PH_1 + \epsilon E_1^T E_{21} \;\; PH_2 \;\; PH_3] \\
\mathcal{P}_{22} &= \begin{bmatrix} -S_1 + \epsilon E_{d1}^T E_{d1} & 0 \\ 0 & -S_2 + \epsilon E_{d2}^T E_{d2} \end{bmatrix} \\
\mathcal{P}_{23} &= \begin{bmatrix} 0 & \epsilon E_{d1}^T E_{22} & 0 \\ 0 & 0 & \epsilon E_{d2}^T E_{23} \end{bmatrix} \\
\mathcal{P}_{33} &= \begin{bmatrix} -\epsilon I_{k_1} + \epsilon E_{21}^T E_{21} & 0 & 0 \\ 0 & -\epsilon I_{k_2} + \epsilon E_{22}^T E_{22} & 0 \\ 0 & 0 & -\epsilon I_{k_3} + \epsilon E_{23}^T E_{23} \end{bmatrix}.
\end{aligned}$$

Moreover, the cost function (3.10) satisfies the following bound:

$$\mathcal{J} \le x^T(0)Px(0) + \int_{-\tau_1}^{0} x^T(\theta)S_1x(\theta)d\theta + \int_{-\tau_2}^{0} x^T(\theta)S_2x(\theta)d\theta, \quad \forall t > 0. \tag{3.12}$$

Proof. Consider the following Lyapunov-Krasovskii functional candidate:

$$V(t, x_t) = x^T(t)Px(t) + \int_{t-\tau_1}^{t} x^T(\theta)S_1x(\theta)d\theta + \int_{t-\tau_2}^{t} x^T(\theta)S_2x(\theta)d\theta \tag{3.13}$$

where $P > 0$, $S_1 > 0$ and $S_2 > 0$.

We can easily verify that

$$\lambda_{min}(P)\|x(t)\|^2 \le V(t, x_t) \le (\lambda_{max}(P) + \tau_1\lambda_{max}(S_1) + \tau_2\lambda_{max}(S_2))\|x_t\|_C^2. \tag{3.14}$$

Denote $\dot{V}(t, x_t)$ the derivative of the Lyapunov-Krasovskii functional $V(t, x_t)$, then the following inequality

$$\dot{V}(t, x_t) + x^T(t)Qx(t) \le 0 \tag{3.15}$$

guarantees both uniform asymptotic stability and the upper bound (3.12) for the cost function (3.10). Indeed, we have

$$\dot{V}(t, x_t) \le -x^T(t)Qx(t) \le 0, \qquad \forall t > 0 \tag{3.16}$$

and $x^T(t)Qx(t) = 0$ if and only if $x(t) = 0$. According to Lyapunov-Krasovskii stability theorem, conditions (3.15) and (3.16) guarantee the uniform asymptotic stability of system (3.1)-(3.4) without constraint (3.5).

On the other hand, integrating (3.15) on $[0,t]$, then it is obvious that (3.15) is a sufficient condition to guarantee

$$\begin{aligned} V(t,x_t) - x^T(0)Px(0) - \int_{-\tau_1}^{0} x^T(\theta)S_1x(\theta)d\theta \\ - \int_{-\tau_2}^{0} x^T(\theta)S_2x(\theta)d\theta + \mathcal{J}(t) \le 0, \\ \forall t > 0. \end{aligned}$$

Since $V(t,x_t) \ge 0,\ \forall t > 0$, we have

$$\mathcal{J}(t) \le x^T(0)Px(0) + \int_{-\tau_1}^{0} x^T(\theta)S_1x(\theta)d\theta + \int_{-\tau_2}^{0} x^T(\theta)S_2x(\theta)d\theta, \qquad \forall t > 0. \tag{3.17}$$

Furthermore, due to the uniform asymptotic stability we have established, $x(t) \to 0$ when $t \to \infty$. Then we have

$$\mathcal{J} \le x^T(0)Px(0) + \int_{-\tau_1}^{0} x^T(\theta)S_1x(\theta)d\theta + \int_{-\tau_2}^{0} x^T(\theta)S_2x(\theta)d\theta.$$

Applying $\mathcal{S}$-procedure to inequality (3.15) with constraint (3.5), we conclude that (3.15) is satisfied under constraint (3.5) if there exists scaling parameter $\epsilon > 0$ such that

$$\dot{V}(t,x_t) + x^T(t)Qx(t) + \epsilon\sum_{i=1}^{3}(\|z_i(t)\|^2 - \|\zeta_i(t)\|^2) \le 0, \tag{3.18}$$

and the $\mathcal{S}$-procedure (3.18) is lossless according to Lemma 3.

Since

$$\begin{aligned} \dot{V}(t,x_t) &= \dot{x}^T(t)Px(t) + x^T(t)P\dot{x}(t) + x^T(t)S_1x(t) - x^T(t-\tau_1)S_1x(t-\tau_1) \\ &\quad + x^T(t)S_2x(t) - x^T(t-\tau_2)S_2x(t-\tau_2), \end{aligned}$$

we can rewrite (3.18) in the following form:

$$\begin{bmatrix} x(t) \\ x(t-\tau_1) \\ x(t-\tau_2) \\ \zeta_1(t) \\ \zeta_2(t) \\ \zeta_3(t) \end{bmatrix}^T \mathcal{L}_1 \begin{bmatrix} x(t) \\ x(t-\tau_1) \\ x(t-\tau_2) \\ \zeta_1(t) \\ \zeta_2(t) \\ \zeta_3(t) \end{bmatrix} \le 0. \tag{3.19}$$

(3.11) is a sufficient condition to guarantee that (3.19) is satisfied for all admissible uncertainty.

In fact, we can require that the following inequality instead of (3.18) holds:

$$\dot{V}(t,x_t) + \delta x^T(t)Qx(t) + \epsilon\sum_{i=1}^{3}(\|z_i(t)\|^2 - \|\zeta_i(t)\|^2) \le 0 \tag{3.20}$$

for any given $\delta > 0$. The physical explanation of (3.20) is that we require

$$\dot{V}(t, x_t) \le -\tilde{\delta}\|x(t)\|^2$$

holds for any given $\tilde{\delta} > 0$ subject to constraint (3.5). However, since $\delta > 0$, we can absorb it into P, S_1, S_2 and ϵ, namely, let

$$\tilde{P} = \frac{1}{\delta}P; \quad \tilde{S}_1 = \frac{1}{\delta}S_1; \quad \tilde{S}_2 = \frac{1}{\delta}S_2; \quad \tilde{\epsilon} = \frac{1}{\delta}\epsilon,$$

then (3.20) takes the same form as (3.18), and this absorbing procedure doesn't invalidate condition 1 of Lyapunov-Krasovskii stability theorem and affect the upper bound for $\mathcal{J}$. Therefore, without loss of generality, we use (3.18). □

Instead of system (3.1)-(3.5), we can alternatively consider the following system with a different uncertainty structure:

$$\dot{x}(t) = Ax(t) + A_d x(t-\tau_1) + B_d x(t-\tau_2) + H_1\zeta(t) \quad (3.21)$$

$$z(t) = E_1 x(t) + E_{21}x(t-\tau_1) + E_{22}x(t-\tau_2) + E_3\zeta(t) \quad (3.22)$$

where the admissible uncertain variables satisfy

$$\|\zeta(t)\|^2 \le \|z(t)\|^2, \quad \forall t \ge t_0. \quad (3.23)$$

We can use the same method as described above to tackle system (3.21)-(3.23). As a direct application, we consider the following uncertain linear system with single state delay:

$$\dot{x}(t) = Ax(t) + A_d x(t-\tau) + H_1\zeta(t) \quad (3.24)$$

$$z(t) = E_1 x(t) + E_{1d}x(t-\tau) + E_2\zeta(t) \quad (3.25)$$

with $z(t) \in \mathbb{R}^k$ and the admissible uncertain variable $\zeta(t) \in \mathbb{R}^k$ satisfying the following quadratic constraint

$$\|\zeta(t)\|^2 \le \|z(t)\|^2, \quad \forall t \ge t_0. \quad (3.26)$$

Then we have the following corollary straightforwardly:

Corollary 6. *Consider the system (3.24)-(3.25) with uncertainty satisfying (3.26), the robust performance analysis problem associated with the uncertain time-delay system (3.24)-(3.26) is solvable if there exist matrices* $P \in \mathbb{R}^{n\times n}$, $S \in \mathbb{R}^{n\times n}$, $P > 0$, $S > 0$ *and a scaling scalar parameter* $\epsilon > 0$ *such that the following LMI is feasible:*

$$\begin{bmatrix} A^TP + PA + \epsilon E_1^TE_1 + Q + S & PA_d + \epsilon E_1^TE_{1d} & PH_1 + \epsilon E_1^TE_2 \\ A_d^TP + \epsilon E_{1d}^TE_1 & \epsilon E_{1d}^TE_{1d} - S & \epsilon E_{1d}^TE_2 \\ H_1^TP + \epsilon E_2^TE_1 & \epsilon E_2^TE_{1d} & -\epsilon I_k + \epsilon E_2^TE_2 \end{bmatrix} < 0. \quad (3.27)$$

Moreover, the cost function (3.10) satisfies the following bound:

$$\mathcal{J}(t) \le x^T(0)Px(0) + \int_{-\tau}^{0} x^T(\theta)Sx(\theta)d\theta, \qquad \forall t > 0. \quad (3.28)$$

Remark 4. The feasibility of LMIs (3.11) and (3.27) can be easily determined using *Matlab LMI Control Toolbox* [5].

In the sequel, we will consider robust guaranteed cost control problem using memoryless static state feedback. We will simplify our derivative steps in order to reduce redundancy.

4 Robust Guaranteed Cost Control – Single State-delay Case

Consider the following uncertain linear time-delay system

$$\dot{x}(t) = Ax(t) + A_d x(t-\tau) + Bu(t) + H_1\zeta(t) \tag{4.1}$$

$$z(t) = E_1 x(t) + E_{1d} x(t-\tau) + E_3 u(t) + E_2\zeta(t) \tag{4.2}$$

with the initial condition

$$x(t_0+\theta) = \phi(\theta), \quad \forall\theta \in [-\tau, 0]; \quad (t_0, \phi) \in \mathbb{R}^+ \times \mathcal{C}^{\nu}_{\tau}$$

where $x(t) \in \mathbb{R}^n$ is the state, $u(t) \in \mathbb{R}^m$ the control input, $z(t) \in \mathbb{R}^k$ the fictitious output, and $\zeta(t) \in \mathbb{R}^k$ the uncertain variable satisfying the following quadratic constraint

$$\|\zeta(t)\|^2 \leq \|z(t)\|^2. \tag{4.3}$$

Again, A, A_d, B, H_1, E_1, E_{1d}, E_2 and E_3 are constant matrices of appropriate dimension.

Similar to the robust performance analysis problem, we take $t_0 = 0$ without loss of generality.

Associated with the system (4.1)-(4.3) is the following quadratic cost function:

$$\mathcal{J}(t) = \int_0^t (x^T(t)Qx(t) + u^T(t)Ru(t))dt,$$

$$Q \in \mathbb{R}^{n\times n}, \; R \in \mathbb{R}^{m\times m}, \; Q > 0, \; R > 0, \quad \forall t > 0. \tag{4.4}$$

We consider the following robust guaranteed cost control problem associated with system (4.1)-(4.3):

Find a controller in the following form:

$$u(t) = Kx(t) \tag{4.5}$$

such that the closed-loop system (4.1)-(4.5) is uniformly asymptotically stable and give an upper bound for the cost function (4.4) for all admissible uncertainty satisfying (4.3).

We have the following theorem for robust guaranteed cost control using memoryless static state feedback:

Theorem 7. *Consider the system (4.1)-(4.2) with uncertainty satisfying (4.3), then there exists a memoryless static state feedback controller (4.5) that solves the addressed robust guaranteed cost control problem if there exist matrices* $X \in \mathbb{R}^{n\times n}$, $Y \in \mathbb{R}^{n\times n}$, $X > 0$, $Y > 0$ *and a scaling scalar parameter* $\epsilon > 0$ *such that the following LMIs are feasible:*

$$\begin{bmatrix} -Y & 0 & YE_{1d}^T \\ 0 & -\epsilon^{-1}I_k & \epsilon^{-1}E_2^T \\ E_{1d}Y & \epsilon^{-1}E_2 & -\epsilon^{-1}I_k \end{bmatrix} < 0 \tag{4.6}$$

$$\mathcal{N}^T \left[\begin{array}{cc|cccc} XA^T + AX & XE_1^T & A_dY & \epsilon^{-1}H_1 & X & X \\ E_1X & -\epsilon^{-1}I_k & E_{1d}Y & \epsilon^{-1}E_2 & 0 & 0 \\ \hline YA_d^T & YE_{1d}^T & -Y & 0 & 0 & 0 \\ \epsilon^{-1}H_1^T & \epsilon^{-1}E_2^T & 0 & -\epsilon^{-1}I_k & 0 & 0 \\ X & 0 & 0 & 0 & -Y & 0 \\ X & 0 & 0 & 0 & 0 & -Q^{-1} \end{array}\right] \mathcal{N} < 0 \tag{4.7}$$

where

$$\mathcal{N} = \left[\begin{array}{c|c} N & 0 \\ \hline 0 & I_{3n+k} \end{array}\right]$$

with N *any matrix whose columns form a basis of the null space of* $[B^T \;\; E_3^T]$. *Moreover, the cost function (4.4) satisfies the following bound:*

$$\mathcal{J}(t) \le x^T(0)X^{-1}x(0) + \int_{-\tau}^{0} x^T(\theta)Y^{-1}x(\theta)d\theta, \qquad \forall t > 0. \tag{4.8}$$

Proof. The closed-loop system of (4.1)-(4.2) with controller (4.5) is the following:

$$\dot{x}(t) = (A + BK)x(t) + A_d x(t-\tau) + H_1\zeta(t) \tag{4.9}$$
$$z(t) = (E_1 + E_3K)x(t) + E_{1d}x(t-\tau) + E_2\zeta(t) \tag{4.10}$$

where the uncertain variable $\zeta(t)$ satisfying

$$\|\zeta(t)\|^2 \le \|z(t)\|^2. \tag{4.11}$$

Applying Corollary 6 to system (4.9)-(4.11), it is an easy exercise of using Schur complements that (4.9)-(4.11) is uniformly asymptotically stable and the guaranteed cost (4.8) is satisfied if there exist $P > 0$, $S > 0$ and $\epsilon > 0$ such that the following LMI is feasible:

$$\begin{bmatrix} \begin{matrix}(A+BK)^TP + P(A+BK) \\ +S + Q + K^TRK\end{matrix} & PA_d & PH_1 & \epsilon(E_1 + E_3K)^T \\ A_d^TP & -S & 0 & \epsilon E_{1d}^T \\ H_1^TP & 0 & -\epsilon I_k & \epsilon E_2^T \\ \epsilon(E_1 + E_3K) & \epsilon E_{1d} & \epsilon E_2 & -\epsilon I_k \end{bmatrix} < 0. \tag{4.12}$$

We then rewrite (4.12) into the following form:

$$\begin{bmatrix} A^TP+PA & PA_d & PH_1 & \epsilon E_1^T & I_n & I_n & 0 \\ A_d^TP & -S & 0 & \epsilon E_{1d}^T & 0 & 0 & 0 \\ H_1^TP & 0 & -\epsilon I_k & \epsilon E_2^T & 0 & 0 & 0 \\ \epsilon E_1 & \epsilon E_{1d} & \epsilon E_2 & -\epsilon I_k & 0 & 0 & 0 \\ I_n & 0 & 0 & 0 & -S^{-1} & 0 & 0 \\ I_n & 0 & 0 & 0 & 0 & -Q^{-1} & 0 \\ 0 & 0 & 0 & 0 & 0 & 0 & R^{-1} \end{bmatrix} +$$

$$+ \begin{bmatrix} I_n \\ 0 \\ 0 \\ 0 \\ 0 \\ 0 \\ 0 \end{bmatrix} K^T \left[B^TP \; 0 \; 0 \; \epsilon E_3^T \; 0 \; 0 \; I_m\right] + \begin{bmatrix} PB \\ 0 \\ 0 \\ \epsilon E_3 \\ 0 \\ 0 \\ I_m \end{bmatrix} K \left[I_n \; 0 \; 0 \; 0 \; 0 \; 0 \; 0\right] < 0. \quad (4.13)$$

Applying Lemma 4 to (4.13), note that

$$\begin{bmatrix} 0 \\ \hline I_{n+2k} \end{bmatrix}$$

is a matrix whose columns form a basis of the null space of $[I \mid 0 \; \dots \; 0]$, and

$$\left[\begin{array}{ccc|c} P^{-1} & & & \\ & \epsilon^{-1} I_m & & \\ & & I_m & \\ \hline & & & I_{3n+k} \end{array}\right] \left[\begin{array}{c|c} N & 0 \\ 0 & 0 \\ \hline 0 & I \end{array}\right]$$

is a matrix whose columns form a basis of the null space of $[B^TP \; \epsilon E_3^T \; I_m \mid 0 \; \dots \; 0]$, further let $X = P^{-1}$ and $Y = S^{-1}$, we obtain (4.6)-(4.7) after some algebraic manipulations. □

Remark 5. When $P > 0$, $S > 0$ and $\epsilon > 0$ are obtained, we can synthesize the controller using LMI (4.13). Indeed, we replace P, S and ϵ in LMI (4.13) with their obtained forms, then the controller gain matrix K is the only unknown variable in LMI (4.13). It then requires some algebraic manipulations to get explicit expression for K. However, we observe that K is not unique due to the non-uniqueness of N. This is in fact one advantage which allows us to explore all possible controllers which solve the addressed robust guaranteed cost control problem.

5 Robust Guaranteed Cost Control – Mixed State and Input Delays

Consider the following uncertain linear time-delay system

$$\begin{aligned}\dot{x}(t) &= Ax(t) + A_d x(t-\tau_1) + Bu(t) + B_d u(t-\tau_2) \\ &\quad + H_1\zeta_1(t) + H_2\zeta_2(t) + H_3\zeta_3(t) &(5.1)\\ z_1(t) &= E_1 x(t) + E_{21}\zeta_1(t) + E_3 u(t) &(5.2)\\ z_2(t) &= E_{d1} x(t-\tau_1) + E_{22}\zeta_2(t) &(5.3)\\ z_3(t) &= E_{d2} u(t-\tau_2) + E_{23}\zeta_3(t) &(5.4)\end{aligned}$$

with the initial condition

$$x(t_0+\theta) = \phi(\theta), \quad \forall\theta \in [-\max\{\tau_1,\tau_2\}, 0]; \quad (t_0,\phi) \in \mathbb{R}^+ \times \mathcal{C}^{\nu}_{\max\{\tau_1,\tau_2\}}$$

where $x(t) \in \mathbb{R}^n$ is the state, $u(t) \in \mathbb{R}^m$ the control input, $z_i(t) \in \mathbb{R}^{k_i}$, $i = 1, 2, 3$, the fictitious outputs, and $\zeta_i(t) \in \mathbb{R}^{k_i}$, $i = 1, 2, 3$, the uncertain variables satisfying the following quadratic constraints

$$\sum_{i=1}^{3} \|\zeta_i(t)\|^2 \leq \sum_{i=1}^{3} \|z_i(t)\|^2, \qquad \forall t \geq t_0. \tag{5.5}$$

Again, A, A_d, B, B_d, H_1, H_2, H_3, E_1, E_{21}, E_{22}, E_{23}, E_{d1}, E_{d2} and E_3 are known constant matrices of appropriate dimension.

Generally, it should be very restrictive to require $\tau_1 = \tau_2$. The two uncertain delays impose on output channels and input channels separately, therefore, it is unlikely for them to have a exact match. For this reason, we would like to treat τ_1 and τ_2 as two independent delays while it is clear that the unified state and input delay is just a special case of our problem.

We also take $t_0 = 0$ without loss of generality.

Associated with the system (5.1)-(5.5) is the following quadratic cost function:

$$\begin{aligned}\mathcal{J} &= \int_0^{\infty} (x^T(t)Qx(t) + u^T(t)Ru(t))dt,\\ &Q \in \mathbb{R}^{n\times n},\ R \in \mathbb{R}^{m\times m},\ Q > 0,\ R > 0, \quad \forall t > 0. &(5.6)\end{aligned}$$

We consider the following robust guaranteed cost control problem associated with system (5.1)-(5.5):

Find a controller in the following form:

$$u(t) = Kx(t) \tag{5.7}$$

such that the closed-loop system (5.1)-(5.5) is uniformly asymptotically stable and give an upper bound for the cost function (5.6) for all admissible uncertainty satisfying (5.5).

We have the following theorem for robust guaranteed cost control using memoryless static state feedback:

Theorem 8. *Consider the system (5.1)-(5.4) with uncertainty satisfying (5.5), then there exists a memoryless static state feedback controller (5.7) that solves the addressed robust guaranteed cost control problem if there exist matrices* $X \in \mathbb{R}^{n\times n}$, $Y_1 \in \mathbb{R}^{n\times n}$, $Y_2 \in \mathbb{R}^{n\times n}$, $X > 0$, $Y_1 > 0$, $Y_2 > 0$ *and scaling scalar parameter* $\epsilon > 0$ *such that the following LMIs are feasible:*

$$\begin{bmatrix} \mathcal{R}_{11} & \mathcal{R}_{12} \\ \mathcal{R}_{12}^T & \mathcal{R}_{22} \end{bmatrix} < 0 \tag{5.8}$$

$$\mathcal{N}^T \left[\begin{array}{c|cc} \mathcal{Q}_{11} & \mathcal{Q}_{12} & \mathcal{Q}_{13} \\ \hline \mathcal{Q}_{12}^T & \mathcal{Q}_{22} & \mathcal{Q}_{23} \\ \mathcal{Q}_{13}^T & \mathcal{Q}_{23}^T & \mathcal{Q}_{33} \end{array}\right] \mathcal{N} < 0 \tag{5.9}$$

where

$$\mathcal{N} = \left[\begin{array}{c|c} N & 0 \\ \hline 0 & I_{4n+k_1+2k_2+k_3} \end{array}\right]$$

with N *any matrix whose columns form a basis of the null space of*

$$\begin{bmatrix} B^T & E_3^T & 0 \\ B_d^T & 0 & E_{d2}^T \end{bmatrix},$$

and

$$\mathcal{R}_{11} = \begin{bmatrix} -Y_1 & 0 & 0 \\ 0 & -\epsilon^{-1} I_{k_1} + \epsilon^{-1} E_{21}^T E_{21} & 0 \\ 0 & 0 & -\epsilon^{-1} I_{k_2} \end{bmatrix}$$

$$\mathcal{R}_{12} = \begin{bmatrix} 0 & Y_1 E_{d1}^T \\ 0 & 0 \\ 0 & \epsilon^{-1} E_{22}^T \end{bmatrix}, \quad \mathcal{R}_{22} = \begin{bmatrix} -\epsilon^{-1} I_{k_3} + \epsilon^{-1} E_{23}^T E_{23} & 0 \\ 0 & -\epsilon^{-1} I_{k_2} \end{bmatrix}$$

$$\mathcal{Q}_{11} = \begin{bmatrix} XA^T + AX & XE_1^T & 0 \\ E_1 X & -\epsilon^{-1} I_{k_1} & 0 \\ 0 & 0 & -\epsilon^{-1} I_{k_3} \end{bmatrix}$$

$$\mathcal{Q}_{12} = \begin{bmatrix} A_d Y_1 & \epsilon^{-1} H_1 & \epsilon^{-1} H_2 \\ 0 & \epsilon^{-1} E_{21} & 0 \\ 0 & 0 & 0 \end{bmatrix}, \quad \mathcal{Q}_{13} = \begin{bmatrix} \epsilon^{-1} H_3 & 0 & X & X & X \\ 0 & 0 & 0 & 0 & 0 \\ \epsilon^{-1} E_{23} & 0 & 0 & 0 & 0 \end{bmatrix}$$

$$\mathcal{Q}_{22} = \begin{bmatrix} -Y_1 & 0 & 0 \\ 0 & -\epsilon^{-1} I_{k_1} & 0 \\ 0 & 0 & -\epsilon^{-1} I_{k_2} \end{bmatrix}, \quad \mathcal{Q}_{23} = \begin{bmatrix} 0 & Y_1 E_{d1}^T & 0 & 0 & 0 \\ 0 & 0 & 0 & 0 & 0 \\ 0 & \epsilon^{-1} E_{22}^T & 0 & 0 & 0 \end{bmatrix}$$

$$\mathcal{Q}_{33} = \begin{bmatrix} -\epsilon^{-1} I_{k_3} & 0 & 0 & 0 & 0 \\ 0 & -\epsilon^{-1} I_{k_2} & 0 & 0 & 0 \\ 0 & 0 & -Q^{-1} & 0 & 0 \\ 0 & 0 & 0 & -Y_1 & 0 \\ 0 & 0 & 0 & 0 & -Y_2 \end{bmatrix}$$

Moreover, the cost function (5.6) satisfies the following bound:

$$\mathcal{J} \leq x^T(0)X^{-1}x(0) + \int_{-\tau_1}^{0} x^T(\theta)Y_1^{-1}x(\theta)d\theta + \int_{-\tau_2}^{0} x^T(\theta)Y_2^{-1}x(\theta)d\theta, \ \forall t > 0. \tag{5.10}$$

Proof. The closed-loop system of (5.1)-(5.4) with controller (5.7) is the following:

$$\begin{aligned}
\dot{x}(t) &= (A+BK)x(t) + A_d x(t-\tau_1) + B_d K x(t-\tau_2) \\
&\quad + H_1\zeta_1(t) + H_2\zeta_2(t) + H_3\zeta_3(t) && (5.11)\\
z_1(t) &= (E_1 + E_3K)x(t) + E_{21}\zeta_1(t) && (5.12)\\
z_2(t) &= E_{d1}x(t-\tau_1) + E_{22}\zeta_2(t) && (5.13)\\
z_3(t) &= E_{d2}Kx(t-\tau_2) + E_{23}\zeta_3(t) && (5.14)
\end{aligned}$$

where the uncertain variables $\zeta_i(t)$, $i = 1, 2, 3$ satisfying

$$\sum_{i=1}^{3} \|\zeta_i(t)\|^2 \leq \sum_{i=1}^{3} \|z_i(t)\|^2, \quad \forall t \geq 0 \tag{5.15}$$

Applying Theorem 5 to system (5.11)-(5.15), it is an easy exercise of using Schur complements that (5.11)-(5.15) is uniformly asymptotically stable and the guaranteed cost (5.10) is satisfied if there exist $P > 0$, $S > 0$ and $\epsilon > 0$ such that the following LMI is feasible:

$$\left[\begin{array}{ccc}
(A+BK)^TP + P(A+BK) + K^TRK & PA_d & PB_dK \\
A_d^TP & -S_1 + \epsilon E_{d1}^TE_{d1} & 0 \\
K^TB_d^TP & 0 & -S_2 + \epsilon K^TE_{d2}^TE_{d2}K \\
H_1^TP + \epsilon E_{21}^T(E_1 + E_3K) & 0 & 0 \\
H_2^TP & \epsilon E_{22}^TE_{d1} & 0 \\
H_3^TP & 0 & \epsilon E_{23}^TE_{d2}K \\
I_n & 0 & 0 \\
I_n & 0 & 0 \\
I_n & 0 & 0
\end{array}\right.$$

$$\leftarrow \begin{array}{ccc}
PH_1 + \epsilon(E_1 + E_3K)^TE_{21} & PH_2 & PH_3 \\
0 & \epsilon E_{d1}^TE_{22} & 0 \\
0 & 0 & \epsilon K^TE_{d2}^TE_{23} \\
-\epsilon I_{k_1} + \epsilon E_{21}^TE_{21} & 0 & 0 \\
0 & -\epsilon I_{k_2} + \epsilon E_{22}^TE_{22} & 0 \\
0 & 0 & -\epsilon I_{k_3} + \epsilon E_{23}^TE_{23} \\
0 & 0 & 0 \\
0 & 0 & 0 \\
0 & 0 & 0
\end{array}$$

$$
\leftarrow \begin{array}{ccc} I_n & I_n & I_n \\ 0 & 0 & 0 \\ 0 & 0 & 0 \\ 0 & 0 & 0 \\ 0 & 0 & 0 \\ 0 & 0 & 0 \\ -Q^{-1} & 0 & 0 \\ 0 & -S_1^{-1} & 0 \\ 0 & 0 & -S_2^{-1} \end{array} \left. \vphantom{\begin{array}{c} 0 \\ 0 \\ 0 \\ 0 \\ 0 \\ 0 \\ 0 \\ 0 \\ 0 \end{array}} \right] < 0. \tag{5.16}
$$

We then rewrite (5.16) into the following form:

$$
\left[\begin{array}{cccccc} A^TP + PA & PA_d & 0 & PH_1 & PH_2 & PH_3 \\ A_d^TP & -S_1 & 0 & 0 & 0 & 0 \\ 0 & 0 & -S_2 & 0 & 0 & 0 \\ H_1^TP & 0 & 0 & -\epsilon I_{k_1} & 0 & 0 \\ H_2^TP & 0 & 0 & 0 & -\epsilon I_{k_2} & 0 \\ H_3^TP & 0 & 0 & 0 & 0 & -\epsilon I_{k_3} \\ \epsilon E_1 & 0 & 0 & \epsilon E_{21} & 0 & 0 \\ 0 & \epsilon E_{d1} & 0 & 0 & \epsilon E_{22} & 0 \\ 0 & 0 & 0 & 0 & 0 & \epsilon E_{23} \\ I_n & 0 & 0 & 0 & 0 & 0 \\ I_n & 0 & 0 & 0 & 0 & 0 \\ I_n & 0 & 0 & 0 & 0 & 0 \\ 0 & 0 & 0 & 0 & 0 & 0 \end{array} \right.
$$

$$
\leftarrow \left. \begin{array}{ccccccc} \epsilon E_1^T & 0 & 0 & I_n & I_n & I_n & 0 \\ 0 & \epsilon E_{d1}^T & 0 & 0 & 0 & 0 & 0 \\ 0 & 0 & 0 & 0 & 0 & 0 & 0 \\ \epsilon E_{21}^T & 0 & 0 & 0 & 0 & 0 & 0 \\ 0 & \epsilon E_{22}^T & 0 & 0 & 0 & 0 & 0 \\ 0 & 0 & \epsilon E_{23}^T & 0 & 0 & 0 & 0 \\ -\epsilon I_{k_1} & 0 & 0 & 0 & 0 & 0 & 0 \\ 0 & -\epsilon I_{k_2} & 0 & 0 & 0 & 0 & 0 \\ 0 & 0 & -\epsilon I_{k_3} & 0 & 0 & 0 & 0 \\ 0 & 0 & 0 & -Q^{-1} & 0 & 0 & 0 \\ 0 & 0 & 0 & 0 & -S_1^{-1} & 0 & 0 \\ 0 & 0 & 0 & 0 & 0 & -S_2^{-1} & 0 \\ 0 & 0 & 0 & 0 & 0 & 0 & -R^{-1} \end{array} \right]
$$

$$
+\begin{bmatrix} I_n & 0 \\ 0 & 0 \\ 0 & I_n \\ 0 & 0 \\ 0 & 0 \\ 0 & 0 \\ 0 & 0 \\ 0 & 0 \\ 0 & 0 \\ 0 & 0 \\ 0 & 0 \\ 0 & 0 \\ 0 & 0 \end{bmatrix} K^T \begin{bmatrix} PB & PB_d \\ 0 & 0 \\ 0 & 0 \\ 0 & 0 \\ 0 & 0 \\ 0 & 0 \\ \epsilon E_3 & 0 \\ 0 & 0 \\ 0 & \epsilon E_{d2} \\ 0 & 0 \\ 0 & 0 \\ 0 & 0 \\ I_m & 0 \end{bmatrix}^T + \begin{bmatrix} PB & PB_d \\ 0 & 0 \\ 0 & 0 \\ 0 & 0 \\ 0 & 0 \\ 0 & 0 \\ \epsilon E_3 & 0 \\ 0 & 0 \\ 0 & \epsilon E_{d2} \\ 0 & 0 \\ 0 & 0 \\ 0 & 0 \\ I_m & 0 \end{bmatrix} K \begin{bmatrix} I_n & 0 \\ 0 & 0 \\ 0 & I_n \\ 0 & 0 \\ 0 & 0 \\ 0 & 0 \\ 0 & 0 \\ 0 & 0 \\ 0 & 0 \\ 0 & 0 \\ 0 & 0 \\ 0 & 0 \\ 0 & 0 \end{bmatrix}^T < 0. \quad (5.17)
$$

Applying Lemma 4 to (5.17), note that

$$
\left[\begin{array}{cc} 0 & 0 \\ \hline I_n & 0 \\ \hline 0 & 0 \\ \hline 0 & I_{4n+2k_1+2k_2+2k_3} \end{array}\right]
$$

is a matrix whose columns form a basis of the null space of

$$
\left[\begin{array}{c|c|c|c} I_n & 0 & 0 & 0 \\ 0 & 0 & I_n & 0 \end{array}\right]
$$

and

$$
\left[\begin{array}{ccc|c|c} P^{-1} & & & & \\ & \epsilon^{-1} I_m & & & \\ & & \epsilon^{-1} I_m & & \\ \hline & & & I_m & \\ \hline & & & & I_{4n+k_1+2k_2+k_3} \end{array}\right] \left[\begin{array}{c|c} N & 0 \\ \hline 0 & 0 \\ \hline 0 & I_{4n+k_1+2k_2+k_3} \end{array}\right]
$$

is a matrix whose columns form a basis of the null space of

$$
\left[\begin{array}{ccc|c|c} B^T P & \epsilon E_3^T & 0 & I_m & 0 \\ B_d^T P & 0 & \epsilon E_{d2}^T & 0 & 0 \end{array}\right],
$$

we obtain (5.8)-(5.9) after some tedious algebraic manipulations. □

6 Illustrative Examples

Example 1. Consider the uncertain system (4.1)-(4.2) with parameters given by:

$$A = \begin{bmatrix} -3 & 0 \\ 0.5 & 1 \end{bmatrix}; \qquad A_d = \begin{bmatrix} -1.4 & 0 \\ -0.5 & -0.6 \end{bmatrix}$$

$$B = \begin{bmatrix} 0 \\ 1 \end{bmatrix}; \quad H_1 = \begin{bmatrix} 0 \\ 1 \end{bmatrix}$$

$$C = \begin{bmatrix} 1 & 0 \\ 0 & 1 \\ 0 & 0 \\ 0 & 0 \end{bmatrix}; \quad C_d = \begin{bmatrix} 0 & 0 \\ 0 & 0 \\ 1 & 0 \\ 0 & 1 \end{bmatrix}$$

$$E_1 = [0.1\ 1]; \quad E_{1d} = [0.05\ 0.6]$$

$$E_2 = 0.9; \quad E_3 = 0.4.$$

We consider the following quadratic cost function:

$$\mathcal{J}(t) = \int_0^\infty (x^T(t)Qx(t) + u^T(t)Ru(t))dt$$

where

$$Q = \begin{bmatrix} 3 & 0 \\ 0 & 2 \end{bmatrix}; \qquad R = 0.1.$$

If the state is measurable, we can possibly design a memoryless static state feedback controller. First of all, we determine if the LMIs (4.6)-(4.7) are feasible for the above given system parameters. Then we can synthesize the feedback gain matrix K based on LMI (4.12) if LMIs (4.6)-(4.7) are feasible. Note however, we have the freedom in selecting the matrix N in LMI (4.7).

First, we select N as a matrix whose columns form an orthonormal basis of the null space of $[B^T\ E_3^T]$.

Simulation using *Matlab LMI Control Toolbox* [5] shows that LMIs (4.6)-(4.7) are feasible for our uncertain system, and we obtain the following variables from the feasibility test:

$$X^{-1} = \begin{bmatrix} 1.4808 & 0.0187 \\ 0.0187 & 1.0885 \end{bmatrix}; \qquad Y^{-1} = \begin{bmatrix} 1.1205 & 0.1195 \\ 0.1195 & 0.9826 \end{bmatrix}$$

$$\epsilon = 0.4181.$$

With the obtained variables, it is straightforward to get a suitable memoryless static state feedback controller as follows:

$$u(t) = [-0.4667\ \ -7.3572]x(t) \tag{6.1}$$

which guarantees that the cost function satisfies (4.8).

Alternatively, we choose

$$N = \begin{bmatrix} 2 & 0 \\ 0 & -0.2 \\ 0 & 0.5 \end{bmatrix}$$

whose columns also form a basis of the null space of $[B^T \; E_3^T]$.

Simulation shows that for memoryless static state feedback control, we now have

$$X^{-1} = \begin{bmatrix} 4.3731 & -0.0241 \\ -0.0241 & 1.2132 \end{bmatrix}; \qquad Y^{-1} = \begin{bmatrix} 2.3136 & 0.1544 \\ 0.1544 & 1.2817 \end{bmatrix}$$
$$\epsilon = 0.4987$$

while the controller is given by

$$u(t) = [-0.0890 \; -8.0439]x(t). \tag{6.2}$$

Example 2. Consider the system (5.1)-(5.4) with the given parameters:

$$A = \begin{bmatrix} -3 & 0 \\ 0.5 & 1 \end{bmatrix}; \quad A_d = \begin{bmatrix} -1.4 & 0 \\ -0.5 & -0.6 \end{bmatrix}; \quad B = \begin{bmatrix} 0 \\ 1 \end{bmatrix}$$
$$B_d = \begin{bmatrix} 0 \\ 1 \end{bmatrix}; \quad H_1 = \begin{bmatrix} 0 \\ 1 \end{bmatrix}; \quad H_2 = \begin{bmatrix} 0 \\ 1 \end{bmatrix}; \quad H_3 = \begin{bmatrix} 0 \\ 1 \end{bmatrix}$$
$$E_1 = [0.1 \; 1]; \quad E_{d1} = [0.05 \; 0.6]; E_{d2} = 0.6$$
$$E_{21} = E_{22} = E_{23} = 0.9; \quad E_3 = 0.4.$$

The following quadratic cost function is given by:

$$\mathcal{J}(t) = \int_0^\infty (x^T(t)Qx(t) + u^T(t)Ru(t))dt$$

where

$$Q = \begin{bmatrix} 3 & 0 \\ 0 & 2 \end{bmatrix}; \qquad R = 0.1.$$

Simulation using shows that LMIs (5.8)-(5.9) are feasible for our uncertain system, and we obtain the following variables from the feasibility test:

$$X^{-1} = \begin{bmatrix} 1.4936 & 0.0019 \\ 0.0019 & 1.0926 \end{bmatrix}$$
$$Y_1^{-1} = \begin{bmatrix} 1.2903 & -0.0823 \\ -0.0823 & 0.9641 \end{bmatrix}; \quad Y_2^{-1} = \begin{bmatrix} 0.1746 & -0.0017 \\ -0.0017 & 0.2054 \end{bmatrix}$$
$$\epsilon = 0.1320.$$

With the obtained variables, it is straightforward to get the controller gain matrix

$$K = [-1.3548 \; -9.8765].$$

Therefore, a suitable memoryless static state feedback controller is given as follows:

$$u(t) = [-1.3548 \quad -9.8765]x(t) \tag{6.3}$$

which guarantees that the cost function satisfies (5.10).

7 Conclusion

We consider the robust guaranteed cost control problem for linear time-delay systems with quadratically constrained uncertainty. Our uncertainty description is more general than norm-bounded time-varying uncertainty and linear fractional transform type uncertainty descriptions. We show that the feasibility of LMI (3.27) guarantees the solvability of the corresponding robust performance analysis problem while the feasibility of a pair of LMIs (4.6)-(4.7) insures the existence of a memoryless static state feedback controller which solves the addressed robust guaranteed control problem. Once the solvability issue is determined, it is then straightforward to construct a family of desired memoryless static state controllers numerically.

References

1. S. Boyd, L. El Ghaoui, E. Feron, and V. Balakrishnan, *Linear Matrix Inequalities in System and Control Theory*, vol. 15. Philadelphia: SIAM Studies in Appl. Math., 1994.
2. J. C. Doyle, K. Glover, P. P. Khargonekar, and B. A. Francis, "State–space solutions to standard $\mathcal{H}_2$ and $\mathcal{H}_\infty$ control problems," *IEEE Trans. Auto. Contr.*, vol. 34, no. 8, pp. 831–847, 1989.
3. E. Feron, V. Balakrishnan and S. Boyd, "A design of stabilizing state feedback for delay systems via convex optimization," *Proc. 31st IEEE Conf. Dec. Contr.*, Tuscon, Arizona, USA, pp. 147–148, 1992.
4. P. Gahinet and P. Apkarian, "A linear matrix inequality approach to $\mathcal{H}_\infty$ control," *Int. J. Robust and Nonlinear Contr.*, vol. 4, pp. 421–428, 1994.
5. P. Gahinet, A. Nemirovskii, A. J. Laub and M. Chilali, *LMI Control Toolbox*, The MathWorks, Inc., 1995.
6. H. Górecki, S. Fuksa, P. Gabrowski and A. Korytowski, *Analysis and Synthesis of Time Delay Systems*, John Wiley & Sons, Warszawa, Poland, 1989.
7. V. B. Kolmanovskii and V. R. Nosov, *Stability of Functional Differential Equations*. New York: Academic Press, 1986.
8. N. N. Krasovskii, *Stability of motion: Application of Lyapunov's Second Method to Defferential Systems and Equations with Delay*. Stanford, California: Stanford University Press, 1963.
9. X. Li and C. E. de Souza, "LMI approach to delay-dependent robust stability and stabilization of uncertain linear delay systems," in *Proc. 34th IEEE CDC*, New Orleans, Louisiana, 1995.
10. J. E. Marshall, H. Górecki, K. Walton and A. Korytowski, *Time-delay systems: Stability and performance criteria with applications*. Ellis Horwood, New York, 1992.

11. S.-I. Niculescu, *On the stability and stabilization of linear systems with delayed-state*, (in French). Ph.D. Thesis, Laboratoire d'Automatique de Grenoble, INPG, February, 1996.
12. S. I. Niculescu, C. E. de Souza, J. -M. Dion and L. Dugard, "Robust stability and stabilization for uncertain linear systems with state delay: Single delay case (I)," *Proc. IFAC Workshop on Robust Control Design*, Rio de Janeiro, Brazil, pp. 469–474, 1994.
13. S.-I. Niculescu, E. I. Verriest, L. Dugard and J.-M. Dion, "Stability and Robust Stability of Time-Delay Systems: A Guided Tour, " this monography (first chapter), LNCIS, Springer-Verlag, London, 1997.
14. I. R. Petersen and D. C. MacFarlane, "Optimal guaranteed cost control and filtering for uncertain linear systems," *IEEE Trans. Automat. Contr.*, vol. 39, pp. 1971–1977, 1994.
15. V. Răsvan, *Absolute stability of automatic control systems with delays* (in Romanian). Eds. Academiei RSR, Bucharest, Romania, 1975.
16. S. O. Reza Moheimani and I. R. Petersen, "Optimal quadratic guaranteed cost control of a class of uncertain time-delay systems," in *Proc. 34th IEEE CDC*, New Orleans, U.S.A., pp. 1513–1518, 1995.
17. A. Thowsen, "Uniform ultimate boundness of the solutions of uncertain dynamic delay systems with state-dependent and memoryless feedback control," *Int. J. Contr.*, vol. 37, pp. 1153–1143, 1983.
18. L. Xie and C. E. de Souza, "Robust stabilization and disturbance attenuation for uncertain delay system," *Proc. 2nd European Contr. Conf.* Groningen, The Netherlands, pp. 667–672, 1993
19. V. A. Yakubovich, "S-procedure in nonlinear control theory," *Vestnik Leningradskogo Universiteta, Ser. Matematika*, pp. 62–77, 1971.
20. K. Zhou, J. Doyle and K. Glover, *Robust and optimal control*, Prentice Hall, New Jersey, 1995.

Local Stabilization of Continuous-Time Delay Systems with Bounded Inputs

Sophie Tarbouriech

L.A.A.S - C.N.R.S.
7, Avenue du Colonel Roche,
31077 Toulouse Cedex 4, France.
E-mail : tarbour@laas.fr - Fax : +33 (0) 561 33 69 69

Abstract. This chapter deals with the stabilization of linear continuous-time systems with time-delay in the state and subject to bounded inputs. A saturated state feedback control law is used. Sufficient conditions addressing the local stabilization of such systems are proposed. The methodology consists in determining some domains of safe admissible states for which the stability of the saturated closed-loop system is guaranteed.

1 Introduction

In practical control problems, many constraints have to be treated in order to design suitable controllers operating in real environment. Hence, controls and states of practical systems are bounded and therefore subject to amplitude saturations.

The stabilization of linear systems with saturating actuators has been widely investigated in the last years: see, for example, [1] and references herein. The problems of local and global stabilization for such a class of systems have been studied. Some of these results have been extended to the case of linear systems with delayed state and then sufficient conditions for state feedback stabilization have been given, for example, in [3], [5], [18] (independent of the size of delay) or in [3], [16] (dependent of the size of delay). The stability conditions presented in these papers are mainly based on the use of matrix measure, complex Lyapunov equations, or still Razumikhin-type theorems. For an outline concerning the last results on the delay systems see, for example, [9] and references herein, or still the different papers on the subject in the 13th World IFAC Congress (San Francisco, USA - July 1996).

In this chapter, we consider a linear continuous-time system with saturating controls and with time-delay in the state. The main objective of this chapter consists in determining some domains of safe admissible states for which the stability of the saturated closed-loop system is guaranteed. The approach is based on a Lyapunov-Krasovskii technique for analysing the uniform asymptotic stability of solutions of functional differential equations. The main results consist in proposing simultaneously delay-independent sufficient conditions for the asymptotic stability of the closed-loop system via memoryless static state feedback and a suitable domain of safe admissible states. These conditions are given in terms of solutions of appropriate finite dimensional algebraic Riccati equations. The suitable associated domain is obtained from an optimization linear program. To obtain these results, the saturated closed-loop system is written as a convex combination of matrices belonging to a convex polyhedron of matrices.

The chapter is organized as follows. Section 2 presents the considered system with its hypotheses and states the objectives. Section 3 proposes a solution when saturation of controls has to be avoided. Section 4 addresses the problem when controls are allowed to saturate. In Section 5, a numerical example, borrowed from the literature, illustrates the results. Finally, Section 6 gives some concluding remarks.

Notations. Throughout this chapter, the following notations are used. Re denotes the set of real numbers, Re^+ is the set of non-negative real numbers, Re^n denotes the n dimensional Euclidean space, and $\mathrm{Re}^{n\times m}$ denotes the set of all $n\times m$ real matrices. The notation $X \geq Y$ (respectively, $X > Y$), where X and Y are symmetric matrices, means that the matrix $X - Y$ is positive semi-definite (respectively, positive definite). For a real matrix A, A^T and $A_{(i)}$ denote the transpose of matrix A and the ith row vector of matrix A respectively. I_n denotes the identity matrix in $\mathrm{Re}^{n\times n}$. $\lambda_{max}(P)$ and $\lambda_{min}(P)$ denote respectively the maximal and minimal eigenvalues of matrix P. $\mathcal{C}_\tau = \mathcal{C}([-\tau, 0], \mathrm{Re}^n)$ denotes the Banach space of continuous vector functions mapping the interval $[-\tau, 0]$ into Re^n with the topology of uniform convergence. The following norms will be used: $\|\cdot\|$ refers to either the Euclidean vector norm or the induced matrix 2-norm ; $\|\phi\|_c = \sup_{-\tau\leq t\leq 0} \|\phi(t)\|$ stands for the norm of a function $\phi \in \mathcal{C}_\tau$. Moreover, we denote by $\mathcal{C}_\tau^v$ the set defined by $\mathcal{C}_\tau^v = \{\phi \in \mathcal{C}_\tau \; ; \; \|\phi\|_c < v\}$, where v is a positive real number.

2 Problem statement

Consider the linear time-delay system described by:

$$\dot{x}(t) = Ax(t) + A_d x(t-\tau) + Bu(t) \tag{2.1}$$

with the initial condition

$$x(t_0+\theta) = \phi(\theta), \; \forall\theta \in [-\tau, 0], \; (t_0, \phi) \in \mathrm{Re}^+ \times \mathcal{C}_\tau^v \tag{2.2}$$

where $x(t) \in \mathrm{Re}^n$ is the state, $u(t) \in \mathrm{Re}^m$ is the control input, τ is the time-delay of the system, A, A_d, B are known real constant matrices of appropriate dimensions. Furthermore, pair (A, B) is assumed to be stabilizable.

The vector $u(t)$ is assumed to take values in the compact set $\Omega \in \mathrm{Re}^m$:

$$\Omega = \{u \in \mathrm{Re}^m; -u_0 \leq u \leq u_0\} \tag{2.3}$$

with u_0 component-wise positive vector of Re^m.

From (2.3), the saturation function $sat(Kx(t))$, $K \in \mathrm{Re}^{m\times n}$, is defined as

$$sat(Kx(t)) = \left[\ sat(K_{(1)}x(t)) \ \ ... \ \ sat(K_{(m)}x(t))\ \right]^T \tag{2.4}$$

with for $i = 1, ..., m$:

$$sat(K_{(i)}x(t)) = \begin{cases} -u_{0(i)} & \text{if } K_{(i)}x < -u_{0(i)} \\ K_{(i)}x & \text{if } -u_{0(i)} \leq K_{(i)}x \leq u_{0(i)} \\ u_{0(i)} & \text{if } K_{(i)}x > u_{0(i)} \end{cases} \tag{2.5}$$

By implementing such a saturated control law, the closed-loop system is :

$$\dot{x}(t) = Ax(t) + A_d x(t-\tau) + Bsat(Kx(t)) \tag{2.6}$$

When the controls do not saturate, that is, for all $x(t) \in \mathcal{S}(K, u_0)$ described as follows:

$$\mathcal{S}(K, u_0) = \{x \in \mathrm{Re}^n; -u_0 \leq Kx \leq u_0\} \tag{2.7}$$

system (2.6) admits the linear model:

$$\dot{x}(t) = (A + BK)x(t) + A_d x(t-\tau) \tag{2.8}$$

The aim of this chapter is to investigate conditions for closed-loop stability of the saturated system (2.6) via memoryless state feedback. The approach developed is mainly based on the Lyapunov-Krasovskii Theorem [4], [8]. However, some results based on the Razumikhin's approach are discussed. No assumption on the stability of the open-loop system is made. When the open-loop system ($u = 0$) is stable the global stabilization can be studied (see [14] and references therein).

First the linear model (2.8) is considered: a state feedback matrix K and a set of safe admissible states (domain of linear behavior) guaranteeing the asymptotic stability of the system are then determined.

Next, considering the saturated system (2.6), a domain of nonlinear behavior is determined in order to guarantee the asymptotic stability of the system.

3 Closed-loop stability without saturations

This section addresses the determination of a local domain of stability in which the control law is not saturated. In other words, a local domain included in $\mathcal{S}(K, u_0)$ and in which the model (2.8) is valid has to be found.

Lemma 1. *Assume that for two Lyapunov functions the inequality $V_1(x) \leq V_2(x) \leq c_0$ holds. Then the set $\mathcal{D}_1$ contains the set $\mathcal{D}_2$, where the sets $\mathcal{D}_i$, $i = 1, 2$ are defined by*

$$\mathcal{D}_i = \{x \in \mathrm{Re}^n; V_i(x) \leq c_0\}$$

Proposition 2. *Given symmetric and positive definite matrices Q and R, if there exist two symmetric and positive definite matrices P and S solutions of*

$$A^T P + PA + PA_d S^{-1} A_d^T P - PBR^{-1} B^T P + S + Q = 0 \tag{3.1}$$

then system (2.8) is asymptotically stabilizable by the state feedback matrix

$$K = -R^{-1} B^T P \tag{3.2}$$

for all initial condition $\phi \in \mathcal{B}(\sigma)$ defined by

$$\begin{aligned} &\mathcal{B}(\sigma) = \{\phi \in C_\tau^v; \| \phi \|_c^2 \leq \sigma\} \\ &\textit{with } \sigma = \tfrac{\mu}{\lambda_{max}(P) + \tau\lambda_{max}(S)} \end{aligned} \tag{3.3}$$

where the scalar μ corresponds to the largest ellipsoid $\mathcal{D}(P, \mu) = \{x \in \mathrm{Re}^n; x^T Px \leq \mu\}$ contained in $\mathcal{S}(K, u_0)$.

Proof. Let us introduce the following Lyapunov functional candidate:

$$V(x_t) = x(t)^T Px(t) + \int_{t-\tau}^{t} x(\theta)^T Sx(\theta)d\theta \tag{3.4}$$

where P and S are solutions of the Riccati equation (3.1). Furthermore, one gets:

$$\beta_1 \| x(t) \|^2 \leq V(x_t) \leq \beta_2 \| x_t \|_c^2 \tag{3.5}$$

where $\beta_1 = \lambda_{min}(P)$ and $\beta_2 = \lambda_{max}(P) + \tau\lambda_{max}(S)$. The time-derivative of $V(x_t)$ is given by:

$$\begin{aligned} \dot{V}(x_t) &= \dot{x}(t)^T Px(t) + x(t)^T P\dot{x}(t) + x(t)^T Sx(t) \\ &\quad -x(t-\tau)^T Sx(t-\tau) \end{aligned}$$

Then from (3.1) and (3.2), along the trajectories of system (2.8) it follows:

$$\begin{gathered} \dot{V}(x_t) = -x(t)^T Qx(t) - x(t)^T PBR^{-1} B^T Px(t) \\ -[x(t-\tau) - S^{-1} A_d^T Px(t)]^T S[x(t-\tau) - S^{-1} A_d^T Px(t)] \end{gathered}$$

Hence, it follows that there exists a positive scalar β_3 such that one gets $\dot{V}(x_t) \leq -\beta_3 \| x(t) \|^2 < 0$, and therefore $V(x_t) < V(x_{t_0})$, provided that the linear model (2.8) remains valid. According to Lemma 1, it is clear from (3.5) that both:

- $\forall \phi \in \mathcal{B}(\sigma)$ it follows $\phi \in \mathcal{D}(V, \mu)$, which is the domain defined by $\mathcal{D}(V, \mu) = \{x \in \mathrm{Re}^n; V(x_t) \leq \mu\}$.
- $\forall x_t \in \mathcal{D}(V, \mu)$ it follows $x(t) \in \mathcal{D}(P, \mu) = \{x \in \mathrm{Re}^n; x^t Px \leq \mu,\ \mu > 0\}$.

Furthermore, since $V(x_t) < V(x_{t_0})$, from (3.5), it follows that $\forall\phi \in \mathcal{B}(\sigma)$ one gets $x(t) \in \mathcal{D}(P,\mu)$. Thus, if we determine the largest ellipsoid $\mathcal{D}(P,\mu) = \{x \in \mathrm{Re}^n; x^t P x \leq \mu,\ \mu > 0\}$ included in $\mathcal{S}(K,u_0)$, it follows that $\forall\phi \in \mathcal{B}(\sigma) = \{\phi \in C^v_\tau; \| \phi \|^2_c \leq \sigma\}$ with $\sigma = \frac{\mu}{\beta_2}$, $x(t) \in S(K,u_0)$. Then, for any initial condition in $\mathcal{B}(\sigma)$ the linear system (2.8) is valid. Hence, using the Lyapunov-Krasovskii functional approach (see the first chapter of this monograph), for any initial condition in $\mathcal{B}(\sigma)$ the local asymptotic stability of system (2.8) is guaranteed.

□

Then the following Algorithm can be stated.

Algorithm 4 *1. Given symmetric and positive definite matrices Q and R, compute solutions P and S of (3.1).*

2. Compute $\mu = \min_i \eta_i$ *where* $\eta_i = \frac{u^2_{0(i)}}{K_{(i)}P^{-1}K^T_{(i)}}$ *is the solution of the optimization program:*

$$\max x^T P x = \eta_i$$
$$\textit{subject to}\, K_{(i)}x \leq u_{0(i)}$$

3. Compute $\sigma = \frac{\mu}{\lambda_{max}(P)+\tau\lambda_{max}(S)}$.

It is worth to notice that the set $\mathcal{D}(V,\mu) = \{x \in \mathrm{Re}^n; V(x_t) \leq \mu\}$ obtained from Step 2 is a positively invariant and strictly contractive set [6] with respect to the trajectories of system (2.8). Nevertheless, a better way to determine a set of safe admissible states would consist in finding the maximal set $\mathcal{D}(V,\mu)$ (defined above) with $\mu = \min_i \zeta_i$ where ζ_i would be the solution of the optimization program:

$$\max V(x_t) = \zeta_i$$
$$\text{subject to}\, K_{(i)}x \leq u_{0(i)}$$

However such a computation is very hard and no simple solutions are really attainable at the current time.

Since one gets:

$$\forall\phi \in \mathcal{B}(\sigma) \text{ then } x_t \in \mathcal{D}(V,\mu) \text{ and } x(t) \in \mathcal{D}(P,\mu) \tag{3.6}$$

it could be interesting to know if the ellipsoid $\mathcal{D}(P,\mu)$ is a positively invariant and contractive set for system (2.8). In this sense, based on the Razumikhin's approach the following proposition can be stated.

Proposition 3. *Given symmetric and positive definite matrices Q and R, if there exist two symmetric and positive definite matrices P and S solutions of (3.1) then system (2.8) is asymptotically stabilizable by the state feedback K, defined in (3.2), for all initial condition* $\phi \in \mathcal{D}(P,\mu) = \{x \in \mathrm{Re}^n; x^T P x \leq \mu\}$, *where* μ *is defined as in Step 2 of Algorithm 4. Hence,* $\mathcal{D}(P,\mu)$ *is a positively invariant and contractive set for system (2.8).*

Proof. Compute the time-derivative of the quadratic Lyapunov function $V(x) = x^T P x$ along the trajectories of system (2.8). Then from (3.1) and (3.2) it follows:

$$\dot{V}(x) = -x(t)^T(Q + PBR^{-1}B^TP + S)x(t) + 2x(t)^T PA_d x(t-\tau) \\ -x(t)^T PA_d S^{-1} A_d^T P x(t)$$

Let us first consider the following term

$$[x(t-\tau) - S^{-1}A_d^T Px(t)]^T S[x(t-\tau) - S^{-1}A_d^T Px(t)] \geq 0$$

which is equal to

$$x(t-\tau)^T Sx(t-\tau) - 2x(t)^T PA_d x(t-\tau) + x(t)^T PA_d S^{-1} A_d^T Px(t) \geq 0$$

therefore it follows:

$$2x(t)^T PA_d x(t-\tau) - x(t)^T PA_d S^{-1} A_d^T Px(t) \leq x(t-\tau)^T Sx(t-\tau)$$

Hence the time-derivative of $V(x)$ satisfies:

$$\dot{V}(x) \leq -x(t)^T(Q + PBR^{-1}B^TP + S)x(t) + x(t-\tau)^T Sx(t-\tau)$$

From the use of the Razumikhin's Theorem [4], it is assumed that there exists a positive number $q > 1$ such that

$$x(t-\tau)^T Px(t-\tau) < q^2 x(t)^T Px(t)$$

which is equivalent to the existence of $\kappa > 1$ such that

$$x(t-\tau)^T Sx(t-\tau) < \kappa^2 x(t)^T Sx(t)$$

Then concerning $\dot{V}(x)$ it follows:

$$\dot{V}(x) \leq (-\lambda_{min}(S^{-\frac{1}{2}}(Q + PBR^{-1}B^TP + S)S^{-\frac{1}{2}}) + \kappa^2)x(t)^T Sx(t)$$

Moreover, remark that $Q + PBR^{-1}B^TP + S > S$ since Q and S are positive definite. Therefore it follows that

$$S^{-\frac{1}{2}}(Q + PBR^{-1}B^TP + S)S^{-\frac{1}{2}} > 1$$

and thus the condition $\lambda_{min}(S^{-\frac{1}{2}}(Q + PBR^{-1}B^TP + S)S^{-\frac{1}{2}}) > 1$ follows. Therefore if $\lambda_{min}(S^{-\frac{1}{2}}(Q + PBR^{-1}B^TP + S)S^{-\frac{1}{2}}) > 1$, there exists κ small enough such that $\dot{V}(x) < 0$. Model (2.8) being valid only in $\mathcal{S}(K, u_0)$, since the choice of $\mu > 0$ follows Step 2 of Algorithm 4, one gets $\mathcal{D}(P, \mu) \subset \mathcal{S}(K, u_0)$ and therefore $\dot{V}(x) < 0$, for any $x \in \mathcal{D}(P, \mu)$. The set $\mathcal{D}(P, \mu)$ is a positively invariant and contractive set for system (2.8). □

Propositions 2 and 3 mean that the function $V(x_t)$, defined in (3.4), and the quadratic function $V(x) = x^T Px$ are both Lyapunov functions for system (2.8). The first one is based on the Lyapunov-Krasovskii approach, whereas the second one uses the Razumikhin's approach. Proposition 2 allows to conclude that $\mathcal{D}(P,\mu)$ is a domain of stability, whereas Proposition 3 allows to conclude that $\mathcal{D}(P,\mu)$ is a positive invariant and contractive set.

In the case $\tau = 0$, one can consider the term $A_d x(t)$ as a norm-bounded uncertain term, that is, $A_d = DFE_1$ where F is the parameter uncertainty satisfying $F^T F \leq 1$. Then to stabilize system (2.1) one can solve equation (3.1) by considering $S = \epsilon^{-1} I_n$, $\epsilon > 0$, that is, one has to determine ϵ and P solutions of [15]:

$$A^T P + PA + \epsilon P A_d A_d^T P - PBR^{-1}B^T P + \epsilon^{-1} I_n + Q = 0$$

Furthermore, in the case $\tau = 0$, the following corollary to Propositions 2 or 3 can be stated.

Corollary 4. *Given symmetric and positive definite matrices Q and R, if there exist two symmetric and positive definite matrices P and S solutions of equation (3.1) then system (2.8) is asymptotically stabilizable by the state feedback K, defined in (3.2), for all initial condition in $\mathcal{D}(P,\mu) = \{x \in \mathrm{Re}^n; x^T Px \leq \mu\}$, where μ is defined as in Step 2 of Algorithm 4. Hence, the set $\mathcal{D}(P,\mu) \subset \mathcal{S}(K,u_0)$ is a positively invariant and contractive set for system (2.8).*

Proof. It suffices to compute the time-derivative of the quadratic Lyapunov function $V(x) = x^T Px$ along the trajectories of system (2.8) in which $\tau = 0$. Then from (3.1) and (3.2) it follows

$$\dot{V}(x) = -x^T(Q + PBR^{-1}B^T P)x$$
$$-x^T(I_n - S^{-1}A_d^T P)^T S(I_n - S^{-1}A_d^T P)x$$

Model (2.8) being valid only in $\mathcal{S}(K,u_0)$, since the choice of $\mu > 0$ follows Step 2 of Algorithm 4, one gets $\mathcal{D}(P,\mu) \subset \mathcal{S}(K,u_0)$ and therefore $\dot{V}(x) < 0$, for any $x \in \mathcal{D}(P,\mu)$. □

4 Closed-loop stability with saturations

To develop the results of this section, the saturated system (2.6) is written under an equivalent form. Let us write the saturation term as:

$$sat(Kx(t)) = D(\alpha(x))Kx(t)\ ;\ D(\alpha(x)) \in \mathrm{Re}^{m\times m} \tag{4.1}$$

where $D(\alpha(x))$ is a diagonal matrix for which the elements $\alpha_i(x)$ satisfy for $i = 1, ..., m$:

$$\alpha_i(x) = \begin{cases} -\frac{u_{0(i)}}{K_{(i)}x} & \text{if } K_{(i)}x < -u_{0(i)} \\ 1 & \text{if } -u_{0(i)} \leq K_{(i)}x \leq u_{0(i)} \\ \frac{u_{0(i)}}{K_{(i)}x} & \text{if } K_{(i)}x > u_{0(i)} \end{cases} \tag{4.2}$$

and

$$0 < \alpha_i(x) \leq 1 \tag{4.3}$$

System (2.6) can then be written in the equivalent form:

$$\dot{x}(t) = (A + BD(\alpha(x))K)x(t) + A_d x(t - \tau) \tag{4.4}$$

Recall that for a given stabilizing state feedback K it is generally not possible to determine analytically the region of attraction of the origin. Nevertheless, the determination of either a positively invariant and contractive set [2] or a domain of stability [10] for system (2.6) may be an interesting way to approximate it.

The determination of such a set for systems (2.6) or (4.4) gives lower bounds for α_i, $i = 1, ..., m$. Thus, if this set is denoted S_0, it follows that for any $x(t)$ belonging to S_0, one may define a lower bound for $\alpha_i(x)$ as:

$$(\alpha_i(x))_{min} = \min\{\alpha_i(x) \; ; \; x \in S_0\} \tag{4.5}$$

Therefore, $\forall x(t) \in S_0$, the scalars $\alpha_i(x)$, $i = 1, ..., m$, satisfy $(\alpha_i(x))_{min} \leq \alpha_i(x) \leq 1$. From a convex linear combination of matrices A_j defined by [12]:

$$A_j = A + BD(\gamma_j)K \tag{4.6}$$

where $D(\gamma_j)$ is a diagonal matrix of positive scalars $\gamma_{j(i)}$, for $i = 1, ..., m$, which arbitrarily take the value 1 or $(\alpha_i(x))_{min}$, system (2.6) may be written, for any $x(t) \in S_0$, as:

$$\dot{x}(t) = \sum_{j=1}^{2^m} \lambda_j(x(t)) A_j x(t) + A_d x(t - \tau) \tag{4.7}$$

with

$$\sum_{j=1}^{2^m} \lambda_j(x(t)) = 1 \; , \; \lambda_j(x(t)) \geq 0 \tag{4.8}$$

Note that the matrices A_j are the vertices of a convex polyhedron of matrices. Note also that $(\alpha_i(x))_{min}$, $i = 1, ..., m$, define the polyhedral set

$$\mathcal{S}(K, u_0^\alpha) = \{x \in \mathrm{Re}^n; -u_0^\alpha \leq Kx \leq u_0^\alpha\} \tag{4.9}$$

where every component of vector u_0^α is defined by $\frac{u_{0(i)}}{(\alpha_i(x))_{min}}$, $i = 1, ..., m$. This set contains S_0 and corresponds to the maximal set in which model (4.7) represents system (2.6) or (4.4).

Proposition 5. *Assume that matrices P and S are solutions of (3.1) and K is given by (3.2). If for all $j = 1, ..., 2^m$, one gets:*

$$- Q + K^T(R - RD(\gamma_j) - D(\gamma_j)R)K < 0 \tag{4.10}$$

then system (2.6) is asymptotically stabilizable by the state feedback K for all initial condition $\phi \in \mathcal{B}(\delta)$ defined by

$$\begin{array}{c} \mathcal{B}(\delta) = \{\phi \in \mathcal{C}_{\tau}^{v}; \| \phi \|_c^2 \leq \delta\} \\ \text{with } \delta = \frac{\rho}{\lambda_{max}(P) + \tau \lambda_{max}(S)} \end{array} \tag{4.11}$$

where the scalar ρ corresponds to the largest ellipsoid $\mathcal{D}(P,\rho) = \{x \in \mathrm{Re}^n; x^T P x \leq \rho\}$ contained in $\mathcal{S}(K, u_0^{\alpha})$.

Proof. Using the same Lyapunov functional as defined in (3.4) and considering its time-derivative along the trajectories of system (4.7) it follows

$$\begin{array}{c} \dot{V}(x_t) = \sum_{j=1}^{2^m} \lambda_j(x) x(t)^T (A_j^T P + P A_j) x(t) \\ +2x(t)^T P A_d x(t-\tau) + x(t)^T S x(t) - x(t-\tau)^T S x(t-\tau) \end{array}$$

From (4.8), (3.1), (3.2) and from the convexity of function $V(x_t)$ it follows that:

$$\dot{V}(x_t) = \sum_{j=1}^{2^m} \lambda_j(x) \dot{V}_j(x_t)$$

where $\dot{V}_j(x_t)$ is defined by:

$$\begin{array}{c} \dot{V}_j(x_t) = -x(t)^T Q x(t) + x(t)^T K^T (R - RD(\gamma_j) - D(\gamma_j)R) K x(t) \\ -[x(t-\tau) - S^{-1} A_d^T P x(t)]^T S [x(t-\tau) - S^{-1} A_d^T P x(t)] \end{array}$$

Hence, if condition (4.10) holds, then there exists $\beta_4 > 0$ such that $\dot{V}(x_t) \leq -\beta_4 \| x(t) \|^2 < 0$, provided that model (4.7) is valid. The end of the proof of Proposition 2 can be mimicked. For any initial condition in the ball $\mathcal{B}(\delta)$, the trajectories remains in $\mathcal{S}(K, u_0^{\alpha})$ and the model (4.7) is valid. Hence, for any initial condition in the ball $\mathcal{B}(\delta)$ the local stability of system (2.6) is guaranteed. □

Notice that if we consider the ellipsoid $\mathcal{D}(P,\rho)$, we can define the resulting $\alpha_i(x)_{min}$ as

$$\alpha_i(x)_{min} = \min\left(\frac{u_{0(i)}}{\sqrt{\rho}\sqrt{K_{(i)} P^{-1} K_{(i)}^T}}, 1\right), \; i = 1, ..., m \tag{4.12}$$

Therefore the definition of vectors γ_j, $j = 1, ..., 2^m$, follows. A way to compute the suitable vectors γ_j and the positive scalar δ is now proposed.

Algorithm 5 *1. From the solution obtained in Algorithm 4 compute $s_i = \frac{u_{0(i)}}{\sqrt{\mu}\sqrt{K_{(i)} P^{-1} K_{(i)}^T}}$, $i = 1, ..., m$. One gets $s_i \geq 1$, $i = 1, ..., m$.*
2. Choose an increment $\Delta\omega$ and iterate ω from 1.

3. Compute the 2^m possible combinations of vector γ_j from $\gamma_{j(i)} = 1$ or $\min(\frac{s_i}{\omega}, 1)$, $i = 1, ..., m$. Test if condition (4.10) holds.

4. If condition (4.10) is verified, ω is a suitable value then increment it and go to the step above. Otherwise stop.

5. Among the suitable values of ω select $\omega_{max} = \max\omega$ and compute $\rho = \omega_{max}^2 \mu$.

6. Compute $\delta = \frac{\rho}{\lambda_{max}(P) + \tau\lambda_{max}(S)}$.

As in section 3, since

$$\forall\phi \in \mathcal{B}(\delta),\ x_t \in \mathcal{D}(V, \rho) \text{ and } x(t) \in \mathcal{D}(P, \rho) \tag{4.13}$$

the following proposition can be stated, based on the Razumikhin's approach, to establish in what case the ellipsoid $\mathcal{D}(P, \rho)$ is a positively invariant and contractive set for system (2.6).

Proposition 6. *Assume that two symmetric and positive definite matrices P and S are solutions of (3.1) and K is given by (3.2). If for all $j = 1, ..., 2^m$ condition (4.10) holds, then system (2.6) is asymptotically stabilizable by K, defined in (3.2), for all initial condition $\phi \in \mathcal{D}(P, \rho) = \{x \in \mathrm{Re}^n; x^T P x \leq \rho\}$, where ρ is defined by Step 5 of Algorithm 5. Furthermore the set $\mathcal{D}(P, \rho)$ is a positively invariant and contractive set for system (2.6).*

Proof. Compute the time-derivative of the quadratic Lyapunov function $V(x) = x^T P x$ along the trajectories of system (2.6). Then from (3.1) and (3.2) it follows:

$$\dot{V}(x(t)) = \sum_{j=1}^{2^m} \lambda_j(x) x(t)^T (A_j^T P + P A_j) x(t) + 2x(t)^T P A_d x(t - \tau)$$

From (4.8), (3.1), (3.2) and from the convexity of function $V(x)$ it follows that :

$$\dot{V}(x(t)) = \sum_{j=1}^{2^m} \lambda_j(x) \dot{V}_j(x(t))$$

where $\dot{V}_j(x(t))$ is defined by:

$$\begin{gathered} \dot{V}_j(x(t)) = -x(t)^T Q x(t) \\ -x(t)^T S x(t) - x(t)^T P A_d S^{-1} A_d^T P x(t) + 2x(t)^T P A_d x(t - \tau) \\ +x(t)^T K^T (R - R D(\gamma_j) - D(\gamma_j) R) K x(t) \end{gathered}$$

Thus, one gets:

$$\begin{aligned} \dot{V}_j(x(t)) \ \leq\ & -x(t)^T Q x(t) - x(t)^T S x(t) + x(t - \tau)^T S x(t - \tau) \\ & +x(t)^T K^T (R - R D(\gamma_j) - D(\gamma_j) R) K x(t) \end{aligned}$$

Then if condition (4.10) holds for all $j = 1, ..., 2^m$, therefore since S is positive definite it follows that

$$Q + S - K^T (R - R D(\gamma_j) - D(\gamma_j) R) K > S$$

therefore

$$S^{-\frac{1}{2}}(Q+S-K^T(R-RD(\gamma_j)-D(\gamma_j)R)K)S^{-\frac{1}{2}}>1$$

and the condition $\lambda_{min}(S^{-\frac{1}{2}}(Q+S-K^T(R-RD(\gamma_j)-D(\gamma_j)R)K)S^{-\frac{1}{2}})>1$, $\forall j=1,...,m$, follows. Furthermore, by mimicking the proof of Proposition 3 one can conclude that by Razumikhin's approach there exists a sufficiently small $\kappa>1$ satisfying $x(t-\tau)^TSx(t-\tau)<\kappa^2x(t)^TSx(t)$ such that $\dot V_j(x)<0$. Moreover, model (2.6) is valid only in $\mathcal{S}(K,u_0^\alpha)$. Since $\rho>0$ is chosen such that $\mathcal{D}(P,\rho)\subset\mathcal{S}(K,u_0^\alpha)$, by using Razumikhin's approach the same type of reasoning can be applied to conclude that $\dot V_j(x)<0$, and therefore $\dot V(x)<0$, for any $x\in\mathcal{D}(P,\rho)$. Thus, $\forall\phi\in\mathcal{D}(P,\rho)$, one gets $x(t)\in\mathcal{D}(P,\rho)$, $\forall t\geq t_0$. The set $\mathcal{D}(P,\rho)$ is a positively invariant and contractive set for system (2.6). □

The objective of Proposition 6 was to express the possible links between the Lyapunov-Krasovkii and the Razumikhin's approaches for a system with saturating controls as system (2.6). Hence, it is clear that other sufficient conditions may be found in order to guarantee the positive invariance of $\mathcal{D}(P,\rho)$ for system (2.6).

In the case $\tau=0$, the following corollary of Proposition 5 can be stated.

Corollary 7. *Assume that matrices P and S are solutions of (3.1) and K is given by (3.2). System (2.6) is asymptotically stabilizable by the state feedback K in $\mathcal{D}(P,\rho)=\{x\in\mathrm{Re}^n; x^TPx\leq\rho\}$ if, for all $j=1,...,2^m$, one gets:*

$$\begin{array}{c}-Q-(I_n-S^{-1}A_d^TP)^TS(I_n-S^{-1}A_d^TP)\\+K^T(R-RD(\gamma_j)-D(\gamma_j)R)K<0\end{array}\tag{4.14}$$

Proof. Using the Lyapunov function $V(x)=x(t)^TPx(t)$ and considering its time-derivative along the trajectories of system (4.7) in which $\tau=0$, it follows:

$$\dot V(x(t))=\sum_{j=1}^{2^m}\lambda_j(x)x(t)^T(A_j^TP+PA_j)x(t)+2x(t)^TPA_dx(t)$$

From (4.8), (3.1), (3.2) and from the convexity of function $V(x)$ it follows that :

$$\dot V(x(t))=\sum_{j=1}^{2^m}\lambda_j(x)\dot V_j(x(t))$$

where $\dot V_j(x(t))$ is defined by:

$$\begin{array}{c}\dot V_j(x(t))=-x(t)^TQx(t)\\-x(t)^T[I_n-S^{-1}A_d^TPx(t)]^TS[I_n-S^{-1}A_d^TP]x(t)]\\+x(t)^TK^T(R-RD(\gamma_j)-D(\gamma_j)R)Kx(t)\end{array}$$

Hence, if condition (4.14) holds, then $\dot V_j(x(t))<0$ and therefore $\dot V(x(t))<0$ for all $x\in\mathcal{D}(P,\rho)$. In this case, the set $\mathcal{D}(P,\rho)$ is positively invariant and contractive for system (4.7). Thus, the local stability of system (2.6) is guaranteed in $\mathcal{D}(P,\rho)$. □

5 Numerical example

Consider the numerical example borrowed from [17]. System (2.1) with constraint (2.3) is described by the following data:

$$A = \begin{bmatrix} 1 & 1.5 \\ 0.3 & -2 \end{bmatrix} ; \; A_d = \begin{bmatrix} 0 & -1 \\ 0 & 0 \end{bmatrix}$$

$$B = \begin{bmatrix} 10 \\ 1 \end{bmatrix} ; \; u_0 = 15 ; \; \tau = 1$$

Then by choosing $R = 1$ and $Q = \begin{bmatrix} 2 & 1 \\ 1 & 3 \end{bmatrix}$ one gets:

$$P = \begin{bmatrix} 0.2387 & -0.0537 \\ -0.0537 & 1.3102 \end{bmatrix} ; \; S = \begin{bmatrix} 2.9810 & 0.0043 \\ 0.0043 & 2.9990 \end{bmatrix}$$

Then from (3.2) the resulting state feedback matrix is

$$K = \begin{bmatrix} -2.3335 & -0.7734 \end{bmatrix}$$

5.1 Closed-loop stability without saturations

By applying algorithm 4, it follows:

$$\mu = 9.3328 ; \; \sigma = 2.1639$$

Thus for any initial condition belonging to $\mathcal{B}(\sigma)$ the resulting trajectories of the closed-loop system are those of system (2.8) since they remain in $\mathcal{S}(K, u_0)$. Furthermore the asymptotic stability of system (2.8) is guaranteed.

From Proposition 3 since

$$\lambda_{min}(S^{-\frac{1}{2}}(Q + S + PBR^{-1}B^T P)S^{-\frac{1}{2}}) < 1$$

we can conclude that the set $\mathcal{D}(P, \mu)$ obtained is a positively invariant set with respect to the trajectories of system (2.8).

Moreover, in the linear case (that is, in the Algorithm 4 case) the results obtained here can be compared to those of Theorem 2 in [7]. The comparison can be made in terms of size of domains of safe admissible states. The results given in [7] are based on the use of the Razumikhin's approach. Then, let us first consider the symmetric and positive definite matrix P_0 solution of

$$(A + BK)^T P_0 + P_0(A + BK) = -I_2$$

One obtains:

$$P_0 = \begin{bmatrix} 0.0247 & -0.0253 \\ -0.0253 & 0.2373 \end{bmatrix}$$

By applying Step 2 of Algorithm 4 in order to compute the maximal ellipsoid $\mathcal{D}(P_0, \mu_0) = \{x \in \mathrm{Re}^2; x^T P_0 x \leq \mu_0\}$ included in $\mathcal{S}(K, u_0)$, one obtains $\mu_0 = 0.8394$. From Theorem 2 in [7], domain $\mathcal{D}(P_0, \mu_0)$ is positively invariant and contractive for system (2.8), that is, the time-derivative of $V_0(x) = x^T P_0 x$ is strictly negative along the trajectories of system (2.8), if:

$$\| P_0 A_d \| < \frac{\lambda_{min}(Q_0)}{2} \sqrt{\frac{\lambda_{min}(P_0)}{\lambda_{max}(P_0)}}$$

In the present case, it follows:

$$\| P_0 A_d \| = 0.0354 \; ; \quad \frac{\lambda_{min}(Q_0)}{2} \sqrt{\frac{\lambda_{min}(P_0)}{\lambda_{max}(P_0)}} = 0.1503$$

It clearly appears that the domains of safe admissible states is smaller than that obtained from Proposition 2 or Proposition 3.

5.2 Closed-loop stability with saturations

Next, by applying Algorithm 5, one obtains:

$$\omega_{max} = 3.1208 \; ; \; \rho = 90.8941 \; ; \; \delta = 21.0751$$

One gets the following corresponding lower bound $\alpha(x)_{min}$ for the saturation term:

$$\alpha(x)_{min} = 0.3204$$

which generates the set

$$\mathcal{S}(K, u_0^\alpha) = \{x \in \mathrm{Re}^2; -46.8115 \leq \begin{bmatrix} -2.3335 & -0.7734 \end{bmatrix} x \leq 46.8115\}$$

We obtain a domain of safe initial conditions $\mathcal{B}(\delta)$, larger than $\mathcal{B}(\sigma)$, such that the trajectories of the closed-loop saturated system (2.6) remain confined in the domain $\mathcal{S}(K, u_0^\alpha)$. The local asymptotic stability of system (2.6) is guaranteed.

Proposition 6 may also be applied in order to verify that $\mathcal{D}(P, \rho)$ is a positively invariant and contractive set for system (2.6).

6 Concluding remarks

- The local stabilization of linear continuous-time systems with saturating controls and time-delay in the state was addressed. The approach was based on a Lyapunov-Krasovskii technique. Some domains of safe initial conditions were determined from the solution of an algebraic Riccati equation. Using the Razumikhin's approach some domains of positive invariance and contractivity were determined. Hence, some links between the Lyapunov-Krasovskii and Razumikhin's approaches were discussed.

- The conservativeness of the results proposed in this chapter is mainly due to the representation chosen for the saturated system. Indeed, all the trajectories of system (2.6) can be represented by those of system (4.7) only in $\mathcal{S}(K, u_0^\alpha)$. Hence, all the conditions obtained from this representation are only sufficient.
- The presented results can be extended to the multiple delays case. Consider the following system

$$\dot{x}(t) = Ax(t) + \sum_{i=1}^{r} A_{d_i} x(t - \tau_i) + Bu(t) \tag{6.1}$$

with the initial condition

$$\begin{array}{l} x(t_0 + \theta) = \phi(\theta),\ \forall \theta \in [-\bar{\tau}, 0] \\ \text{with } \bar{\tau} = \max\limits_{i=1,\dots,r} \tau_i(t_0, \phi) \in \mathrm{Re}^+ \times \mathcal{C}_\tau^v \end{array} \tag{6.2}$$

Hence, the following Riccati equation would be considered:

$$A^T P + PA - PBR^{-1}B^T P + \sum_{i=1}^{r} (PA_{d_i} S_i^{-1} A_{d_i}^T P + S_i) + Q = 0 \tag{6.3}$$

which is associated to the following Lyapunov functional:

$$V(x_t) = x(t)^T P x(t) + \sum_{i=1}^{r} \int_{t-\tau_i}^{t} x(\theta)^T S_i x(\theta) d\theta \tag{6.4}$$

- In the time-varying delay case, that is, in the case where the delay satisfies

$$0 \le \tau(t) \le \tau_{max} \text{ and } \dot{\tau} \le \zeta < 1$$

the algebraic Riccati equation (3.1) becomes:

$$A^T P + PA - PBR^{-1}B^T P + PA_d S^{-1} A_d^T P + \frac{1}{1-\zeta} S + Q = 0 \tag{6.5}$$

Then the following Lyapunov function can be used:

$$V(x_t) = x(t)^T P x(t) + \frac{1}{1-\zeta} \int_{t-\tau(t)}^{t} x(\theta)^T S x(\theta) d\theta \tag{6.6}$$

In [11], the authors study the quadratic stabilization of continuous-time systems with time-varying delay and norm-bounded time varying uncertainties but without control constraints. They use a similar algebraic equation to that defined in (6.5).

- In this chapter, the considered control law was memoryless. Nevertheless, the desired control law may be expressed under the form:

$$u(t) = sat(Kx(t) + K_d x(t - \tau)) \tag{6.7}$$

In this case the considered Riccati equation would be formulated as:

$$\begin{aligned} &A^T P + PA - PBR^{-1}B^T P + S + Q \\ &+P(A_d + BK_d)S^{-1}(A_d + BK_d)^T P = 0 \end{aligned} \tag{6.8}$$

- Some results can be obtained by considering the quadratic Lyapunov function $V(x(t)) = x(t)^T Px(t)$ where matrix P is solution of

$$(A + A_d)^T P + P(A + A_d) - PBR^{-1}B^T P + Q = 0 \tag{6.9}$$

for given symmetric and positive definite matrices Q and R. In this last case, the Razumikhin's approach has to be used [13]. Some positively invariant and contractive domains for system (2.6) can be obtained from the solution P.
- When the open-loop stability properties allow it, the global asymptotic stability of the saturated closed-loop system (2.6) can be investigated. Hence, one can show that [14]: Given a symmetric and positive definite matrix Q, if there exist two symmetric and positive definite matrices P and S solutions of

$$A^T P + PA + PA_d S^{-1} A_d^T P + S + Q = 0 \tag{6.10}$$

then system (2.6) is globally asymptotically stabilizable by the state feedback

$$K = -D(\gamma)B^T P$$

where $D(\gamma)$ is a diagonal matrix with positive diagonal elements. One can prove that the time-derivative of the Lyapunov functional $V(x_t)$ defined in (3.4) is negative for all $x \in \text{Re}^n$, $x \neq 0$.

References

1. D.S. Bernstein and A.N. Michel, *A chronological bibliography on saturating actuators*, Int. J. of Robust and Nonlinear Control, vol.5, pp.375-380, 1995.
2. C. Burgat and S. Tarbouriech, *Nonlinear Systems*, Vol.2, Chapter 4, pp.113-197, Appendices C, D, E, pp.217-239, Chapman & Hall, London, 1996.
3. B.S. Chen, S.S. Wang, H.C. Lu, *Stabilization of time-delay systems containing saturating actuators*, Int. J. of Control, vol.47, pp.867-881, 1988.
4. J.K. Hale, *Theory of functional differential equations*, Springer-Verlag, New York, 1977.
5. J.C. Hennet and S. Tarbouriech, *Stability and stabilization of delay differential systems*, Proc. of 13th World IFAC Congress, San Francisco (USA), pp.159-164, July 1996.
6. H.K. Khalil, *Nonlinear system*, Macmillan Ed., 1992.
7. M. Klai, S. Tarbouriech, C. Burgat, *Some independent-time-delay stabilization of linear systems with saturating controls*, Proc. of IEE Control'94, Coventry (U.K), pp.1358-1363, March 1994.

8. V.B. Kolmanovskii and V.R. Nosov, *Stability of functional differential equations*, Academic Press, New York, 1986.
9. J.F. Lafay and G. Conte, *Analysis and design methods for delay systems*, Invited Session, Proc. of 34th IEEE-CDC, New Orleans (USA), pp.2035-2069, December 1995.
10. Z. Lin and A. Saberi, *Semi-global exponential stabilization of linear systems subject to input saturations*, Systems & Control Letters, vol.21, no.3, pp.225-239, 1993.
11. M.S. Mahmound and N.F. Al-Muthairi, *Quadratic stabilization of continuous-time systems with state-delay and norm-bounded time-varying uncertainties*, IEEE Trans. Autom. Control, vol.39, no.10, pp.2135-2139, 1994.
12. A.P. Molchanov and E.S. Pyatnitskii, *Criteria of asymptotic stability of differential and difference inclusions encountered in control theory*, Systems & Control Letters, vol.13, pp.59-64, 1989.
13. S-I. Niculescu, J-M. Dion, L. Dugard, *Robust stabilization for uncertain time-delay systems containing saturating actuators*, IEEE Trans. Autom. Control, vol.41, no.5, pp.742-747, 1996.
14. S-I. Niculescu, S. Tarbouriech, J-M. Dion, L. Dugard: *Stability criteria for bilinear systems with delayed state and saturating actuators*, Proc. of the 34th IEEE-CDC, New Orleans (USA), pp.2064-2069, December 1995.
15. I.R. Petersen, *Stabilization algorithm for a class of uncertain linear systems*, Systems & Control Letters, vol.8, no.4, pp.181-188, 1987.
16. T.J. Su, P.L. Liu, J.T. Tsay, *Stabilization of delay-dependence for saturating actuator systems*, Proc. of 30th IEEE-CDC, Brighton (U.K), pp.2891-2892, December 1991.
17. A. Thowsen, *Stable sampled-data feedback control of dynamic systems with time-delays*, Int. J. Systems Science, vol.13, no.12, pp.1379-1384, 1982.
18. E. Tissir and A. Hmamed, *Further results on the stabilization of time delay systems containing saturating actuators*, Int. J. Systems Science, vol.23, pp.615-622, 1992.

Lecture Notes in Control and Information Sciences

Edited by M. Thoma

1993–1997 Published Titles:

Vol. 186: Sreenath, N.
Systems Representation of Global Climate Change Models. Foundation for a Systems Science Approach.
288 pp. 1993 [3-540-19824-5]

Vol. 187: Morecki, A.; Bianchi, G.; Jaworeck, K. (Eds)
RoManSy 9: Proceedings of the Ninth CISM-IFToMM Symposium on Theory and Practice of Robots and Manipulators.
476 pp. 1993 [3-540-19834-2]

Vol. 188: Naidu, D. Subbaram
Aeroassisted Orbital Transfer: Guidance and Control Strategies
192 pp. 1993 [3-540-19819-9]

Vol. 189: Ilchmann, A.
Non-Identifier-Based High-Gain Adaptive Control
220 pp. 1993 [3-540-19845-8]

Vol. 190: Chatila, R.; Hirzinger, G. (Eds)
Experimental Robotics II: The 2nd International Symposium, Toulouse, France, June 25-27 1991
580 pp. 1993 [3-540-19851-2]

Vol. 191: Blondel, V.
Simultaneous Stabilization of Linear Systems
212 pp. 1993 [3-540-19862-8]

Vol. 192: Smith, R.S.; Dahleh, M. (Eds)
The Modeling of Uncertainty in Control Systems
412 pp. 1993 [3-540-19870-9]

Vol. 193: Zinober, A.S.I. (Ed.)
Variable Structure and Lyapunov Control
428 pp. 1993 [3-540-19869-5]

Vol. 194: Cao, Xi-Ren
Realization Probabilities: The Dynamics of Queuing Systems
336 pp. 1993 [3-540-19872-5]

Vol. 195: Liu, D.; Michel, A.N.
Dynamical Systems with Saturation Nonlinearities: Analysis and Design
212 pp. 1994 [3-540-19888-1]

Vol. 196: Battilotti, S.
Noninteracting Control with Stability for Nonlinear Systems
196 pp. 1994 [3-540-19891-1]

Vol. 197: Henry, J.; Yvon, J.P. (Eds)
System Modelling and Optimization
975 pp approx. 1994 [3-540-19893-8]

Vol. 198: Winter, H.; Nüßer, H.-G. (Eds)
Advanced Technologies for Air Traffic Flow Management
225 pp approx. 1994 [3-540-19895-4]

Vol. 199: Cohen, G.; Quadrat, J.-P. (Eds)
11th International Conference on Analysis and Optimization of Systems – Discrete Event Systems: Sophia-Antipolis, June 15–16–17, 1994
648 pp. 1994 [3-540-19896-2]

Vol. 200: Yoshikawa, T.; Miyazaki, F. (Eds)
Experimental Robotics III: The 3rd International Symposium, Kyoto, Japan, October 28-30, 1993
624 pp. 1994 [3-540-19905-5]

Vol. 201: Kogan, J.
Robust Stability and Convexity
192 pp. 1994 [3-540-19919-5]

Vol. 202: Francis, B.A.; Tannenbaum, A.R. (Eds)
Feedback Control, Nonlinear Systems, and Complexity
288 pp. 1995 [3-540-19943-8]

Vol. 203: Popkov, Y.S.
Macrosystems Theory and its Applications: Equilibrium Models
344 pp. 1995 [3-540-19955-1]

Vol. 204: Takahashi, S.; Takahara, Y.
Logical Approach to Systems Theory
192 pp. 1995 [3-540-19956-X]

Vol. 205: Kotta, U.
Inversion Method in the Discrete-time Nonlinear Control Systems Synthesis Problems
168 pp. 1995 [3-540-19966-7]

Vol. 206: Aganovic, Z.;.Gajic, Z.
Linear Optimal Control of Bilinear Systems with Applications to Singular Perturbations and Weak Coupling
133 pp. 1995 [3-540-19976-4]

Vol. 207: Gabasov, R.; Kirillova, F.M.; Prischepova, S.V.
Optimal Feedback Control
224 pp. 1995 [3-540-19991-8]

Vol. 208: Khalil, H.K.; Chow, J.H.; Ioannou, P.A. (Eds)
Proceedings of Workshop on Advances inControl and its Applications
300 pp. 1995 [3-540-19993-4]

Vol. 209: Foias, C.; Özbay, H.; Tannenbaum, A.
Robust Control of Infinite Dimensional Systems: Frequency Domain Methods
230 pp. 1995 [3-540-19994-2]

Vol. 210: De Wilde, P.
Neural Network Models: An Analysis
164 pp. 1996 [3-540-19995-0]

Vol. 211: Gawronski, W.
Balanced Control of Flexible Structures
280 pp. 1996 [3-540-76017-2]

Vol. 212: Sanchez, A.
Formal Specification and Synthesis of Procedural Controllers for Process Systems
248 pp. 1996 [3-540-76021-0]

Vol. 213: Patra, A.; Rao, G.P.
General Hybrid Orthogonal Functions and their Applications in Systems and Control
144 pp. 1996 [3-540-76039-3]

Vol. 214: Yin, G.; Zhang, Q. (Eds)
Recent Advances in Control and Optimization of Manufacturing Systems
240 pp. 1996 [3-540-76055-5]

Vol. 215: Bonivento, C.; Marro, G.; Zanasi, R. (Eds)
Colloquium on Automatic Control
240 pp. 1996 [3-540-76060-1]

Vol. 216: Kulhavý, R.
Recursive Nonlinear Estimation: A Geometric Approach
244 pp. 1996 [3-540-76063-6]

Vol. 217: Garofalo, F.; Glielmo, L. (Eds)
Robust Control via Variable Structure and Lyapunov Techniques
336 pp. 1996 [3-540-76067-9]

Vol. 218: van der Schaft, A.
L_2 Gain and Passivity Techniques in Nonlinear Control
176 pp. 1996 [3-540-76074-1]

Vol. 219: Berger, M.-O.; Deriche, R.; Herlin, I.; Jaffré, J.; Morel, J.-M. (Eds)
ICAOS '96: 12th International Conference on Analysis and Optimization of Systems - Images, Wavelets and PDEs:
Paris, June 26-28 1996
378 pp. 1996 [3-540-76076-8]

Vol. 220: Brogliato, B.
Nonsmooth Impact Mechanics: Models, Dynamics and Control
420 pp. 1996 [3-540-76079-2]

Vol. 221: Kelkar, A.; Joshi, S.
Control of Nonlinear Multibody Flexible Space Structures
160 pp. 1996 [3-540-76093-8]

Vol. 222: Morse, A.S.
Control Using Logic-Based Switching
288 pp. 1997 [3-540-76097-0]

Vol. 223: Khatib, O.; Salisbury, J.K.
Experimental Robotics IV: The 4th International Symposium, Stanford, California, June 30 - July 2, 1995
596 pp. 1997 [3-540-76133-0]

Vol. 224: Magni, J.-F.; Bennani, S.; Terlouw, J. (Eds)
Robust Flight Control: A Design Challenge
664 pp. 1997 [3-540-76151-9]

Vol. 225: Poznyak, A.S.; Najim, K.
Learning Automata and Stochastic Optimization
219 pp. 1997 [3-540-76154-3]

Vol. 226: Cooperman, G.; Michler, G.; Vinck, H. (Eds)
Workshop on High Performance Computing and Gigabit Local Area Networks
248 pp. 1997 [3-540-76169-1]

Vol. 227: Tarbouriech, S.; Garcia, G. (Eds)
Control of Uncertain Systems with Bounded Inputs
203 pp. 1997 [3-540-76183-7]

GPSR Compliance

The European Union's (EU) General Product Safety Regulation (GPSR) is a set of rules that requires consumer products to be safe and our obligations to ensure this.

If you have any concerns about our products, you can contact us on ProductSafety@springernature.com

In case Publisher is established outside the EU, the EU authorized representative is:

Springer Nature Customer Service Center GmbH
Europaplatz 3
69115 Heidelberg, Germany

Batch number: 10704711

Printed by Printforce, the Netherlands